Mycoplasma Diseases of Crops

Karl Maramorosch   S.P. Raychaudhuri
Editors

Mycoplasma Diseases of Crops
Basic and Applied Aspects

With 54 Illustrations

Springer-Verlag
New York Berlin Heidelberg
London Paris Tokyo

Karl Maramorosch
Department of Entomology and Economic Zoology
Cook College
Rutgers—The State University
New Brunswick, New Jersey 08903, USA

S.P. Raychaudhuri
IUFRO Working Party on Mycoplasma Diseases
A-61 Alkananda
Kalkaji, New Delhi 110019, India

Library of Congress Cataloging-in-Publication Data
Mycoplasma diseases of crops.
 Bibliography: p.
 Includes index.
 1. Mycoplasma diseases in plants.   I. Maramorosch,
Karl.   II. Raychaudhuri, S.P. (Syama Prasad)
SB737.M93   1988   632′.32   87-26435

Camera-ready copy prepared by the editors.
Printed and bound by Arcata Graphics/Halliday, West Hanover, Massachusetts.
Printed in the United States of America.

9 8 7 6 5 4 3 2 1

ISBN 0-387-96646-3 Springer-Verlag New York Berlin Heidelberg
ISBN 3-540-96646-3 Springer-Verlag Berlin Heidelberg New York

PREFACE

Mycoplasmas are placed in a separate class, Mollicutes, which removes them from bacteria. Their main characteristics are lack of a cellular wall and inability to synthesize the peptidoglycan polymer. The lack of a cell wall accounts for the pleomorphism, osmotic sensitivity, sensitivity to antibiotics that inhibit peptidoglycan polymerization and synthesis, susceptibility to lysis by alcohol and detergents, and the ability to grow on agar gel. At present, three families are placed in the class Mollicutes: Mycotaceae, Acholetaceae, and Spiroplasmataceae.

The first pathogenic mycoplasmas were discovered in Pasteur's laboratory nearly 90 years ago as the causative agents of a sheep disease. They were first named PPLO, pleuropneumonia-like organisms. In 1928, Nocard in France coined the name mycoplasma for PPLO, but his publication and the new name remained practically unnoticed until Leonard Hayflick and Robert Channock succeeded in culturing the "PPLO" of human "atypical virus pneumonia" in the United States in 1960. Hayflick resurrected the name given by Nocard and since then, the causative agent of human "atypical virus pneumonia" is known as *Mycoplasma pneumoniae*. Other mycoplasmas cause diseases in dogs, sheep, birds, cattle, pigs, etc.

A breakthrough was made in Japan in 1967, when several plant diseases, earlier considered to be caused by viruses, were found to be associated with mycoplasma-like organisms (MLO). Soon the Japanese finds were confirmed by others in the United States and subsequently, in many other countries. Now more than 100 different plant diseases, previously categorized as caused by viruses, are classified as plant mycoplasma diseases. At present, a dozen plant and insect-associated mycoplasmas have been found to produce helical forms that distinguish them from typical mycoplasmas and places them in the family called Spiroplasmataceae. Most of these have been grown on artificial media. The remaining large number of MLOs have not yet been grown in cell-free media, and this formidable difficulty has prevented their detailed characterization.

Many reviews and several books have been published on basic aspects of Mycoplasmas, including plant and insect spiroplasmas, but only one, limited to mycoplasma diseases of trees and shrubs, has appeared until now (*Mycoplasma Diseases of Trees and Shrubs,* K. Maramorosch and S.P. Raychaudhuri, eds.,

Academic Press, New York, 1981). The purpose of the present volume is to bring together basic and applied information concerning several plant mycoplasma diseases of crops. In addition to information pertaining to new means of detection, characterization, and cultivation of plant mycoplasmas (Part I), interactions between insect vectors, plant mycoplasmas, and viruses in mixed mycoplasma-virus infections are presented (Part II). Part III includes mycoplasma diseases of rice, potato, corn, citrus, and other cultivated plants. A historical account of the earliest description of a plant mycoplasma disease, made nearly 1000 years ago in China, is published here for the first time. Four chapters on current control technologies constitute the last part of the volume (Part IV).

The contributors to this book are active workers from the United States, Canada, India, The Netherlands, Yugoslavia, and The Peoples' Republic of China.

The subject of this treatise should be of considerable interest and importance, and one that will appeal to an audience representing plant pathologists, entomologists, horticulturists, as well as other branches of agriculture.

Karl Maramorosch
S.P. Raychaudhuri

CONTENTS

Part II: Interactions with Plants, Insects, and Viruses

Part III: Diseases of Rice, Potato, Corn, Citrus, Eggplant and Other Plants

Part IV: Chemotherapy and Other Methods of Control

CONTRIBUTORS

Y.K. Arora
Chemistry and Biology Research
  Institute
Agriculture Canada, Research Branch
Ottawa, Ontario
Canada K1A OC6

Ernest E. Banttari
Department of Plant Pathology
University of Minnesota
St. Paul, MN 55108, USA

R.E. Davis
Microbiology and Plant
  Pathology Laboratory
ARS-USDA
Beltsville, MD 20705, USA

G.T.N. de Leeuw
Willie Commelin Scholten
  Phytopathological Laboratory
Baarn, The Netherlands

Z. Eric
Department of Biology
Faculty of Science
University of Sarajevo
Vojvode Putnika 43A
Sarajevo 71000, Yugoslavia

Donald E. Gardner
National Park Service
Cooperative Park Studies Unit
Department of Botany
University of Hawaii at Manoa
Honolulu, HA 96822, USA

Bijan K. Ghosh
Department of Physiology
  and Biophysics
University of Medicine and Dentistry
Robert Wood Johnson Medical
  School
Piscataway, NJ 08554, USA

David J. Gumpf
Department of Plant Pathology
University of California–Riverside
Riverside, CA 92521, USA

Chuji Hiruki
Department of Plant Science
University of Alberta
Edmonton, Alberta
Canada T6G 2P5

D.M. Kaley
Central Potato Research Station
Rajgurunagar 410 505
Pune, Maharashtra, India

S.M. Paul Khurana
Central Potato Research Institute
Simla-171001, India

K. Krivokapic
Department of Biology
Faculty of Science
University of Sarajevo
Vojvode Putnika 43A
Sarajevo 71000, Yugoslavia

Ing-Mee Lee
Department of Botany
University of Maryland
College Park, MD 20742, USA

Alan Liss
Department of Biological Sciences
State University of New York
University Center at Binghamton
Binghamton, NY 13901, USA

Karl Maramorosch
Department of Entomology and
  Economic Zoology
Cook College
Rutgers–The State University
New Brunswick, NJ 08903, USA

S. Misra
Department of Botany
University of Rajasthan
Jaipur, India

D.K. Mitra
Nuclear Research Laboratory
Indian Agricultural Research Institute
New Delhi 110012, India

V. Muniyappa
Department of Plant Pathology
University of Agricultural Sciences
Hebbal, Bangalore 560024, India

V.M.G. Nair
College of Environmental Sciences
University of Wisconsin at Green Bay
Green Bay, WI 54301, USA

George N. Oldfield
USDA, ARS
Boyden Fruit and Vegetable Insects
  Laboratory
University of California–Riverside,
Riverside CA 92521, USA

Biljana Plavsic
Department of Biology
Faculty of Science
University of Sarajevo
Vojvode Putnika 43A
Sarajevo 71000, Yugoslavia

A.A. Polak-Vogelzang
National Institute of Public Health
  and Environmental Hygiene
Bilthoven, The Netherlands

S.P. Raychaudhuri
IUFRO Working Party on
  Mycoplasma Diseases
A-61 Alkananda, Kalkaji
New Delhi 110019, India

Narayan Rishi
Department of Plant Pathology
Haryana Agricultural University
Hisar 125005, India

Sushma Rishi
Department of Pharmacology
Haryana Agricultural University
Hisar 125005, India

R.A. Samson
Centraal Bureau voor Schimmelcultures
Baarn, The Netherlands

R.A. Singh
Central Potato Research Institute
Simla-17001, India

Ramish C. Sinha
Chemistry and Biology
  Research Institute
Agriculture Canada, Research Branch
Ottawa, Ontario
Canada K1A OC6

James H. Tsai
Fort Lauderdale Research and
   Education Center
University of Florida
3205 College Avenue
Fort Lauderdale, FL 33314, USA

P.A.M. van Vught
Willie Commelin Scholten
   Phytopathological Laboratory
Baarn, The Netherlands

M.Q. Wang
Department of Biology
Fudan University, Shanghai
The People's Republic of China

Part I: Detection, Characterization, and Cultivation of Plant Mycoplasmas

1

PLANT PATHOGENIC MYCOPLASMAS:

MORPHOLOGICAL AND BIOCHEMICAL

CHARACTERISTICS

by

Y.K. Arora and R.C. Sinha

Chemistry and Biology Research Institute

Agriculture Canada, Research Branch

Ottawa, Ontario, Canada   K1A OC6

INTRODUCTION

Association of wall-less prokaryotes with plants was first reported by Japanese workers in 1967 (13,23). This association has since been reported for more than 100 diseases inflicting damage to a variety of crops of agricultural importance throughout the world (29). Almost all such diseases are disseminated in nature by means of insect vectors, the leafhoppers (Cicadellids), in a persistent manner. The pathogen undergoes multiplication in both vectors and host plants (51). Most of these diseases were presumed to be caused by viruses in spite of the fact that the causal agents were never morphologically identified. Several of these organisms have not yet been cultured in cell-free medium and are termed as mycoplasma-like organisms (MLOs). These microbes are pleomorphic and wall-less, morphologically and ultrastructurally resembling non-helical mycoplasmas. These are located in the sieve tubes of plants showing symptoms such as leaf yellows, leaf dwarfing, leaf mottling, flower dwarfing, flower virescence, witches' broom, shortening of internodes, stunting etc. MLOs fulfill the first of Koch's postulates in being constantly associated with diseased plants and never present in healthy plants. Since they have never been cultured *in vitro*, Koch's remaining postulates have not yet been fulfilled.

The term MLO does not include a) sieve-tube restricted, helical mycoplasmas, b) sieve-tube restricted bacterial-like agents such as organisms associated with citrus greening (4), and clover club leaf diseases (57), c) xylem-limited bacteria such as Pierce's disease agent (20,21) and d) organisms such as *Mycoplasma* sp., *Spiroplasma* sp. and *Acholeplasma* sp. which occur on the surface of some plants. Some of the plant pathogenic sieve-tube restricted mycoplasmas have been cultured *in vitro*, Koch's postulates satisfied and placed in a genus *Spiroplasma* because they produce helical and motile filaments (14,44,45). Rapid progress has since been made on immunological, microbiological, biochemical, and ultrastructural characteristics of these mycoplasmas especially *Spiroplasma citri* -- the causal agent of citrus stubborn disease (3,7,11,35,58,59).

Attempts to obtain continuous cultures of MLOs in cell-free medium have failed so far and this has impeded research on their characterization. As an alternative, MLOs have been purified from diseased plants (46,50) in order to determine the properties of these microbes. The major difficulty with this approach is the low yield of the purified preparations from diseased plants (49). However, these preparations have been subjected to morphological, serologiical and biochemical characterization. MLOs are morphologically similar to pleomorphic animal or human mycoplasmas. Immunological techniques have played an important role in establishing the antigenic differences among individual mycoplasmas. Biochemical characterization has proved useful in understanding and comparing the molecular properties of MLOs with other mycoplasmas. In this article we present and evaluate the various approaches undertaken for characterization of plant pathogenic mycoplasmas, emphasizing the research on MLOs which should unravel the fundamental properties of these organisms.

ISOLATION AND PURIFICATION OF THE MLOs: A STEP LEADING TO CHARACTERIZATION

The yield of the purified MLOs (48) is usually 2-4 mg (freeze-dried weight) per 100 g leaves. This estimate is based on the yields from aster (*Callistephus chinensis* Nees) leaves infected with aster yellows or clover phyllody, and celery (*Apium graveolans* L.) leaves infected with eastern peach-X disease. The loss of MLO during purification, determined through serological titer, was estimated to be about 75%. Various host plants infected with MLOs may contain different concentration of the pathogen, and therefore the yield of the purified MLOs from various hosts can differ drastically (52). The MLO concentration may also vary in various parts of the host plant. Studies conducted to determine the concentration of aster yellows-MLO in various parts of the aster plants revealed the highest MLO-antigen concentration in infected leaves determined by ELISA (Table 1). The concentration in stem and petiole of diseased plants was negligible. MLO-antigens in roots of the diseased aster plants were also detectable but healthy root extracts showed a higher absorbance value as compared with control samples from other tissues. The reason for this is not known.

PLANT PATHOGENIC MLOs AND SPIROPLASMAS

Morphology

Electron microscopic examination of purified clover phyllody-MLO suspensions stained
with phosphotungstic acid (PTA) has shown numerous pleomorphic bodies (46). Many spherical
forms of the MLO cells showed small vesicles,
giving a budding appearance. The spherical and
oblong cells ranged from 175-400 nm in diameter
while filamentous forms were about 1700 nm
long. Scanning electron microscopy of freeze
fractured root and petiole tissues of clover
phyllody and aster yellows-infected plants (16)
has provided proof that MLOs are non-helical
and such structural features observed *in vitro* are not artifacts due to PTA staining
(61). Studies on the untrastructure of purified MLO cells (47) have shown that they have a
well defined trilaminar unit membrane with ribosome-like granules in the cytoplasm, and some
cells show a nuclear area with DNA-like fibrils
(Fig. 1). The symmetrical unit membrane is composed of two electron-dense layers and a less
dense intermediate layer. The average thickness of the membrane was estimated to be about
7 nm.

Morphologically, spiroplasmas can be described as motile, wall-less prokaryotes with
helical morphology (3,12). *S. citri* and
corn stunt spiroplasma strains are morphologically very similar to each other with the exception of a non-helical strain of *S. citri*. Morphology of spiroplasmas is influenced
by the age of the culture. In early log-phase
increase in broth culture, spiroplasma cells
are helical filaments ranging between 2 to 5 m
in length, 0.15 to 0.20 m in diameter and
pointed at one end (9,10,60). The helix is generally uniform in wave (gyre) length and ampli-
tude in a given filament although this as well
as filament length can vary with changes in the
composition of culture medium. In late log-

phase and stationary phase, spiroplasma cells elongate, lose helicity and deteriorate in shape. The structural changes with the age of the culture and pH have been described in detail elsewhere (3,9,10).

For morphological and etiological studies of microbes *in vitro*, it is desirable to preserve the structure of the mycoplasmas and to use live and unfixed organisms. To fulfill this objective, a techniques was developed in which sieve elements of periwinkle plants infected *S. citri* or MLOs were separated from other cell types by partial digestion of the plant tissues with cellulase and macerozyme (26). The sieve-tube elements of the diseased plants showed the presence of microorganisms through the cell walls by dark field microscopical examination. Pleomorphic non-helical bodies were observed in sieve elements of plants infected with aster yellows, and helical bodies in preparations from plants infected with *S. citri.* MLO morphology has also been studied by rupturing the sieve elements by osmotic shock. The MLOs observed exhibited the same morphology as described above. These results along with the scanning electron microscopical examination of MLOs (16), show that these microbes are non-helical in shape.

Immunology

Morphological characteristics of the MLOs found associated with different diseases are very similar. The application of immunological techniques has proved to be useful in analysing the antigens of an MLO and to develop rapid disease diagnostic methods (6,8,46,52). Immunodiffusion tests were first used to determine the antigenicity of the membrane and cytoplasmic contents of the MLO cells purified from aster yellows-infected plants. The presence of at least two cytoplasmic and two membrane antigens was demonstrated. The minimum concentration of the MLO required for detecting the slow moving cytoplasmic antigen in the agar gel was about 20 mg/ml and for the fast moving one was 2.5 mg/ml (50). Membrane preparations required the solubilization with the mild detergents to release the antigens bound with lipids of the membrane as has also been demonstrated for animal mycoplasmas (19,25,42).

Enzyme linked immunosorbent assay (ELISA) and immunosorbent electron microscopy (ISEM) techniques have recently been used to detect MLO antigens in preparations from infected plants. In ELISA tests, preparations containing concentrations of aster yellows-MLO as low as 4µg/ml gave a positive reaction whereas in ISEM tests, the minimum concentration needed to trap the MLO cells on the antiserum-coated grids was about 12.5 g/ml (51). Both these tests have also been employed to detect eastern peach X-MLO antigens in preparations from infected celery plants (53).

The conventional polyclonal antisera used, however, do not fully elucidate the chemically and structurally complex antigens through serological analysis. The quality, quantity and affinity of antibodies specific for the epitope of an antigen often vary from animal to animal and even from different bleedings from the same animal. The hybridoma technique has the potential of producing monospecific immunoglobulins (monoclonal antibodies) against a single epitope and are ideal serological reagents for structural, immunological and biochemical analyses of the antigens. Monoclonal antibodies have been prepared against *S. citri* (32) to characterize spiroplasma protein antigens. The specific monoclonal antibodies were directed against spiralin, the main membrane protein. MLO antigens have been detected in preparations from diseased lettuce and periwinkle plants using monoclonal antibodies against aster yellows agent (27). These were highly specific and did not react with MLOs from ash yellows, elm yellows, maize bushy stunt, paulownia witches' broom and peanut rosette infected plants.   Diagnostic and toxonomic applications of monoclonal antibodies should prove useful in defining antigenic relationships among the pathogenic mycoplasmas as has been demonstrated for viruses (17, 18).

Biochemistry

Amino Acid Composition

Amino acid composition of freeze-dried MLO preparations from clover phyllody and aster yellows-infected plants has been determined (54). The concentration of various amino acids in both MLOs was very similar. Glycine concentration was the highest followed closely by alanine.   Isoleucine, leucine, valine and arginine were also fairly good amounts. The concentrations of phenylalanine and tyrosine were low. The high concentration of glycine in the MLOs was probably due to glycine-magnesium chloride solution used in the purification procedure. This was indicated by the incorporation of label into the cells from C-glycine-magnesium chloride used in the purification procedure. The amino acid composition of *S. citri* appears to be different from that of MLOs. *S. citri* has been shown to contain the highest concentration of leucine and is characterized by the absence of arginine, histidine and proline (2).   The concentration of amino sugars was high in aster yellows-MLO and clover phyllody-MLO which contained 2.5% and 4.5%, respectively, as compared with 1-2% in *S. citri* (40).

Proteins

Electrophoretic profile of proteins of a cell provide essential information about the characteristics of the genome. The number and pattern of protein bands of animal mycoplasmas obtained through sodium dodecyl sulfate (SDS) - polyacrylamide gel electrophoresis (PAGE) has been widely used as one of the criteria for species differentiation and characterization (55). So far there is only one report on the protein pattern of plant MLOs (54). SDS-electrophoresis of purified aster yellows-MLO and clover phyllody-MLO resolved proteins ranged from about 38 to 170 x 10^3 daltons for aster yellows-MLO and 33 to 170 x 10^3 daltons for clover phyllody-MLO.

Gel electrophoresis of *S. citri* proteins was first carried out by the disk technique in the presence of 5M urea and 35% acetic acid (38,39). *S. citri* has been shown to produce 18 bands while corn stunt spiroplasma produced only 12 bands (35). Studies on two strains of *S. citri* have revealed identical cell protein patterns but these were very distinct from the patterns of other mycoplasmas (45). Taxonomic implications have been drawn from the protein pattern differences between these two spiroplasmas. The protein pattern of *S. citri* exhibited more than 50 bands by gradient slab gel electrophoresis in the presence of SDS (3). The pattern shows a difference in one protein with acidic isoelectric point but with the same molecular weight.

Enzyme Distribution

Recently, the distribution of enzymes has been described for the non-helical, non-culturable microbes parasitic to plants (1). MLOs purified from aster yellows-infected plants were osmotically lysed and the membranes (Fig. 2) separated from the cytoplasmic fractions through differential centrifugation. NADH oxidase activity was markedly higher in the membrane fraction than in either whole cells or the cytoplasmic fraction (Table 2). The specific activity of the enzyme in the whole cells was 21.5% of the total activity in three preparations. The high value in whole cells was partly ascribed to the leaky nature of the MLO membranes. NADPH oxidase on the other hand was extremely low in the cytoplasmic and membrane fraction and was not detectable in the whole cells. The location of NADH oxidase in the membranes was similar to that reported for saprophytic *Mycoplasma laidlawii* (37). However, this distribution differed from that of *S. citri* (33) causing citrus stubborn disease (22,44) and of parasitic *Mycoplasma* strains (37) in being associated with the cytoplasmic fraction. This represented an important difference in the membrane composition between parasitic MLOs and *Mycoplasma* species. High NADH oxidase and low NADPH oxidase in cell fractions implied that the NADH-specific form may be a dominant form in the MLOs.

ATPase and n-nitrophenyl phosphatase have been found associated with the trilaminar membrane of the MLOs (Table 3 and 4). Similar disposition has been found in *Mycoplasma gallisepticum* and other *Mollicutes* (34). However, whereas the transport of sodium and potassium through the animal cell membrane is mediated by a Na /K concentration-activated ATPase system located in the membrane, the mycoplasma ATPase resembles the mitochondrial and microbial ATPase insofar as it is not stimulated by sodium and potassium (43). Membrane ATPase acts as a monovalent cation pump and regulates intracellular osmolarity of the organism similar to the Na /K -activated ATPase of eucaryotic cells (24). RNase and DNase activities were detected in the MLO-cells and fractions (Table 4). RNase exhibited highest level of activity in membranes while DNase was mainly cytoplasmic. It is speculated that membrane-associated hydrolytic enzymes are involved in degrading macromolecules into smaller molecules capable of permeating the cells. Nucleases may induce and contribute to host pathology (15,36, 56) by hydrolysing host nucleic acids to produce nucleotides and nucleosides (30,31).

DNA ANALYSIS

Protein synthesis of an organism is genetically directed and properties of the proteins reflect the nature and complexity of the genome of the organism. The electrophoretic patterns of the proteins may be considered partly as 'finger prints' of the genetic constitution of the microorganisms. Protein analysis, however, is the indirect approach to characterize a genome since it is the genome-directed products which are analysed and not the genome itself. Restriction enzyme analysis offers a means for studying a genome directly. These enzymes which are endonucleases in nature split DNA only at highly specific points yielding a number of DNA fragments. The profiles of the resulting fragments obtained by PAGE and subsequent ethidium bromide staining were found to be reproducible for a given DNA and have been used to characterize DNA of *Mycoplasma* sps. (MW 5 x 10 daltons), *S. citri* and *Acholeplasma* sps. (MW 10 x 10 daltons) (3,5,41). Two groups of *S. citri* have been distinguished by employing restriction enzyme analysis. One group was found to exhibit only one band having similar restriction fragment patterns. The other group exhibited one or several additional DNA bands with different restriction fragment patterns. The technique has not yet been applied to MLOs because attempts to isolate DNA from purified MLOs have so far been unsuccessful. Biochemical properties of MLOs closely resembled those of animal *Mycoplasma* sp. and *Acholeplasma* sp. and differed from *S. citri* (Table 5). The genome size and the disposition of enzymes are important properties in defining the phylogenetic and taxonomic positions.

SUMMARY AND GENERAL REMARKS

Sieve-tube specific yellows disease-associated MLOs and pathogenic spiroplasmas are agents responsible for plant diseases. Unlike *S. citri* and corn stunt spiroplasma, plant pathogenic MLOs have not been successfully cultivated in pure culture. However, they have been isolated and purified from plants infected with aster yellows, clover phyllody, eastern peach-X, ash yellows and elm yellows diseases. The purification technique led to the morphological, immunological and biochemical characterization of some of these non-helical and non-culturable plant parasitic microbes. MLOs are wall-less non-helical prokaryotes, pleomorphic in nature having well-defined trilaminar membrane while spiroplasmas are motile wall-less prokaryotes with helical cell shape. MLOs associated with plant diseases are morphologically indistinguishable from each other. The purified MLO preparations have been used to prepare antisera against these microbes. MLO-antigens have been detected by immunodiffusion, ELISA and ISEM. Serological relatedness and differences have been established for some of the MLO diseases. The use of monoclonal antibodies is ideal for structural, immunological and biochemical analyses of plant pathogenic organisms. They have been used to differentiate *S. citri* within the same and different serogroups.

Amino acid composition and concentration of various amino acids has been found to be very similar for MLOs purified from aster yellows- and clover phyllody-infected plants. The protein pattern of these MLOs exhibited almost the same number of bands. The distribution of enzymes in MLO cells purified from aster yellows-infected plants resembles that of some mycoplasmas that infect animals but differed from *S. citri*, a plant pathogenic mycoplasma. *S. citri* DNA has been characterized and two groups distinguished by restriction enzyme analysis. The close relationship between *S. citri* and corn stunt spiroplasma has also been speculated from their genome size and DNA G + C content. There is no evidence to suggest that plant pathogenic spiroplasmas are serologically related to MLOs.

The properties of plant pathogenic MLOs have not been extensively elucidated. Studies on proteins that play structural and catalytic roles might contribute to our understanding of the functional relationship of membrane proteins to the molecular organization of the MLOs. Attempts should also be made to obtain pure MLOs not only for future characterization but also for satisfying Koch's postulates. Also, the metabolic changes in different host-parasite systems should be studied to provide the biochemical basis of plant-mycoplasma pathogenicity.

REFERENCES

1. Arora, Y.K. and Sinha, R.C. 1985. Enzymatic activities in cell fractions of mycoplasma-like organisms purified from aster yellows-infected plants. J. Bacteriol. 164: In Press.

2. Bebear, C.H., Latrille, J., Fleck, J., Roy, B. and Bové, J.M. 1974. *Spiroplasma citri*: un mollicute. pps. 33-52. In: Mycoplasma of man, animals, plants and insects. INSERM. J.M. Bové and J.F. Duplan, eds. Vol. 33. Inst. Natl. Sante, Rech. Med. Paris.

3. Bové, J.M. and Saillard, C. 1979. Cell Biology of spiroplasmas. pps. 83-153. In: The Mycoplasmas. Vol. 3. Plant and Insect Mycoplasmas. R.F. Whitcomb and J.G. Tully, eds. Academic Press, New York.

4. Bové, J.M., Bonnet, P., Garnier, M. and Aubert, B. 1980. Penicillin and tetracycline treatments of greening disease-affected citrus plants in the glasshouse, and the bacterial nature of prokaryote associated with greening. pps. 91-102. In: Proceedings 8th Conference IOCV, E.C. Calavan,

Garnsey and L.W. Timmer, eds. IOCV, Dept.
Plant Pathol. Univ. Calif. Riverside, California.

5. Bové, J.M., Saillard, C., Junca, P.,
Degorce-Dumas, J.R., Ricard, B., Nhami, A.,
Whitcomb, R.F., Williamson, D. and Tully,
J.G. 1982. Guanine-plus-cytosine content,
hybridization percentages, and EcoRI re-
striction enzyme profiles of spiroplasmal
DNA. Rev. Inf. Dis. 4: S129-136.

6. Caudwell, A., Meignoz, R., Kuszala, C.,
Larrue, J., Fleury, A. and Boudon, E.
1982. Purification serologique et observation ultra-microscopique de l'agent patho-
gene (MLO) de la flavascence doree de la
vigne dans les extrais liquides de feves
(*Vicia faba L.*) malades. C.R. Seances
Soc. Biol. 176: 723-729.

7. Chen, T.A. and Davis, R.E. 1979. Cultivation of spiroplasmas. pps. 65-82. In: The
Mycoplasmas. Vol. 3. Plant and Insect
Mycoplasmas. R.F. Whitcomb and J.G. Tully,
eds. Academic Press, New York.

8. Clark, M.F., Barbara, D.J. and Davis, D.L.
1983. Production and characteristics of
antisera to *Spiroplasma citri* and clover phyllody-associated antigens derived
from plants. Ann. Appl. Biol. 103: 251-259.

9. Cole, R.M., Tully, J.G., Popkin, T.J. and
Bové, J.M. 1973. Ultrastructure of the
agent of citrus stubborn disease. Ann.
N.Y. Acad. Sci. 225: 491-493.

10. Cole, R.M., Tully, J.G., Popkin, T.J. and
Bové, J.M. 1973. Morphology, ultrastructure and bacteriophage infection of the
helical mycoplasma-like organism (*Spiroplasma citri* gen. nov., sp. nov.) cultured from stubborn disease of citrus. J.

Bacteriol. 115: 367-386.

11. Daniels, M.J. 1979. Mechanism of spiroplasma pathogenicity. pps. 209-277. In: The Mycoplasmas. Vol. 3. Plant and Insect Mycoplasmas. R.F. Whitcomb and J.G. Tully, eds. Academic Press, New York.

12. Davis, R.E. and Worley, J.F. 1973. Spiroplasma: Motile helical microorganism associated with corn stunt disease. Phytopathology. 63: 403-408.

13. Doi, Y., Teranaka, M., Yora, K. and Asuyama, H. 1967. Mycoplasma or PLT group-like microorganisms found in the phloem elements of plants infected with mulberry dwarf, potato witches' broom, aster yellows, or Paulownia witches' broom. Ann. Phytopathol. Soc. Japan. 33: 259:266.

14. Fudl-Allah, A., Calavan, E.C. and Igwegbe, E.C.K. 1972. Culture of a mycoplasma-like organism associated with stubborn disease of citrus. Phytopathology. 62: 729-731.

15. Gabridge, M.G. and Stahl, Y.D.B. 1978. Role of adenine in the pathogenesis of *Mycoplasma pneumoniae* infections of trachael epithelium. Med. Microbiol. Immuno. 165: 43-55.

16. Haggis, G.H. and Sinha, R.C. 1978. Scanning electron microscopy of mycoplasma-like organisms after freeze fracture of plant tissues affected with clover phyllody and aster yellows. Phytopathology. 68: 677-680.

17. Halk, E.L. 1984. Production of monoclonal antibodies to three ilarviruses: Use as serotyping and diagnostic reagents. pps. 236-238. In: Hybridoma Technology in Agricultural and Veterinary Research. N. Stern

and H.R. Gamble, eds. Rowman and Allenheld, New Jersey.

18. Halk, E.L., Hsu, H.T., Aebig, J. and Franke, J. 1984. Production of monoclonal antibodies against three ilarviruses and alfalfa mosaic virus and their use as serotyping reagents. Phytopathology. 74: 367-372.

19. Hollingdale, M.R. and Lemcke, R.M. 1969. The antigens of *Mycoplasma hominis*. J. Hyg. Comb. 67: 585-602.

20. Hopkins, D.L. 1977. Diseases caused by leafhopper-borne, rickettsia-like bacteria. Ann. Rev. Phytopathol. 15: 277-294.

21. Hopkins, D.L. 1983. Gram-negative, xylem limited bacteria in plant disease. Phytopathology 73: 347-350.

22. Igwegbe, E.C.K. and Calavan, E.C. 1970. Occurrence of mycoplasma-like bodies in phloem of stubborn-infected citrus seedlings. Phytopathology. 60: 1525-1526.

23. Ishiie, T., Doi, Y., Yora, K. and Asuyama, H. 1967. Suppressive effects of antibiotics of tetracycline group on symptom development of mulberry dwarf disease. Ann. Phytopathol. Soc. Japan. 33: 267-275.

24. Jinks, D.C., Silvius, J.R. and McElhaney, R.N. 1978. Physiological role and membrane lipid modulation of the membrane-bound (Mg^{2+}, Na^{+})-adenosine triphosphatase activity in *Acholeplasma laidlawii*. J. Bacteriol. 136: 1027-1036.

25. Kahane, I. and Razin, S. 1969. Immunological analysis of mycoplasma membranes. J. Bacteriol. 100: 187-194.

26. Lee, I.M. and Davis, R.E. 1983. Phloem limited prokaryotes in sieve elements isolated by enzyme treatment of diseased plant tissue. Phytopathology. 73: 1540-1543.

27. Lin, C.-P and Chen, T.A. 1985. Monoclonal antibodies against aster yellows agent. Science 227: 1233-1235.

28. Maniloff, J. 1983. Evolution of wall-less prokaryotes. Ann. Rev. Microbiol. 37: 477-499.

29. McCoy, R.E. 1979. Mycoplasma and yellows diseases. pps 229-264. In: The Mycoplasmas. Vol. 3. Plant and Insect Mycoplasmas. R.F. Whitcomb and J.G. Tully, eds. Academic Press, New York.

30. McIvor, R.S. and Kenny, G.E. 1978. Differences in incorporation of nucleic acid bases and nucleosides by various *Mycoplasma* and *Acholeplasma* species. J. Bacteriol. 135: 483-489.

31. Mitchell, A. and Finch, L.R. 1977. Pathways of nucleotide biosynthesis in *Mycoplasma mycoides* sub sp. *mycoides*. J. Bacteriol. 139: 1047-1054.

32. Mouches, C., Candresse, T., McGarrity, G.J. and Bové, J.M. 1983. Analysis of spiroplasma proteins: Contribution to the taxonomy of group IV spiroplasmas and the characterization of spiroplasma protein antigens. Yale J. Biol. Med. 56: 431-437.

33. Mudd, J.B., Ittig, M., Roy, B., Latrille, J. and Bové, J.M. 1977. Composition and enzyme activities of *Spiroplasma citri*. J. Bacteriol. 129: 1250-1256.

34. Munkers, M. and Wachtel, A. 1967. Histochemical localization of phosphatase in

Mycoplasma gallisepticum. J. Bacteriol. 129: 1096-1103.

35. Padhi, S.B., McIntosh, A.H. and Maramorosch, K. 1977. Characterization and identification of spiroplasmas by polyacrylamide gel electrophoresis. Phytopath. Z. 90: 268-272.

36. Pollack, J.D. and Hoffman, P.J. 1982. Properties of the nucleases of *Mollicutes*. J. Bacteriol. 152: 538-541.

37. Pollack, J.D., Razin, S. and Cleverdon, R. C. 1965. Localization of enzymes in *Mycoplasma*. J. Bacteriol. 152: 538-541.

38. Razin, S. 1968. *Mycoplasma* taxonomy studied by electrophoresis of cell proteins. J. Bacteriol. 96: 687-694.

39. Razin, S.R. and Rottem, S. 1967. Identification of *Mycoplasma* and other microorganisms by polyacrylamide-gel electrophoresis of cell proteins. J. Bacteriol. 94: 1807-1810.

40. Razin, S., Hasin, M., Neeman, Z. and Rottem, S. 1973. Isolation, chemical composition and ultrastructural features of the cell membrane of the mycoplasma-like organism *Spiroplasma citri*. J. Bacteriol. 116: 1421-1435.

41. Razin, S., Tully, J.G., Rose, D.L. and Barile, M.F. 1983. DNA cleavage pattern as indicators of genotype heterogeneity among strains of *Acholeplasma* and *Mycoplasma* species. J. Gen. Microbiol. 129: 1935-1944.

42. Rodwell, A.W., Razin, S., Rottem, S. and Argaman, M. 1967. Association of protein and lipid in *Mycoplasma laidlawii* mem-

branes disaggregated by detergents. Arch. Biochem. Biophys. 122: 621-628.

43. Rottem, S. and Razin, S. 1966. Adenosine triphosphatase activity of *Mycoplasma* membranes. J. Bacteriol. 92: 714-722.

44. Saglio, P., Laflèche, D., Bonissol, C. and Bové, J.M. 1971. Isolement, culture et observation au microscope electronique des structure de type mycoplasma associees au stubborn des agrumes et leur comparison avec les structures observees dans le cas de la maladie du greening des agrumes. Physiol. Ve. 9: 569-582.

45. Saglio, P., L'Hospital, M., Laflèche, D., Dupont, G., Bové, J.M., Tully, J.G. and Freundt, E.A. 1973. *Spiroplasma citri* gen. and sp. L.: A mycoplasma-like organism associated with stubborn disease of citrus. Intl. J. Syst. Bacteriol. 23: 191-204.

46. Sinha, R.C. 1974. Purification of mycoplasma-like organisms from china aster plants affected with clover phyllody. Phytopathology. 64: 1156-1158.

47. Sinha, R.C. 1976. Ultrastructure of mycoplasma-like organisms purified from clover phyllody-affected plants. J. Ultrastruct. Res. 54: 183-189.

48. Sinha, R.C. 1979. Purification and serology of mycoplasma-like organisms from aster yellows-infected plants. Can. J. Pl. Pathol. 1: 65-70.

49. Sinha, R.C. 1983. Physical techniques for purification of mycoplasmas from plant tissues. pps. 243-247. In: Methods in Mycoplasmology. Vol. 2. Diagnostic Mycoplasmology, J.G. Tully and S. Razin, eds. Academic Press, New York.

50. Sinha, R.C. 1983. Relative concentration of mycoplasma-like organisms in plants at various times after infection with aster yellows. Can. J. Pl. Pathol. 5: 7-10.

51. Sinha, R.C. 1984. Transmission mechanism of mycoplasma-like organism by leafhopper vectors. pps. 93-109. In: Current Topics in Vector Research. Vol. 2. K.F. Harris, ed. Praeger Scientific, New York.

52. Sinha, R.C. and Benhamou, N. 1983. Detection of mycoplasma-like organism antigens from aster yellow-diseased plants by two serological methods. Phytopathology. 73: 1199-1202.

53. Sinha, R.C. and Chiykowski, L.N. 1984. Purification and serological detection of mycoplasma-like organisms from plants affected by peach eastern X-disease. Can. J. Pl. Pathol. 6: 200-205.

54. Sinha, R.C. and Madhosingh, C. 1980. Proteins of mycoplasma-like organisms purified from clover phyllody and aster-yellows affected plants. Phytopathol. Z. 99: 294-300.

55. Tully, J.G. 1973. Biological and serological characteristics of the acholeplasmas. Ann. N.Y. Acad. Sci. 225: 74-93.

56. Upchurch, S. and Gabridge, M.G. 1981. Role of host cell metabolism in the pathogenesis of *Mycoplasma pneumoniae* infection. Infect. and Immun. 31: 174-181.

57. Whitcomb, R.F. and Black, L.M. 1982. Plant and anthropod mycoplasmas: A Historical Perspective. pps. 40-81. In: Plant and Insect Mycoplasma Techniques, M.J. Daniels and P.G. Markham, eds. Croom Helm, London.

58. Williamson, D.L. 1982. Serological characterization of spiroplasmas and other mycoplasmas. pps. 240-267. In: Plant and Insect Mycoplasma Techniques. M.J. Daniels and P.G. Markham, eds. Croom Helm, London.

59. Williamson, D.L. and Whitcomb, R.F. 1975. Plant mycoplasmas: A cultivable spiroplasma causes corn stunt disease. Science. 188: 1018-1020.

60. Williamson, D.L. and Whitcomb, R.F. 1983. Special serological tests for spiroplasma identification. pps. 249-259. In: Methods in Mycoplasmology. Vol. 2. Diagnostic mycoplasmology. J.G. Tully and S. Razin, Eds. Academic Press, New York.

61. Wolanski, B. and Maramorosch, K. 1970. Negatively stained mycoplasmas: fact or artifact. Virology 42: 319-327.

TABLE 1. Relative Concentration of MLO antigens Determined by ELISA in Partially
Purified Preparations from Different Parts of Aster Plants Infected with Aster Yellows.[a]

Source of Preparation	Absorbance at 405 nm (mean ± SD)	
	Diseased	Healthy
Leaves	1.274 ± 0.074	0.072 ± 0.009
Petioles	0.134 ± 0.018	0.075 ± 0.008
Stems	0.070 ± 0.023	0.063 ± 0.018
Roots	0.712 ± 0.138	0.272 ± 0.006

[a]Preparations from different parts of diseased and healthy plants tested by ELISA using a 1/2 dilution. The dilution is calculated from the fresh weight of the tissue used for obtaining the pellet to the volume of the resuspending buffer. ELISA tests were conducted using enzyme congugate at 1/400, gamma glubulin at 1 µg/ml and absorbance measured after 45 min. with p-nitrophenyl phosphate as substrate.

[b]Values are from two experiments, each with duplicate determinations.

23

TABLE 2. Enzymatic Activities in MLO cells and Fractions

Preparation	Mean Protein concentration (mg/ml) ±SD	Activity (Mean + SD)			
		NADH oxidase[a]		NADPH oxidase[a]	
		Total	Specific	Total	Specific
Whole cells	0.77 ± 0.01	0.20 ± 0.01	2.64 ± 0.21	0.00	0.00
Membrane fraction	0.75 ± 0.11	0.68 ± 0.15	9.07 ± 0.69	0.002 ± 0.00	0.02 ± 0.004
Cytoplasmic fraction	2.86 ± 0.78	0.15 ± 0.02	0.54 ± 0.07	0.018 ± 0.002	0.07 ± 0.007

[a]Decrease in A_{340nm} per minute (total) per minute per milligram of protein (specific)

(From Arora and Sinha, 1985).

TABLE 3. Enzymatic Activities in MLO Cells and Fractions

Preparations	Mean Protein Concentration (mg/ml) ±SD	Activity (Mean ± SD)			
		ATPase[a]		Glucose-6-Phosphate Dehydrogenase[b]	
		Total	Specific	Total	Specific
Whole cells	0.77 ± 0.01	0.13 ± 0.01	1.74 ± 0.10	ND	ND
Membrane fraction	0.75 ± 0.11	0.88 ± 0.14	11.74 ± 0.15	0.02 ± 0.00	0.19 ± 0.02
Cytoplasmic fraction	2.86 ± 0.78	0.27 ± 0.12	0.92 ± 0.16	0.65 ± 0.08	2.32 ± 0.32

[a]Micromoles of inorganic phosphorus released per hour (total) or per hour per milligram of protein (specific).

[b]Increase in $A_{340\ nm}$ per minute (total) or per minute per milligram of protein (specific).

ND = Not Determined

(From Arora and Sinha, 1985).

TABLE 4. Enzymatic Activities in MLO Cells and Fractions

Preparations	Acticity (Mean ± SD)					
	RNase[a]		DNase[a]		P-Nitrophenyl Phosphatase[b]	
	Total	Specific	Total	Specific	Total	Specific
Whole cells	0.95 ± 0.01	12.42 ± 0.16	0.12 ± 0.01	1.56 ± 0.11	0.15 ± 0.01	1.94 ± 0.14
Membrane fraction	1.33 ± 0.32	17.78 ± 2.19	0.05 1 0.02	0.73 ± 0.14	0.18 ± 0.05	2.36 ± 0.39
Cytoplasmic fraction	0.64 ± 0.09	2.28 ± 0.29	0.81 ± 0.35	2.76 ± 0.45	0.06 ± 0.02	0.21 ± 0.02

[a]Increase in A_{260nm} per hour (total) on per hour per milligram of protein (specific).

[b]Units are micromoles of p-nitrophenol released per hour (total) or per hour per milligram of protein (specific).

(From Arora and Sinha, 1985).

TABLE 5. Enzyme Distribution and Genome Sizes of Wall-less Prokaryotes[a]

Organism	ATPase		NADH oxidase		Genome	
	Cytoplasm	Membrane	Cytoplasm	Membrane	G + C Content (mol %)	Size (x mega daltons)
Spiroplasma citri	-	+	+	-	25-57	900-1210
MLO	-	+	-	+	Not known	Not known
Mycoplasma sp.	-	+	-	+	23-41	400-530
Acholeplasma sp.	-	+	-	+	31-34	950-1110

[a]Summarized from Mudd *et al.*, 1977; Maniloff, 1983; Arora and Sinha, 1985.

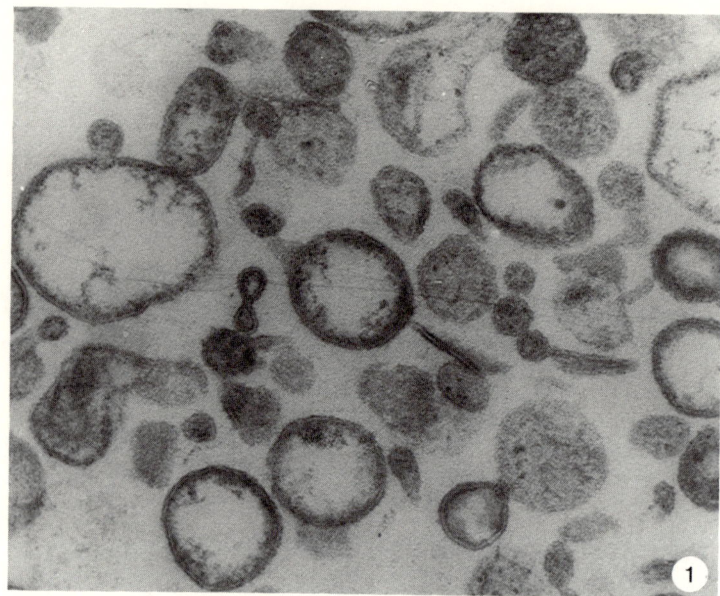

Figure 1. Ultrastructure of purified MLO cells, showing trilaminar unit membrane, ribosome-like granules in the cytoplasm, and DNA-like fibrils.

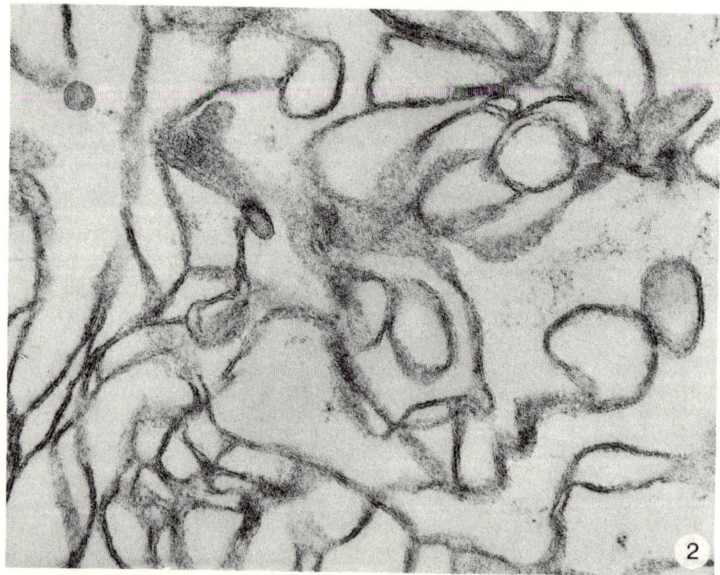

Figure 2. Membranes separated from cytoplasmic fractions through differential centrifugation of purified aster yellows MLOs.

2

PURIFICATION AND PROPERTIES OF

MYCOPLASMA-LIKE ORGANISMS

FROM DISEASED PLANTS

by

R.C. Sinha

Chemistry and Biology Research Institute

Agriculture Canada, Ottawa, Ontario

K1A OC6 Canada, CBRI Contribution

No. 1512

INTRODUCTION

Organisms resembling *Mycoplasma* (Class Mollicutes), first implicated by Japanese workers as the etiological agents of some "yellows" group of plant diseases (6,10), have been shown to be associated with many diseases that inflict damage to a variety of crops of economic importance (11,12). Most of these pleomorphic microbes are non-helical and are commonly referred to as mycoplasma-like organism(s) (MLO) because they have not yet been characterized through cultivation in a cell-free medium. Mycoplasmal etiology of about 100 plant diseases still remains to be demonstrated because Koch's postulates cannot be satisfied. However, a few plant pathogenic mycoplasmas have been cultured *in vitro*, Koch's poltulates fulfilled, and placed in a new genus *Spiroplasma* because they produce helical and motile filaments (7, 14,15). Rapid progress has since been made on immunological, microbiological, biochemical, biophysical, and ultrastructural characterization of these mycoplasmas, especially of *S. citri* -- the etiological agent of citrus stubborn disease (1,3,5,13,29,30).

Failure to obtain continuous cultures of MLOs *in vitro* has impeded research on the properties of these microbes. We took an alternative approach to obtain MLOs in order to determine some of their properties. A procedure was first developed to purify MLOs from diseased plants (16) and then their ultrastructural characteristics (17) as well as some biochemical properties (18,26) were determined. Also antisera against MLOs found associated with aster yellows (AY), clover phyllody (CP), and eastern peach-X (EPX)-infected plants were prepared, and serodiagnostic methods applied for disease identification (24,25). This communication outlines the procedure we have used for purifying MLOs from diseased plants and summarizes our research on the structural, biological, biochemical and immunological characteristics of these non-helical microbes.

PURIFICATION PROCEDURE

The procedure outlined below is based on a method described earlier for purification of the MLO from CP-infected plants (16). Since then, MLOs from AY (19,22), EPX (25), ash yellows and elm yellows (2)-infected plants have also been purified using the same protocol. For detailed description of MLO purification procedure, readers are referred to a recent review (20). All purification procedures should be carried out, whenever possible at 4°C.

1. Cut 50 g of young infected leaves into about 3 cm long pieces and infiltrate with Mg-glycine buffer (0.1M MgCl and 0.3M glycine, pH 8.0) under vacuum for 30 minutes.

This buffer was used throughout the purification procedure.

2. Macerate the infiltrated leaves in a meat grinder, squeeze the juice through two layers of cheese cloth, adjust the volume to 150 ml with the buffer, and centrifuge the juice at 3,000 g for 15 minutes. Pool the supernatants from different tubes, adjust the pH to 8.0 with 1N NaOH, and centrifuge again at 3,000 g for 15 minutes.

3. Divide the supernatant into two equal fractions. To each fraction add 4 g of Darco G60 activated carbon (Atlas Chemical Industries), mix thoroughly for 45-60 seconds on a magnetic stirrer, then add 4 g Celite 545 (Johns Manville Co.) and stir the mixture for another 45-60 seconds.

4. Pour each of the suspension fractions onto separate 1.5 cm thick Celite pads, each on a 12-cm diameter Buchner funnel previously washed and prepared (20). Rinse each flask with 50 ml of the buffer and pour onto the Celite pads. Apply light vacuum until the pads are almost dry and discard the filtrates. Add 200 ml of buffer to each Celite pad, apply gentle vacuum, and combine the eluted filtrates containing the MLO.

5. Centrifuge the filtrate in two bottles at 25,000 g for 30 minutes, discard the supernatants, and soak the pellets in a small volume of buffer overnight. Resuspend the pellets gently with a pasteur pipet, pool the suspensions and adjust the final volume to 5 ml.

6. Layer the 5 ml suspension on top of a Sepharose 2B (Pharmacia Fine Chemicals) column (30 x 2.5 cm) which has been washed and prepared earlier (20). Add buffer to the column after all the MLO suspension has enter-

ed the column, collect 1 ml aliquots, and combine the turbid fractions which usually correspond to aliquot numbers 36-56.

7. Centrifuge the turbid suspension at 25,000 g for 30 minutes, discard the supernatant, and soak the pellet in 3 ml buffer for three hours.

8. Layer the 3 ml suspension on a sucrose density gradient (6.2) ml each of 5,10,15 and 20% sucrose, w/w in buffer), and centrifuge at 10,170 g for 45 minutes. Remove the diffuse whitish zone, visible between 4.0 and 5.0 cm from the bottom of the tube. The total volume of the zone removed is about 10 ml.

9. Add an equal volume of buffer to the suspension and dialyze against the buffer overnight. Centrifuge the dialyzed suspension at 25,000 g for 30 minutes. The light gold-colored pellet contains purified MLO (Fig. 1).

The yield of purified MLOs is usually 1-2 mg (freeze-dried weight) per 50 g of leaves. This estimate is based on yields from aster (*Callistephus chinensis Nees*) leaves infected with either AY or CP, and celery (*Apium graveolans L.*) leaves infected with EPX. Various host plants of a disease may contain different MLO concentration and, therefore, the yield of purified MLOs from different hosts can differ drastically. For example, the concentration of AY-MLO per gram of leaf tissue was estimated, through serological means, to be 80 g for aster and celery; 40 g for periwinkle (*Catharanthus roseus L.*), oat (*Avena sativa L.* 'Random'), and wheat (*Triticum durum Desf.* 'Ramsey'); and 20 g for ladino clover (*Trifolium repens L.*) (24). The time of harvesting the infected leaves also can affect MLO concentration within a plant species.

The relative concentration of AY-MLO in aster plants was estimated to increase 16-fold between the 2nd and 8th week after infection, remained at the same level for another two weeks, and then declined to a low level by the 14th week (21). It has been estimated that about 75% of AY-MLO is lost during the purification procedure, as determined by the serological titer of the preparations obtained after steps 5 and 9 (19). Therefore, the success of MLO purification depends on the selection of a suitable host plant as well as on the time of harvesting the infected tissues. Environmental conditions, such as temperature, light intensity, and plant age at the time of inoculation also may effect the eventual MLO concentration in plants. The MLO concentration in various parts of an infected plant may also vary, and although infected leaves have been used for purification of these organisms, roots may serve as a better source. Such studies, however, have not yet been carried out with MLO-infected plants.

CHARACTERIZATION OF MYCOPLASMA-LIKE ORGANISMS

A. Morphological and ultrastructural

Electron microscopic examination of purified CP-MLO suspensions stained with phosphotungstic acid (PTA) showed numerous pleomorphic bodies (16). Many spherical forms of the MLO-cells showed small vesicles, giving the appearance of budding structures. The size of the spherical or oblong cells ranged from 175-400 nm in diameter. Filamentous forms of the MLO were sometimes as long as 1,700 nm. Scanning electron microscopy of freeze fractured root and petiole tissue of CP and AY-infected plants (8) demonstrated that such morphological features of the MLO are not artifacts of PTA-staining (31). Polymorphic bodies, such as, budding, spherical, dumbbell-shaped cells, and filamentous branching forms were observed in the sieve elements of infected plants only. The size and polymorphic forms of MLOs found in AY and CP-infected plants were very similar.

The ultrastructure of CP-MLO cells has been studied by examining thin sections of purified pellets (17). Small to large, spherical to oblong, and filamentous forms of the MLO, similar in morphology as described above, were observed. The MLO-cells showed a well defined trilaminar unit membrane with ribosome-like granules and, in a few cells, DNA-like fibrils in the cytoplasm. The symmetrical trilaminar unit membrane was composed of two electron-dense layers with a less dense intermediate layer. The average thickness of the membrane was about 7nm. Some fragmented membranes also were observed which were about 12 nm thick, presumably due to swelling and penetration of the stain from the broken ends. Filaments of varying lengths protruding from large spherical cells were often found. Chains of cells connected by small filaments, and dumbbell-shaped cells were observed quite frequently. It is hypothesized that MLOs reproduce by budding and/or binary fission (9), as do other Mollicutes. The various morphotypes of CP-MLO seen in the purified preparations were basically similar to those found *in situ* in diseased plants (27). It should be noted that the morphological and ultrastructural characteristics of MLOs found associated with different diseases are very similar.

Biological

Almost all known MLO-diseases are transmitted to healthy plants by means of leafhopper (Cicadellids) vectors in a persistant manner (23). Infectivity of CP-MLO has been tested by injecting the purified preparations into healthy vector leafhoppers *Macrosteles fascifrons* Stal and then testing the insects for their ability to transmit the disease to aster seedlings. A total of 500 injected leafhoppers were tested in groups of 10 per plant but only

5 plants became infected (16). The low infectivity of purified CP-MLO is not surprising because extracts from plants affected with "yellows"-type diseases are known to lose most of their biological activity within a day, and purification of MLOs involves more than 2 days. Similar results were obtained when infectivity of purified AY-MLO was tested. Only 3% of the vector leafhoppers *M. fascifrons* injected with purified preparations transmitted AY to aster plants (19). However, our recent results have shown that when partially purified preparations of AY-MLO, obtained after step number 5 of the purification procedure, were tested for their infectivity, about 30% of the injected leafhoppers transmitted AY to aster seedlings (R.C.Sinha and N. Benhamou, unpublished data). Hence, it appears that the infectivity of AY-MLO is lost gradually during the purification procedure. In contrast, EPX-MLO loses its infectivity more rapidly during the purification procedure, as none of the vector leafhoppers *Paraphlepsius irroratus* (Say) that were injected with preparations obtained after various stages of purification transmitted the disease to celery plants (25).

Biochemical

Proteins

The protein snythesis of an organism is genetically directed and its analysis provides useful, although indirect, information about the characteristics of its genome. The number and pattern of proteins of animal mycoplasmas, obtained through sodium dodecyl sulphate (SDS)-polyacrylamide gel electrophoresis (PAGE), have been widely used as one of the criteria for species differentiation (28). So far, however, there is only one report dealing with the electrophoretic patterns of MLO-proteins (26). SDS-PAGE (in tube gels) purified CP-MLO and AY-MLO produced 11 and 13 bands, respectively. Molecular weights of different protein bands ranged from about 33,000 to 170,000 daltons for CP-MLO and 38,000 to 170,000 daltons for AY-MLO. Although PAGE of AY-MLO showed two more protein bands than that of CP-MLO, this criteria alone does not appear to be significant enough to clearly differentiate the two MLOs. Other studies showed that CP-MLO and AY-MLO contain 4.6% and 2.5% amino sugars, respectively.

The number of protein bands visible in PAGE most likely represent a small number of major proteins, and each band may comprise several different proteins with similar electrophoretic migration patterns. The use of polyacralamide slab gels or application of two-dimensional PAGE would probably reveal many more MLO proteins with different isoelectric properties, as has been shown for *S. citri* (1).

The amino acid composition of CP-MLO and AY-MLO has also been determined (26). The concentration of various amino acids in both MLOs was very similar, the concentration of glycine being the highest followed closely by alanine (Table 1). It should be noted that the buffer used throughout the purification procedure of MLOs contained glycine. Preliminary experiments have shown that when infected leaves were infiltrated with the buffer containing C^{14}-glycine, traces of radioactivity were detected in the purified MLO-pellets. It is likely, therefore, that the levels of glycine given in Table 1 are higher than naturally present in the organism due to absorption of this amino acid by the MLO-cells.

Our recent work on the isolation and characterization of enzymes of AY-MLO has shown the presence of NADH oxidase, ATPase, ribonuclease, deoxyribonuclease and p-nitrophenyl phosphatase (Y.K. Arora and R.C. Sinha, unpublished data). Further work is in progress to determine the localication of these enzymes in the MLO-cells.

Lipids

The lipid and fatty acid composition of MLOs purified from CP and AY-infected plants have been reported (18). Lipid compromised about 30% and 40% of the total dry weight of CP-MLO and AY-MLO, respectively. The fatty acid composition of AY-MLO and CP-MLO was very similar with the exception that the proportion of palmitate was a little higher in AY-MLO (Table 2). Further analysis of various lipid classes of AY-MLO showed that each had a characteristic fatty acid composition (Table 3). The phospholipids of AY-MLO constituted the major class of acyl lipids accounting for about 57% of the total fatty acids. Chromatographic analysis of the phospholipids of AY-MLO revealed the following types in decreasing concentration: phosphatidic acid, phosphatidyl inositol, N-methyl phosphatidyl ethanolamine, diphosphatidyl glycerol, phosphatidyl ethanolamine, phosphatidyl glycerol, phosphatidyl choline, lyso-phosphatidyl choline, and lyso-phosphatidyl etanolamine.

The levels of lipids present in CP-MLO and AY-MLO are too high to be attributed to host plant contamination and, therefore, it is reasonable to consider the lipids to be an intrinsic part of these organisms. Whether the MLOs are capable of synthesizing their own lipids or utilize the ones present in host plants is not known. Such information will only be obtained when the MLOs are cultured *in vitro*.

DNA

So far our repeated attempts to isolate DNA from purified AY-MLO have been unsuccessful. Examination of ultrathin sections of the purified MLO-pellet suggests that many of the large spherical cells lose their cytoplasmic contents including DNA-strands during the purification procedure, presumably due to the osmotic fragility of the organism (19). This might also be the reason for the loss in biological activity of the purified MLOs.

Immunological

Applications of immulological techniques can be very helpful in establishing antigenic differences among individual mycoplasmas, but they have not yet been fully utilized for MLOs. Only a few reports have appeared dealing with detection of MLO-antigens in extracts of diseased plants. Antisera against AY-MLO (10), EX-MLO (25), and CP-MLO (R.C. Sinha, unpublished results) have been prepared using purified preparations. AY-MLO antigens have been detected by agar gel double diffusion, enzyme-linked immunosorbent assay (ELISA), and immunosorbent electron microscopy (IEM) tests in purified as well as in partially purified preparations from diseased plants. Gel diffusion tests have revealed that AY-MLO contains at least two cytoplasmic and two membrane antigens (19). The minimum concentration of MLO required for detecting the slow moving cytoplasmic antigen in the gel was about 20 mg/ml, and for the fast moving one 2.5 mg/ml (21). In ELISA tests, preparations containing concentrations of AY-MLO as low as 4 g/ml gave a positive reaction, whereas in IEM tests, the minimum concentration needed to trap the MLO cells on the antiserum-coated grids was about 12.5 g/ml (24). Recently, EXP-MLO antigens have been detected by ELISA and IEM in preparations from infected celery plants (25).
The specificity of ELISA and IEM tests for detection of MLO-antigens has been determined. The preparations of MLOs from EPX and clover yellow edge (CYE)-diseased plants gave negative reactions when tested against AY-MLO antiserum (24). The preparations of MLOs from AY, CP and CYE-infected plants gave negative reactions when tested against EPX-MLO antiserum (25). However, preparations of the MLOs from AY and

CP-diseased aster plants gave identical serological reactions when tested against AY-MLO antiserum (24). These results suggest that although it is possible to distinguish serologically EPX disease from AY, CP and CYE diseases, AY and CP diseases cannot be differentiated from each other by ELISA or IEM. To differentiate these two diseases, it will still be necessary to depend on symptomatology, host range and vector-MLO relationships.

The various antisera used in the above studies were prepared using purified preparations of MLOs from diseased plants. Purifying MLOs from plant tissues is time consuming, but once the antisera are produced, time can be saved by using partially purified preparations as test antigens which can be obtained within a few hours. Also, such antigenic preparations can be obtained from only 5 g of infected leaves through slight modification in the early steps of the purification procedure (24). The antisera against MLOs can also be prepared using partially purified preparations as has been demonstrated for the European isolate of CP disease (4). But the antiserum thus obtained had to be cross absorbed with preparations from healthy leaves to remove non-specific antibodies. The absorbed antiserum was then used to detect by indirect ELISA (a more sensitive method) the MLO-associated antigens in the extracts of infected plants (4). Standard, direct ELISA method could not be used due to relatively low specific antiserum titer.

SUMMARY AND CONCLUDING REMARKS

Mycoplasma-like organism(s) (MLO) from plants infected with clover phyllody (CP), aster yellows (AY), eastern peach-X (EPX), ash yellows and elm yellows diseases have been purified. The purification procedure involves the following steps: 1) infiltration of infected leaves with a buffer containing $MgCl$ and glycine; 2) treatment of extract sap with charcoal powder and Celite, and then passing the mixture through a Celite pad; 3) elution of MLO-cells absorbed on the pad, and high speed centrifugation of the opalescent fractions; 4) fractionation of the concentrated suspension through a Sepharose column; 5) sucrose gradient and high speed centrifugation to obtain a pellet of purified MLO. However, it is difficult to assess the degree of purity of MLOs thus obtained. In control experiments, when healthy aster leaves were subjected to the same purification procedure as the leaves infected with AY or CP, no pellets were produced, but healthy celery leaves, a host used for EXP-MLO purification, produced a very minute pellet. It appears that the degree of purity of a MLO preparation may depend on the host plant used for the purification purposes. The yields of purified MLOs have been estimated to range between 2-4 mg (freeze-dried weight) per 100 g of infected leaves. Higher yields can be obtained if appropriate host plants at an infection age with maximum MLO concentration are used as has been demonstrated for AY-MLO. However, about 75% of the MLO contained in AY-infected aster leaves is lost during the purification procedure. Similar losses occur when MLOs from CP and EPX-infected leaves are purified. Some improvements in the purification procedure that might reduce such losses of the MLOs are needed.

The purified preparations from CP-infected plants show spherical to oblong, and filamentous organisms, of different sizes similar to those found *in situ* in diseased plants. All morphotypes of the MLO-cells are bound by a trilaminar unit membrane, and contain ribosome-like granules and, in a few cells, DNA-like fibrils in the cytoplasm. The average thickness of the membrane is about 7nm. Almost all biological activity of the purified MLOs is lost during the purification procedure. Attempts to isolate DNA from purified AY-MLO have been unsuccessful. Nevertheless, purified preparations of MLOs can be used for determining some other biochemical properties and for preparing the antisera against these microbes. Polyacrylamine gel electrophoresis patterns of purified CL-MLO and AY-MLO as well as their amino acid composition are very similar. Both MLOs are also found to contain a low concentration of amino sugars. Lipid from CP-MLO and AY-MLO constitute about 30% and 40% of the total dry weight of the organisms, respectively. The total fatty acid compositions of the two MLOs are very similar, and main lipid classes found in both are phospholipids, glycolipids, sterols, glycerides, and free fatty acids. Antisera gainst AY-MLO, CP-MLO and EPX-MLO have been prepared using purified preparations of the microbes from infected plants. The MLO-antigens were detected, in purified as well as in partially purified preparations from diseased plants, by immunodiffusion, enzyme-linked immunosorbent assay (ELISA) and immunosorbent electron microscopy (IEM) tests. Heterologus serological reactions showed that it is possible to differentiate serologically the EPX disease from AY, CP and clover yellow edge diseases and vice versa. It appears that ELISA and IEM tests could be useful in the rapid diagnosis of these diseases. However, AY and CP-diseases could not be differentiated from each other because the MLOs from plants affected with the two diseases are closely related serologically.

The results discussed in this review suggest that certain biochemical and immunological characteristics of MLOs can be determined by using purified preparations of the organisms from infected plants. The purification of MLOs from plants is time consuming and to obtain these microbes in sufficient quantity in order to study their properties is indeed a slow process. However, this is the only method available at the present time since attempts in various laboratories to obtain continuous cultures of these organisms *in vitro* have failed so far.

REFERENCES

1. Bové, J.M., and Saillard, C. 1979. Cell biology of spiroplasmas. pps. 83-153. In: The Mycoplasmas. Vol. 3. Plant and Insect Mycoplasmas. R.F. Whitcomb and J.G. Tully, eds. Academic Press, New York.

2. Castello, J.R., Shiel, P., Austin, J.A., Jones, F., Craft, C., and Delgado, G. 1984. Partial purification of ash yellows and elm yellows mycoplasma-like organisms from infected symptomatic periwinkle plants. Phytopathology 74: 804, Abst.

3. Chen, T.A., and Davis, R.E. 1979. Cultivation of spiroplasmas. pps. 65-82. In: The Mycoplasmas. Vol. 3. Plant and Insect Mycoplasmas. R.F. Whitcomb and J.G. Tully, eds. Academic Press, New York.

4. Clark, M.F., Barbara, D.J. and Davis, D.L. 1983. Production and characteristics of antisera to *Spiroplasma citri* and clover phyllody-associated antigens derived from plants. Ann. Appl. Biol. 103: 251-259.

5. Daniels, M.J. 1979. Mechanisms of spiroplasma pathogenicity. Pages 209-277. In: The Mycoplasmas. Vol. 3. Plant and Insect Mycoplasmas. R.F. Whitcomb and J.G. Tully, eds. Academic Press, New York.

6. Doi, Y., Terananka, M., Yora, K. and Asuyama, H. 1967. Mycoplasma- or PLT-like microorganisms found in the phloem elements of plants infected with mulberry dwarf, potato witches' broom, aster yellows or Paulownia witches' broom. Ann. Phytopathol. Soc. Japan 33: 259-266.

7. Fudl-Allah, A., Calavan, E.C. and Igwegbe, E.C.K. 1972. Culture of a mycoplasma-like organism associated with stubborn disease of citrus. Phytopathology 62: 729-731.

8. Haggis, G.H., and Sinha, R.C. 1978. Scanning electron microscopy of mycoplasma-like organisms after freeze fracture of plant tissues affected with clover phyllody and aster yellows. Phytopathology 68: 677-680.

9. Hirumi, H., and Maramorosch, K. 1973. Mycoplasma and mycoplasma-like agents of human, animal and plant diseases. Ann. N.Y. Acad. Sci. 225: 201-222.

10. Ishiie, T., Doi, Y., Yora, K. and Asuyama, H. 1967. Suppressive effects of antibiotics of tetracycline group on symptom development of mulberry dwarf disease. Ann. Phytopathol. Soc. Japan 33: 267-275.

11. Maramorosch, K., Granados, R.R. and Hirumi, H. 1970. Mycoplasma disease of plants and insects. Adv. Virus. Res. 16: 135-193.

12. McCoy, R.E. 1979. Mycoplasmas and yellows diseases. Pages 229-259. In: The Mycoplasmas. Vol. 3. Plant and Insect Mycoplasmas. R.F. Whitcomb and J.G. Tully, eds. Academic Press, New York.

13. Padhi, S.B., McIntosh, A.H. and Maramorosch, K. 1977. Characterization and identification of spiroplasmas by polyacrylamide gel electrophoresis. Phytopath Z 90: 268-272.

14. Saglio, P., L'Hospital, M., Laflèche, D., Dupont, G., Bové, J.M., Tully, G. and Freundt, E.A. 1973. *Spiroplasma citri* gen. and sp. n.: a mycoplasma-like organism associated with "stubborn" disease of citrus. Int. J. Syst. Bacteriol. 23: 191-204.

15. Saglio, P., Laflèuche, D., Bonissol, C. and
    Bové, J.M. 1971. Isolement, culture et ob-
    servation au microscope electronique des
    structure de type mycoplasma associees a la
    maladie du Stubborn des agrumes et leur com-
    paraison avec les structures observees dans
    le cas de la maladie du Greening des agru-
    mes. Physiol. Veg. 9: 569-582.

16. Sinha, R.C. 1974. Purification of myco-
    plasma-like organisms from China aster
    plants affected with clover phyllody.
    Phytopathology 64: 1156-1158.

17. Sinha, R.C. 1976. Ultrastructure of myco-
    plasma-like organisms purified from clover
    phyllody-affected plants. J. Ultrastruct.
    Res. 54: 183-189.

18. Sinha, R.C. 1979. Lipid composition of
    mycoplasma-like organisms purified from
    clover phyllody and aster yellows-affected
    plants. Phytopath. Z. 96: 132-139.

19. Sinha, R.C. 1979. Purification and sero-
    logy of mycoplasma-like organisms from
    aster yellows-infected plants. Can. J.
    Plant Pathol. 1: 65-70.

20. Sinha, R.C. 1983. Physical techniques for
    purification of mycoplasmas from plant tis-
    sues. pps. 243-247. In: Methods in Myco-
    plasmology. Vol. 2. Diagnostic mycoplas-
    mology. J.G. Tully and S. Razin, eds.
    Academic Press, New York.

21. Sinha, R.C. 1983. Relative concentration
    of mycoplasma-like organisms in plants at
    various times after infection with aster
    yellows. Can. J. Plant Pathol. 5: 7-10.

22. Sinha, R.C. 1983. The aster yellows
    controversy. Current status. The Yale J.
    Biol. Med. 56: 737-743..

23. Sinha, R.C. 1984. Transmission mechanisms of mycoplasma-like organisms by leafhopper vectors. pps. 93-109. In: Current Topics in Vector Research, Vol. 2. K.F. Harris, ed. Praeger Scientific, New York.

24. Sinha, R.C. and Benhamou, N. 1983. Detection of mycoplasma-like organism antigens from aster yellow-diseased plants by two serological methods. Phytopathology 73: 1199-1202.

25. Sinha, R.C. and Chiykowski, L.N. 1984. Purification and serological detection of mycoplasma-like organisms from plants affected by peach eastern X-disease. Can. J. Plant Pathol. 6: 200-205.

26. Sinha, R.C., and Madhosingh, C. 1980. Proteins of mycoplasma-like organisms purified from clover phyllody and aster yellows-affected plants. Phytopath. Z. 99: 294-300.

27. Sinha, R.C. and Paliwal, Y.C. 1969. Association, development and growth cycle of mycoplasma-like organisms in plants affected with clover phyllody. Virology 39: 759-767.

28. Tully, J.G. 1973. Biological and serological characteristics of the acholeplasmas. Ann. N.Y. Acad. Sci. 225: 74-93.

29. Williamson, D.L. 1982. Serological characterization of spiroplasmas and other mycoplasmas. pps. 240-267. In: Plant and Insect Mycoplasma Techniques. M.J. Daniels and P.G. Markhan, eds. John Wiley and Sons, New York.

30. Williamson, D.L. and Whitcomb, R.F. 1983. Special serological tests for spiroplasma identification. pps. 249-259. In: Methods in Mycoplasmology. Vol. 2. Diag-

nostic mycoplasmology. J.G. Tully and S. Razin, eds., Academic Press, New York.

31. Wolanski, B. and Maramorosch, K. 1970. Negatively stained mycoplasmas: fact or artifact. Virology 42: 319-327.

TABLE 1.   Amino acid composition of mycoplasma-like
            organisms (MLO) from clover phyllody (CP)
            and aster yellows (AY) - infected plants

| | Relative molar ratio** | |
Amino acid*	CP-MLO	AY-MLO
Alanine	10.15	8.80
Arginine	5.46	4.53
Aspartic acid	3.00	2.53
Glutamic acid	2.77	2.53
Glycine	11.77	10.73
Histidine	1.46	1.20
Isoleucine	6.46	4.90
Leucine	6.54	5.70
Lysine	3.85	3.30
Methionine	1.00	1.00
Phenylalanine	2.92	2.53
Proline	2.08	1.87
Serine	5.70	4.93
Therenine	4.92	4.33
Tyrosine	1.46	1.27
Valine	7.00	5.90

*Cysteine and tryptophan not determined.

**Molar ratios are relative to methionine as 1.

From Sinha and Madhosingh (26).

TABLE 2. Fatty acid composition of mycoplasma-like organisms (MLO)
from clover phyllody (CP) and aster yellows (AY) - infected plants

Lipid Source	Total fatty acid composition*				
	16:0	18:0	18:1	18:2	18:3
CP-MLO	32.5	2.8	3.5	42.9	18.3
AY-MLO	40.2	2.2	2.5	40.1	15.0

*Results expressed as percentage by weight. Palmitate (16:0), stearate (18:0), oleate (18:1, linoleate (18:2), and linolenate (18:3).

From Sinha (18).

TABLE 3. Fatty acid composition of various lipid classes of mycoplasma-like organism purified from aster yellows -infected plants

Lipid class	Percentage by weight	Fatty acid composition*				
		16:0	18:0	18:1	18:2	18:3
Triglyceride	1.9	44.7	4.6	14.0	32.9	3.8
Free fatty acid	17.6	53.7	6.3	5.1	25.0	9.9
Monoglyceride/ Diglyceride	7.7	46.8	2.8	5.1	39.5	5.8
Monogalactosyl Diglyceride	6.8	39.4	1.5	3.8	33.3	22.0
Digalactosyl Diglyceride	8.6	43.7	2.2	3.5	21.4	29.2
Phospholipid	57.4	37.9	1.9	2.8	41.5	15.9

*Results expressed as percentage by weight. Palmitate (16:0), stearate (18:0), oleate (18:1), linoleate (18:2), and linolenate (18:3).

From Sinha (18).

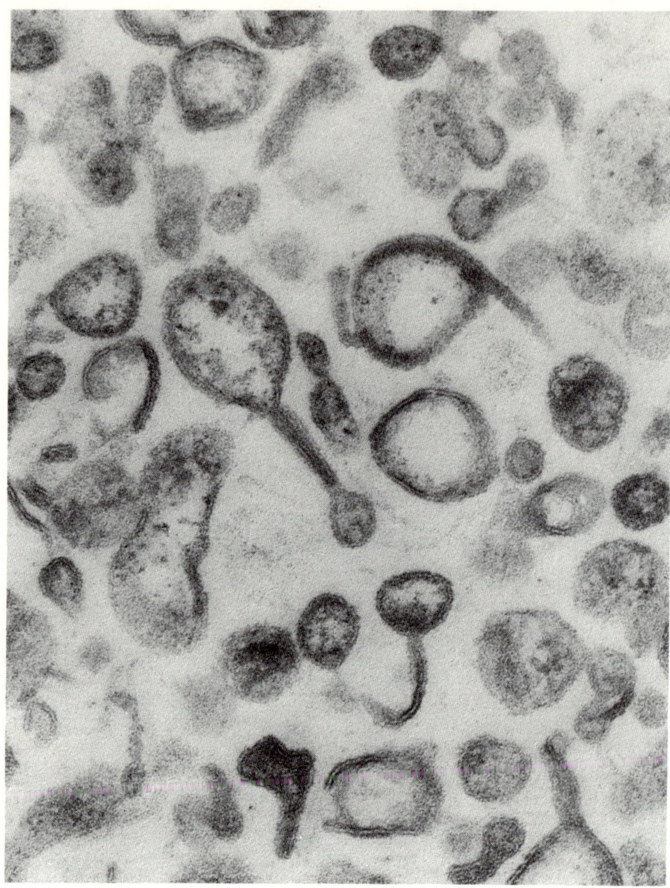

Figure 1. Section of the purified MLO pellet from clover phyllody-affected plants showing pleomorphic bodies. Magnification X 132,000.

3

FLUORESCENCE MICROSCOPY OF YELLOWS DISEASES ASSOCIATED WITH PLANT MYCOPLASMA-LIKE ORGANISMS

by

C. Hiruki

Department of Plant Science

University of Alberta

Edmonton, Alberta, Canada

T6G 2P5

INTRODUCTION

In diagnosis of yellows diseases associated with plant mycoplasma-like organisms (MLOs), a rapid histochemical test of suspected specimens by means of fluorescence microscopy is useful not only for newly recorded diseases from which culture specimens of Mollicutes are not yet available, but also for practical application such as mass-screening of laboratory or field specimens for selection and establishment of MLO-disease-free plants.

The advantages of using fluorescence microscopy in histochemistry are, firstly, its ability to detect target substances in very low concentration, and secondly, its applicability in wide-ranging substances which may be detected, either directly or after certain chemical treatments.

    This chapter will deal with fluorochromy, induced fluorescence, immunofluorescence and combined use of fluorescence and electron microscopy with special reference to their application to yellows diseases associated with MLOs.

PRINCIPLES

Fluorescence Microscope

    A fluorescence microscope is, in addition to a usual optical system, equipped with a light source such as a mercury arc lamp (200, 100 or 50 watts) and a pair of complementary filters. A primary (excitation) filter is inserted in the light pathway to supply mono-chromatic illumination for the excitation of fluorescence, and the other (barrier) filter in the observation pathway to protect the observer's eye from the excitation light, while transmitting the light of longer wavelengths emitted by the fluorescent specimen.

Types of Fluorescence

    Fluorescence is basically a physical phenomenon. A substance (known as a fluorophore) becomes excited as a result of absorption of light and its surplus energy is emitted as fluorescence, light of a longer wavelength.
    Fluorescence in plant tissues infected by MLOs can be found as different types such as autofluorescence, induced fluorescence, fluorescence of dyes and immunofluorescence (Table 1). Unstained plant tissues contain substances which exhibit some degree of fluorescence that is called autofluorescence. In healthy tissues, xylem elements have a strong autofluorescence. Walls of sclerenchyma and collenchyma cells are also known to fluoresce without staining. Additional autofluorescence may be seen in depository materials in the cytoplasm or on the walls of older cells of certain plant species,

or in those of cells under stress including infections by viruses (Namba *et al.*, 1981), MLO (Marwitz and Petzold, 1980; Namba *et al.*, 1981), bacteria or fungi. Accordingly, the use of autofluorescence for disease diagnosis is relatively limited due to its low specificity.

Fluorescent stains used as dyes are referred to as fluorochromes to distinguish them from those used in ordinary diachromes. Many dyes can be used both as diachromes and as fluorochromes. Selected examples are Aniline Blue, Congo Red, Eosin, Fast Green, and basic Fuchsin (pararosaniline). With these dyes, fluorescence microscopy is an alternative way of observing the result of staining. Such a fluorescence may be used for MLO diagnosis, provided that it is known to occur in association with MLO infection. Relatively high sensitivity in MLO diagnosis can be achieved by using certain dyes that are known to combine specifically with MLO-DNA or RNA or both in infected plants. For example, 4', 6-diamidino-2-phenylindole.2HCl (DAPI) is known to bind preferentially to AT-rich DNA forming a complex which exhibits a strong fluorescence. DAPI has been used as a specific fluorescent probe at both molecular and individual cell levels. The Feulgen test is more cumbersome than DAPI but is also useful in MLO diagnosis.

Immunofluorescence test is a serological method that has been used in animal mycoplasmology (Barile and Grabowski, 1983; Gardella *et al.*, 1983), plant spiroplasmology (Williamson, 1982) and especially in virology. It is an advanced technique of histochemistry, which provides a means of observing an antigen-antibody reaction by chemically linking a fluorescent dye, such as fluorescein isothiocyanate (FITC) to specific antibody molecules. Such labeled antibodies retain the ability to react specifically with their respective antigens and when viewed in a fluorescence microscope, the reaction site is usually detected by its fluorescence. Maximum absorption by FITC-IgG occurs at approximately 495 nm and emission occurs at 525 nm. There are two ways of performing the immunofluorescence test, the 'direct' and the 'indirect' methods. In the direct method the antibody produced against the protein to be detected is conjugated with a fluorochrome, such as FITC, and the staining is done simply by exposing test specimens to a solution of the labeled antibody. The indirect method is based on the fact that antibodies (immunoglobulins) are themselves capable of serving as immunogens; thus, for example, if a goat is injected with rabbit immunoglobulins, goat antibody against rabbit immmunoglobulins can be produced and conjugated with a fluorochrome. The staining is then performed in two stages. First, the test specimens are exposed to a solution of unlabeled rabbit antibody produced against the protein to be detected, and unbound antibody is removed by washing. In the second stage, the sites where the rabbit antibodies have attached to the protein are labeled by specific binding with goat fluorescent antibody prepared against rabbit immunoglobulins. The specificity of antigen-antibody reactions, coupled with the relatively short time required to prepare and observe the specimen, makes the immunofluorescence procedure an ideal diagnostic tool for plant diseases associated with MLOs.

MATERIALS AND METHODS

MLO and MLO Infected Plants

All MLO isolates used in this investigation were maintained on *Catharanthus roseus*. All plants were maintained in autoclaved soil (sand:peat:loam/1:1:3 (v/v/v)) in 10 cm pots in insect-proof greenhouses at 20 -25°C. Transmission of MLOs from infected plants to *C. roseus* was done by dodder (*Cuscuta subinclusa*) or by grafting.

Histochemistry

Samples were taken immediately before examination from petioles of young leaves or from young internode regions of diseased and healthy plants maintained in the greenhouses. Free-hand sections, about 20 to 30 µm thick, were made with a razor blade within 5 to 10 min. after excising shoots. For Aniline Blue (AB) staining, the sections were fixed immediately by boiling in tap water for 2-3 min. and then stained in 0.01% AB in 1/15 M K_2HPO_4, pH 8.0 for 20 min. (Hiruki and Shukla, 1973; Hiruki and Dijkstra, 1973; Hiruki *et al.*, 1974). For DAPI staining, the sections were fixed in 5% glutaraldehyde in 0.1 M phosphate buffer pH 7.0 for 2 hr. After rinsing, they were stained in a solution of DAPI (1µg/ml) for 20 min. (Hiruki, 1981. 1982; Hiruki and da Rocha,1986; da Rocha, 1985).

Immunofluorescence (IF)

Preparation of immunogen. The 'membrane fraction' (MF) from healthy *C. roseus* and from *C. roseus* infected with four isolates of MLOs were prepared for use as immunogens as described by Clark *et al.*, (1983). In this investigation, the following MLO isolates were used: clover proliferation (CP, Chen and Hiruki, 1975, 1978a), potato witches' broom (PWB, Hiruki and Chen, 1984), an Alberta isolate of aster yellows (AY27, Hiruki and Chen, 1984) and a New York isolate of Aster Yellows (EAY, obtained from T.A. Chen, Rutgers University, U.S.A.). All procedures were performed at 0-4 C.

Leaves of *C. roseus* were washed for several hours with running tap water to remove soil and other gross contaminants. The tissue (20 g) was shaken dry, passed through a meat grinder, extracted into ice-cold buffer (1 g/5 ml buffer) and squeezed through cheesecloth. The buffer used was 0.3 M glycine-sodium hydroxide, pH 8.0, with 0.02 M magnesium chloride.

The extract was clarified by centrifugation at 2,000 g for 20 min. The supernatant liquid was then centrifuged at 39,000 g for 45 min. and the large green pellet was resuspended in the original volume of buffer overnight. The suspension was clarified by passing the suspension through Whatman No. 2 filter paper in a Buchner funnel. It was subjected to a further cycle of differential centrifugation and the resulting pellet was resuspended in 1/20 volume of buffer. The suspended material was stored in 0.7 ml aliquots at -18 C and was used as immunogen and antigen in serological tests.

Preparation of antiserum. Antisera to MF preparations of CP, PWB, EAY and AY27 MLO-infected and healthy *C. roseus* plants were produced in rabbits. Rabbits were injected at 2-week intervals with a 1:1 (v/v) mixture of the immunogen and either Freund's complete (first injection) or incomplete adjuvant (subsequent injections). One milliliter of the homogenized mixture was administered at multiple sites by subcutaneous, intra-dermal and intramuscular injections. Animals were bled (10 ml) at 2-week intervals, beginning six weeks after the primary injection. The rabbit was exsanguinated when titers of antisera, determined by double immunodiffusion (Kenny, 1983), reached 1/4 - 1/16. As control, antigen from healthy plants replaced the test antigen in the center well. After 2-3 days at room temperature, plates were observed for precipitin lines.

To remove the antibodies to healthy plant proteins, the antisera were cross absorbed with acetone-extracted powder of healthy leaf tissue of *C. roseus*. The immunoglobulin fraction of the antiserum was precipitated with ammonium sulphate (Clark and Adams, 1977).

The concentrations of the purified immunoglobulins were determined by measuring the absorbance at 280 nm (A_{280}) and using an extinction coefficient of 1.35 $(mg/ml)^{-1} cm^{-1}$ (Johnstone and Thorpe, 1982. The immunoglobulins were then diluted with 0.005 M phosphate buffered saline (PBS), pH 7.5, to 10 mg/ml and stored in 1 ml aliquots at -18°C.

Indirect IF. The IF staining procedure described by Kawamura (1977) was modified to provide a simple and rapid method which is suitable for use with thick plant sections. Freehand sections from the internode region, 30- 60 µ m were cut and placed in a 1.5 ml microcentrifuge tube. Two hundred microliters of 95% ethanol were added and the sections were fixed for 5 min. The sections were then washed three times with 200 µ l PBS. The immunoglobulin, 50 l (1 mg/ml) was placed in the tube for 15 min. at room temperature. The sections were washed as above. Labeled antirabbit goat immunoglobulins (50µl [100 µg/ml]) conjugated with FITC (product No. F-0382, Lot. No. 34F-8865, Sigma Chemical Co., St. Louis, MO 63178, USA) were added to the sections in the tube and incubated for 15 min. at room temperature. The sections were washed as above. The final washing was replaced by 200 µl of the mounting medium, 1:1 PBS and glycerol. The sections were mounted on glass slides with coverslips and viewed under a fluorescence microscope.

Immunoglobulins obtained against the membrane fraction of PWB, CP, EAY and AY27 diseased plants were used to test sections from plants with the four diseases as well as sections from healthy control plants.

Fluorescence Microscopy

The sections were examined in a Leitz Orthoplan fluorescence microscope with a mercury arc lamp (HBO200) and with a barrier filter K430 and exciter filters BG3 and UG1 or a Zeiss universal fluorescence microscope (HBO200) with barrier filters 41 and 50 and exciter filters BG3 and UG1 or BG12.

Electron Microscopy

The solution of 12% tannic acid buffered in 0.2 M cacodylate, pH 7.4, was mixed in an equal volume ratio immediately before fixation with a mixture of 4% paraformaldehyde and 5% glutaraldehyde in 0.2 M cacodylate buffer, pH 7.4. The final pH was adjusted to 7.2 with 10 N NaOH. Small pieces of stem internodes from both young infected plants and comparable healthy plants were fixed with the fixative containing tannic acid for 3 hr as well as with conventional fixative as reported previously (Chen and Hiruki, 1977). They were then post-fixed with 2% osmium tetroxide with 1% sucrose in 0.2 M cacodylate buffer for 4 hr. After dehydration, the specimens were embedded in Araldite. Thin sections were stained with uranyl acetate and lead citrate. Micrographs were taken with a Philips Model 201 electron microscope operated at 60 or 80 kV.

RESULTS AND DISCUSSION

AB Test

A time course study was conducted using *C. roseus* plants graft-inoculated with apple proliferation and clover phyllody-1. The purpose was to monitor the appearance of abnormal fluorescence in the phloem elements during the course of disease development and to relate fluorescence appearance to the severity of symptoms. Upon examination of sections from three *Catharanthus* plants that had been in graft contact with apple proliferation MLO for 14 days, slightly abnormal fluorescence was detected in one of them. However, the infected plant did not show any external symptoms at the time. All of the three plants showed abnormal fluorescence and two of them showed external symptoms when examined 26 days after graft-inoculation. Distinctive fluorescence and symptoms developed in all three plants 78 days after graft-inoculation,

With clover phyllody MLO, neither symptoms nor abnormal fluorescence were found 12 days after graft-inoculation. On the 28th day, abnormal fluorescence was detected in three of five plants tested and only two plants showed external symptoms. On the 74th day, four of five plants had both symptoms and abnormal fluorescence. The intensity and distribution of fluorescence in the secondary phloem region increased as external symptoms developed fully. Similar results were obtained with sections from stems of *C. roseus* that had been graft-inoculated with 17 MLO isolates (Table 2). In a separate test, a series of microscopic examinations was also made on sections of host plants in which MLOs were originally found. Typical examples include citrus with citrus stubborn (Hiruki, 1981), sandal tree with sandal spike (Dijkstra and Hiruki, 1974), and potato and tomato with potato witches' broom (Hiruki, unpublished). In testing sections of stem samples of these plants, abnormal fluorescence was present in the primary phloem in early infections up to about a month. In advanced infections of 2-3 months, a strong fluorescence was located in the zone of secondary phloem, whereas some pathological effects were found in the primary phloem region. In the sections from control tissue, less intense fluorescence spots, which are normal to healthy plants, were limited to the primary phloem area. Similarly, abnormal fluorescence in the phloem elements was clearly recognized by comparing the pedicel sections from MLO-infected and healthy onions.

In using the AB test as a diagnostic procedure, care must be taken in sampling tissue specimens from MLO-infected sandal trees and possibly also from other tree species as symptom development and MLO concentration in the tissues appear to be substantially irregular and considerable differences occur even within a single diseased tree (Dijkstra and

Hiruki. 1974).

In evaluating the results obtained with these tests, the understanding of cyto- and histopathology by MLO infection would be of great value. as the intensity and pattern of fluorescence distribution were closely related to internal and external symptom development.

DAPI Test

Diagnosis of MLO diseases by the DAPI test is based on the specific staining of intracellular MLO-DNA. (Figs. 1A, 1B) The results of a time course study with four MLO isolates tested by the DAPI test are shown in Table 3. After graft-inoculation with MLO infected scions, internode portions of stems were examined periodically by the DAPI test. In two week-infection with western aster yellows MLO (AY27, Alberta isolate) and potato witches' broom MLO (PWB, Alberta Isolate), a moderate amount of MLO specific fluorescence was observed. The amount and intensity of the fluorescence was relative to the severity of symptoms. When sections from *C. roseus* plants infected with AY27 for three weeks were compared to those infected for seven weeks, the latter showed more widely distributed fluorescence with high intensity. When a mild MLO disease such as clover proliferation on *C. roseus* and a severe MLO disease such as aster yellows (AY27, Alberta isolate) were compared for MLO-specific fluorescence, the latter developed strong fluorescence sooner after graft- inoculation than did the former (Table 3).

DAPI, a benzimidole derivative, has been shown to possess specific binding properties with adenine-thymine-rich DNA of various origins (Russell *et al.,*). In examination of sections from herbaceous plants (Hiruki and da Rocha, 1984), as well as tree species infected with MLO (Hiruki, 1981) the use of DAPI as a fluorochrome provided a highly sensitive technique for detection of MLOs and confirmed the

results reported from different laboratories
(Seemüller, 1976; deLeeuw, 1977; Samyn and
Welvaert, 1979; Petzold and Marwitz, 1980;
Schaper and Seemüller, 1982; Shaper and Converse, 1985). Some modification of the DAPI
staining procedure such as combined use of ber-
berine sulphate (Petzold and Marwitz, 1978;
1979a) or the pretreatment of plant tissue with
pectinase to induce tissue separation and increase staining efficiency before the application of DAPI has been reported (Cazelles,1978).

Immunofluorescence

MLO infections were detected by the indirect IF method (Fig. 2A, 2B). The use of
'membrane fraction' preparations from
MLO-infected plants as immunogens produced
antisera with titers ranging from 1/4 to 1/16
in the double immunodiffusion test.
Immunofluorescence was positive with
MLO-infected plant sections and negative with
healthy plant sections (Tables 3,4). Four
antisera against EAY, AY27, PWB and CP gave
positive results with all the MLO-infected
plant sections infected with EAY, AY27, PWB and
CP-MLOs (Table 4). Stronger fluorescence was
observed with EAY and AY27 plant sections. This
corresponds with the observations that EAY and
AY27 have stronger symptoms than CP and PWB.
Healthy plant sections tested with antisera
against EAY, AY27, PWB and CP showed no fluorescence in the phloem region.

Combined Use of Fluorescence and Electron
Microscopy

In the study of MLOs by fluorescence microscopy, it is important to establish first the
identity of the fluorescing materials in question. The detection of fluorescence alone does
not necessarily prove that the yellows disease
under investigation is caused by MLOs if MLO
antiserum is not used.
MLOs were observed in thin sections of

diseased plant tissue (Fig. 3, Table 5) from stems that had given positive results in the AB test, DAPI test (Fig. 1A) and indirect immuno-fluorescent test (Fig. 2A). No MLOs were observed in the healthy control.

Reliability of MLO diagnosis based on histochemistry can be increased when sections for fluorescence microscopy are prepared in such a manner that permits histological counter references between light and electron microscopy by using a single block for sectioning (Albert, 1973). The advantages of the combined use of fluorescence and electron microscopy are three-fold. First, when the number of phloem cells infected by MLOs is extremely limited (Hiruki and Shukla,1973), a preliminary test of samples by fluorescence microscopy will ensure that such cells containing MLOs are positively located in electron microscopy by using fluorescent spots as a guide. Second, when the presence of MLOs in a given plant species is established by electron microscopy, fluorescence microscopy will provide an efficient screening procedure by which disease-free stock material can be established readily and quickly as the procedure is less expensive and highly sensitive in detecting MLOs in plant tissue. Third, in basic research, the combined use of fluorescence and electron microscopy permits in-depth study of cytopathology of MLO infected plants and host-parasite interaction. Exact correlation of intracellular fluorescence and MLO distribution can be established as the basis for such a study (Hiruki and Dijkstra, 1973).

FUTURE PROSPECTS AND
CONCLUDING REMARKS

The techniques of histochemistry and cyto-chemistry are constantly being subjected to further improvements and continue to be widely applied. As a result of an expanded search for new fluorochromes and increased sensitivity, highly efficient methods of MLO diagnosis are available. For diagnosis of plant MLO-disease, some of the recently reported improvements include N,N'-diethylpseudoisocyanine chloride as a sensitive DNA fluorochrome (Petzold and Marwitz, 1979c) and sensitivity improvement of the conventional Feulgen staining (Kartha *et al.*, 1976) by the use of auramine O as Schiff's reagent (Petzold and Marwitz, 1979b).

The combined use of fluorescence and electron microscopy should attract more attention in the study of plant MLOs in the future, as etiology still cannot be established for over 100 MLOs that have been documented since 1967 due to the lack of pure cultures of MLOs *in vitro*. For examination of the same section by both fluorescence and electron microscopy using different stains, certain procedures are available for section preparation (Albert, 1973). Immunofluorescence study of MLOs in plant tissue has been non-existent until recently, mainly due to difficulties associated with isolation and purification of MLOs and production of antiserum against MLOs. Although their precipitin titers were relatively low and the antisera contained fractions reactive to host components,(Clark *et al.*, 1983; Sinha, 1979; Sinha and Chiykowski, 1984) the procedure presented in this paper should be applicable to many other yellows diseases that are suspected to be of MLO origin. With the advent of monoclonal antibody technology, it is possible to obtain monoclonal antibodies with high reactivity and specificity for immunofluorescence microscopy as a diagnostic technique (Lin and Chen, 1985, 1986). The recent development of a technique using the avidin (or streptavidin) - biotin complex has proved that it is more sensitive and more specific than conventional ELISA. This is due to the fact that the introduction of biotin groups into antibodies followed by avidin (or streptavidin) - fluorescein conjugates, based on their strong interactions, leads to amplified detectability of antigen in low concentrations (Wilchek and Bayer, 1984). The use of the avidin-biotin-peroxidase complex has also been adapted successfully for histochemical localization studies (Hsu and Raine,1981). It is clear that with the avidin (or streptavidin) - biotin system sophisticated approaches will eventually find ways into fluorescence microscope studies of MLO in the near future.

EXPLANATION OF FIGURES

Fig. 1.   Micrographs showing transverse sections of stem internodes of *Catharanthus roseus* stained with DAPI.

A.   Infected with Western aster yellows MLO (Alberta isolate).

B.   Uninoculated *C. roseus*.

X, xylem; arrows, phloem; bars represent 100 µm. (da Rocha, 1985).

Fig. 2.   Micrographs showing the results of indirect immunofluorescence tests on transverse sections of stem internodes of *Catharanthus roseus*.

A.   Western aster yellows MLO (Alberta isolate) reacted with anti-Eastern aster yellows serum.

B.   Uninoculated specimen reacted with anti-Eastern aster yellows serum.

X, xylem; arrows, phloem; bars represent 100 µm. (da Rocha, 1985).

Fig. 3.   Electron micrograph of a thin section showing MLOs in the phloem of *Catharanthus roseus* affected by Western aster yellows (Alberta isolate). The

DAPI test was positive for MLO-DNA. MLO, mycoplasma-like organism: CW, cell wall; bar represents 1 m. (da Rocha, 1985).

SUMMARY

The advantages of using fluorescence microscopy in the histochemistry of plant infections by mycoplasma-like organisms (MLOs) are its high sensitivity and its wide-ranging applicability to cellular substances which may be detected either directly or after certain chemical treatments. Aniline Blue test is useful in locating MLO-infected cells with abnormal deposition of callose. The specific binding of a fluorochrome, 4', 6-diamidino-2-phenylindole.2HCl, with MLO-DNA is detected by staining hand sections of stems and petioles of MLO-infected plants after glutaraldehyde fixation. The simple and practical method is useful for rapid diagnosis of MLO infection as well as selecting disease-free stock materials from a large number of samples suspected to be infected by MLOs. Antisera prepared against four MLO isolates reacted specifically with the mycoplasma antigens in the hand sections in indirect immunofluorescence microscopy. This method permits specific detection of MLOs in infected tissue specimens and is thus useful for field disease diagnosis.

REFERENCES

Albert, E.N., 1973. A combined fluorescent and electron microscopic stain for elastic tissue. Histochem. J. 5, 157-167.

Barile, M.F. and Grabowski, M.W., 1983. Detection and identification of mycoplasmas in infected cell cultures by direct immunofluorescence staining. In: Methods in Mycoplasmology. Razin, S. and Tully, J.G., (eds.) Vol. 2, pp. 173-181. Academic Press, New York.

Cazelles, O., 1978. Mise en evidence, par fluorescence, des mycoplasmes dans les tubes cribles intact isoles des plantes infectees. Phytopath. Z. 91: 314-319.

Chen, M.H. and Hiruki, C., 1975. Electron microscopy of mycoplasma-like bodies associated with clover proliferation disease. Proc. Amer. Phytopathol. Soc. 2: 52.

Chen, M.H. and Hiruki, C., 1977. Effects of dark treatment on the ultrastructure of the aster yellows agent *in situ*. Phytopathology 67: 321-324.

Chen, M.H. and Hiruki, C., 1978. Occurrence of tubular structures in *Vinca rosea* infected with Alberta isolate of the aster yellows agent. Protoplasma 95, 207-216.

Clark, M.F., Barbara, D.J. and Davis, D.L., 1983. Production and characteristics of antisera to *Spiroplasma citri* and clover phyllody-associated antigens derived from plants. Ann. Appl. Biol. 103: 251-259.

da Rocha, A., 1985. A histochemical study of plant diseases associated with mycoplasma-like organisms. M. Sc. Thesis, University of Alberta, pp. 108.

da Rocha, A., Ohki, S.T. and Hiruki, C., 1986. Detection of mycoplasma-like organisms *in situ* by indirect immunofluorescence microscopy. Phytopathology 76: 864-868.

de Leeuw, G.I.N., 1977. Veld infectie van Aster tripolium L. met mycoplasma's. Grewasbescherming 8: 175-177.

Dijkstra, J. and Hiruki, C., 1974. A histochemical study of sandal (*Santalum album*) affected with spike disease and its diagnostic value. Neth. J.Pl. Pathol. 80:37-47.

Gardella, R.S., DelGuidice, R.A. and Tully, J.G., 1983. Immunofluorescence. In: Methods in Mycoplasmology. Razin, S. and Tully, J.G., (eds.) Vol. 1. pp. 431-439. Academic Press, N.Y.

Hiruki, C., 1981. Fluorescence microscopy in diagnosis of tree diseases associated with mycoplasma-like organisms (MLO). Proc. XVII IUFRO World Cong. Div. 2, 317-322.

Hiruki, C., 1982. Histochemistry of tree mycoplasma disease with special reference to mulberry dwarf and paulownia witches' broom. Proc. 5th IUFRO Conf. Mycoplasma Diseases, Green Bay, 1982.

Hiruki, C. and Chen, M.H., 1984. Plant mycoplasma diseases occurring in Alberta. 7th IUFRO Conference on Mycoplasma Diseases. July 10-12, 1984, p.7.

Hiruki, C. and da Rocha, A., 1986. Histochemical diagnosis of mycoplasma infection in *Catharanthus roseus* by means of a fluorescent DNA-binding agent 4', 6-diamidino-2-phenylindole-2HCL (DAPI). Can. J. Plant Pathol., 8: (In Press).

Hiruki, C. and da Rocha, A., 1984. Histochemical studies of plant mycoplasma diseases. 7th IUFRO Conference on Mycoplasma diseases. July 10-12, 1984, p. 6.

Hiruki, C., Giannotti, J. and Dijkstra, J., 1974. A comparative study of fluorescence in stems of *Vinca rosea* infected with mycoplasmas of different plant origins. Neth. J. Pl. Pathol., 80: 145-153.

Hiruki, C. and Shukla, P., 1973. Mycoplasma-like bodies associated with witches' broom of bleeding heart. Phytopathology 63: 88-92.

Hsu, S.M. and Raine, L., 1981. Protein A, avidin, and biotin in immunohistochemistry.
   J. Histochem. Cytochem. 29: 1349-1353.

Johnstone, A. and Thorpe, R., 1982. Immuno-
   chemistry in Practice. Blackwell Scientific Publ., Oxford. pp. 298.

Kartha, K.K., Cousin, M.T. and Ruegg, E.F.,
   1976. A light microscope detection of
   plant mycoplasma infection by Feulgen
   staining procedure. Indian Phytopathol.
   28: 51-56.

Kawamurua, A., Jr. (ed.),1977. Fluorescent
   antibody techniques and their applications,
   2nd ed. Univ. Tokyo Press: Tokyo, pp. 292.

Kenny, G.E., 1983. Agar precipitin and immuno-
   electrophoretic methods for detection of
   mycoplasmic antigens. In: Methods in Mycoplasmology. Razin, S. and Tully, J.G.,
   (eds.) Vol. 1, pp. 441-456. Academic
   Press, N.Y.

Lin, C.P. and Chen, T.A., 1985. Monoclonal
   antibodies against the aster yellows
   agent. Science 227, 1233-1235.

Lin, C.P. and Chen, T.A., 1986. Comparison of
   monoclonal antibodies and polyclonal antibodies in detection of the aster yellows
   mycoplasma-like organism. Phytopathology
   76: 45-50.

Marwitz, R. and Petzold, H., 1980. Eine einfa-
   che fluoreszenzmikroskopische Nachweis-
   methode für die Infektion von Pflanzen mit
   mykoplasmaahnlichen Organismen. Phytopathol. Z. 97: 302-306.

Namba, S., Yamashita, S., Doi, Y. and Yora, K.,
   1981. Direct fluorescence detection method
   (DFD method) for diagnosing yellows-type

virus diseases and mycoplasma diseases of plants. Ann. Phytopathol. Soc. Japan 47, 258-263.

Petzold, H. and Marwitz, R., 1978. Light microscopical investigations for a simplified detection of plant infection by mycoplasma-like organisms. 3rd Int. Congr. Plant Pathol., Munich, 1978, 63: (Abstr.).

Petzold, H. and Marwitz, R., 1979a. Fluoreszenzmikroskopische Untersuchungen zum Nachweis von mykoplasmaähnlichen Organismen mittels Berberinsulfat. Z. Pflanzenkrankheiten Pflanzenschutz 86: 475-750.

Petzold, H. and Marwitz, R., 1980. Ein verbesserter fluoreszenzmikroskopischer Nachweis fur mykoplasma-ähnliche Organismen in Pflanzengeweben. Phytopathol. Z. 97: 327-331.

Petzold, H. and Marwitz, R., 1979c. Über die Verwendbarkeit des Fluoreszenzfarbstoffes N, N'Diethylpseudoisocyaninchlorid zum Nachweis von mykoplamaähnlichen Organismen. Z. Pflanzenkraukheiten Pflanzenschultz 86: 670-674.

Russel, W.C., Newman, C. and Williamson, D.H., 1975. A simple cytochemical technique for demonstration of DNA in cells infected with mycoplasma and viruses. Nature, 253: 461-461.

Samyn, G. and Welvaert, W., 1979. The use of 4',6-diamidino-2-phenyl-indol.2HCl (DAPI) compared with other stains for a quick diagnosis of mycoplasma infections in ornamental plants. Med. Fac. Landbouww. Rijksuniv. Gent. 44: 623-628.

Schaper, U. and Converse, R.H., 1985. Detection of mycoplasma-like organisms in infected blueberry cultivars by the DAPI technique. Plant Disease, 69: 193-196.

Schaper, H. and Seemüller, E., 1982. Condition of the phloem and the persistance of mycoplasma-like organisms associated with apple proliferation and pear decline. Phytopathology, 72: 736-742.

Seemüller, E., 1976. Fluoreszenzoptischer Direktnachweis von mykoplasmaähnlichen Organismen in Phloem pea-decline-und triebsuchtkranker Bäume. Phytopathol. Z. 85: 368-372.

Sinha, R.C., 1979. Purification and serology of mycoplasma-like organisms from aster yellow-infected plants. Can. J. Plant Pathol. 1: 65-70.

Sinha, R.C. and Chiykowski, L.N., 1984. Purification and serological detection of mycoplasma-like organisms from plants affected by peach eastern X-disease. Can. J. Plant Pathol. 6: 200-205.

Wilchek, M. and Bayer, E.A., 1984. The avidin-biotin complex in immunology. Immunology Today 5, 39-43.

Williamson, D.L., 1982. Serological characterization of spiroplasmas and other mycoplasmas. In: Plant and Insect Mycoplasma Techniques Daniels, M.J., and Markham, P.G., (eds.) pp. 240-267. Croom Helm: London.

TABLE 1. Types of fluorescence in plant tissue and
their specificity in MLO diagnosis

Type	Characteristics	Stain (example)	Specificity
Autofluorescence	Natural fluorescence(s) in tissue	Nil	Very Low
Induced	Substance in tissue converted to fluorophore	Aniline Blue	Low
Fluorescence of dye	A. Simple stain w/o pretreatment	DAPI	High
	B. Chemical pretreatment followed by staining	Feulgen	High
Immunofluorescence	Direct or Indirect fluorochromy	FITC	Very High

TABLE 2. Diagnosis of MLO-infected *Catharanthus roseus* by the Aniline
Blue test after graft inoculation at 20-25 C.

MLO isolate	Symptoms[1]	Fluorescence[1]
Apple proliferation	C,sS,sY	++
Apricot chlorotic leafroll	G,→Y,S	+ → ++
Aster yellows (Alberta)	C,E,Y,y	++
Cabbage yellows	C,S.Y,y	+
Carrot yellows	C,P,S,Y	+
Clover phyllodyl-l	C,P,S,Y	+
Clover proliferation	C,G,P,S	+
Cotton phyllody	C,P,S,Y	+
Croton witches' broom	G,S	±
Lavender yellows	C,P,S,Y	+
Plantago yellows	C,Y,S	+ → ++
Potato witches' broom	C,G,P,S	+
Rape yellows	C,S,Y,y	+
Sandal spike	C,P,S,Y	+
Tomato stolbur 'C'	C,sS,sY	++
Tomato stolbur 'SM'	E,G	++
Vinca yellows	C,sS,sY	++

[1]C, chlorosis; E, elongation of internodes; G, greening; P, phyllody;
S, stunting; Y, yellows; y, distortion of young stem, petiole and
veins; s, severe.

[2]Visual rating of fluorescence intensity and distribution in phloem elements. ±,
weak fluorescence; +, strong fluorescence; ++, very strong fluorescence.

(Hiruki *et al.*,1974).

TABLE 3. Diagnosis by the DAPI test of MLO infection in *Catharanthus roseus*

| Sample[b] | Time after graft-inoculation (weeks)[a] | | | | | | | | | | |
| | 2 | | 3 | | 4 | | 5 | | 6 | | 7 |
	S[c]	D[d]	S	D	S	D	S	D	S	D	S	D
Healthy	0	0	0	0	0	0	0	0	0	0	0	0
EAY	0	0	22	22	50	50	80	60	88	75	86	100
AY27	0	8	27	64	33	83	60	60	50	75	100	100
PWB	0	9	8	0	10	40	57	43	17	67	50	83
CP	0	0	0	16	0	80	28	86	20	100	75	100

[a]For graft inoculation, scions from diseased plants were wedge-grafted onto stocks of healthy plants. As controls, healthy scions were grafted onto healthy stocks.

[b]EAY, Eastern aster yellows (New York isolate); AY27, Western aster yellows (Alberta isolate); PWB, potato witches' broom (Alberta isolate); CP, clover proliferation (Alberta isolate).

[c]Symptoms, percentage positive (out of 3 to 14).

[d]DAPI, percentage positive (out of 3 to 14). (Hiruki and da Rocha, 1986).

TABLE 4. Indirect immunofluorescence observed with transverse sections
          of *Catharanthus roseus* stems infected with MLOs of Eastern aster
          yellows (EAY, New York isolate), Western aster yellows (AY27,
          Alberta isolate), potato witches' broom (PWB) and clover pro-
          liferation (CP).

Plant Tissue [a]	EAY	AY27	PWB	CP
Healthy	–[b]	–	–	–
EAY	++++[c]	++++	++++	++++
AY27	++++	++++	++++	++++
PWB	+[α]	+	+	+
CP	+	+	+	+

[a]Three stem sections were used in each test.

[b]–, no fluorescence.

[c]++++, very prominent

[α]+, weak but positive fluorescence (da Rocha *et al.*, 1986).

TABLE 5. Diagnosis of MLO infection by the DAPI test and transmission electron microscopy

Plant[a] Tissue	Weeks After Grafting	Symptoms	DAPI[b]	TEM[c]
Healthy	3	−	0/6	−
EAY	3	−	0/6	−
EAY	4	+	6/6	+
AY27	2	−	0/6	+
AY27	3	−	6/6	+
PWB	7	−	0/6	−
PWB	8	−	6/6	+
CP	3	−	0/6	+
CP	4	−	6/6	+

[a]EAY, Eastern aster yellows (New York isolate); AY27, Western aster yellows (Alberta isolate); PWB, potato witches' broom (Alberta isolate); CP, clover proliferation (Alberta isolate). As control, healthy scions were grafted onto healthy stocks.

[b]Number of sections positive in the DAPI test/Number of sections examined. DAPI fluorescence intensity was high with EAY (4 wk) and AY27 (3 wk).

[c]Transmission electron microscopy. Two ultrathin sections were examined. (Hiruki and da Rocha, 1986).

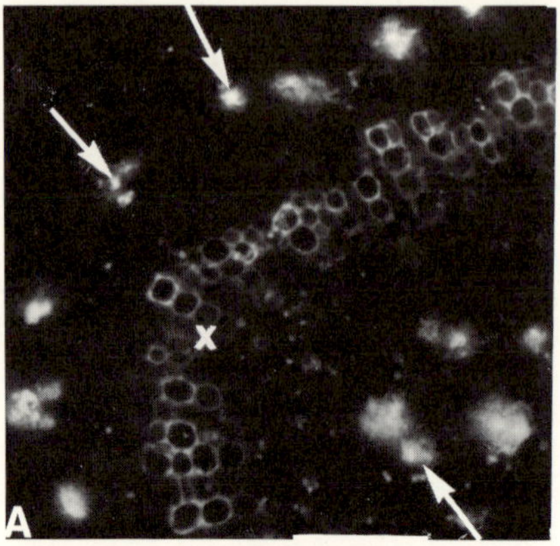

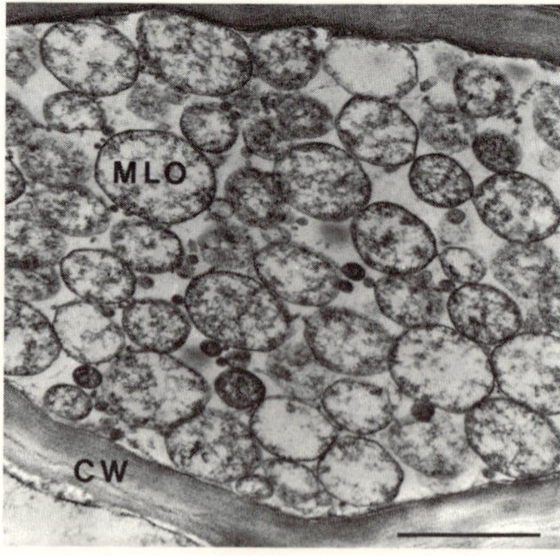

Figure 1A.

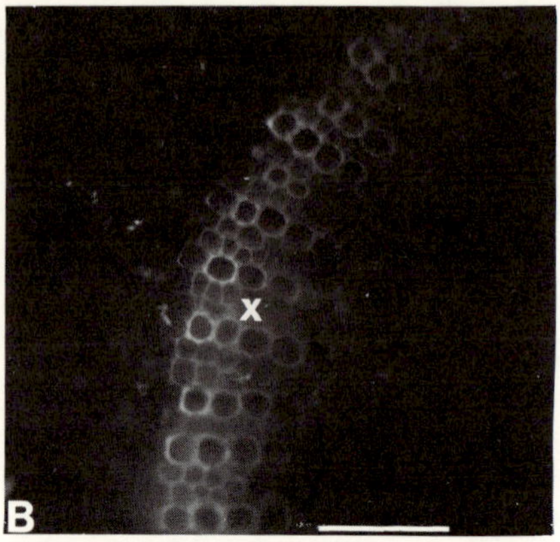

Figure 1B.

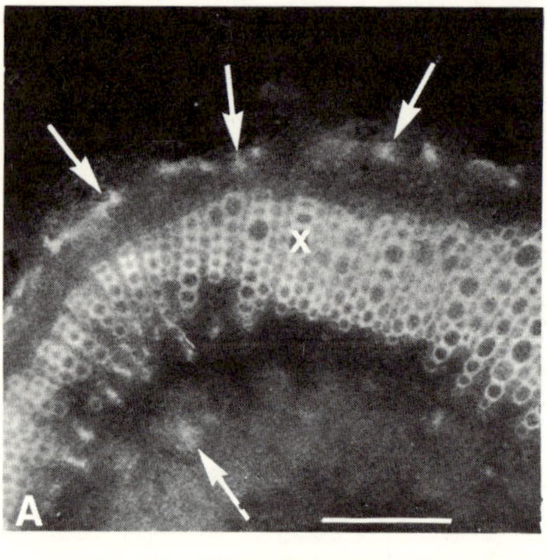

Figure 2A.

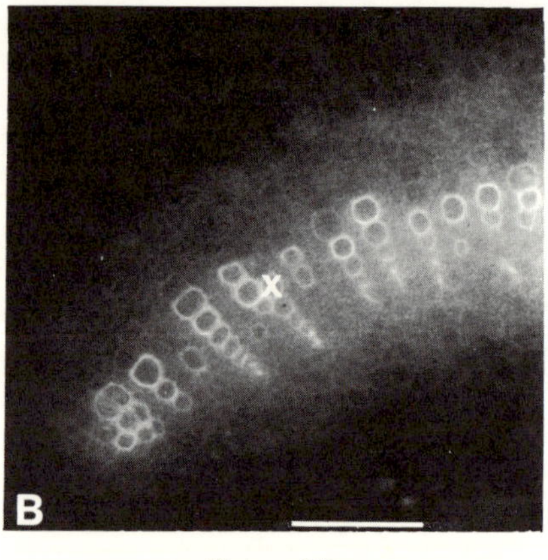

Figure 2B.

Figure 3.

Note: Legends for these figures can be found on page 64.

4

RAPID AND SPECIFIC DETECTION METHODS

FOR PLANT MYCOPLASMAS

by

C. Hiruki

Department of Plant Science

University of Alberta

Edmonton, Alberta T6G 2P5

Canada

INTRODUCTION

Mycoplasmas (class, Mollicutes; order, Mycoplasmatales) are considered to be causal agents of yellows diseases of plants, formerly thought to be of viral origin because of close resemblance in symptomatology and mode of transmission. In plant pathology, however, since the initial discovery by Doi *et al.*, (1967), the term "mycoplasma-like organisms" (MLO) has been in use to describe a group of prokaryotic microorganisms that morphologically resemble members of the Mycoplasmatales but lack proof of Koch's postulates. Most attempts to isolate and culture these microorganisms, except those belonging to the genus *Spiroplasma* (Whitcomb, 1980), have been unsuccessful.

MLO and spiroplasma infections can spread widely throughout many cultivated plants and wild plant species. Infected plants do not normally recover from disease conditions. These organisms perpetuate themselves in perennial plants and vegetatively propagated crops. Our current understanding of yellows diseases and their causal agents is based largely on laboratory and greenhouse tests which are laborious, time-consuming and not highly species-specific. For example, biological assays, such as grafting from MLO-infected plants onto indicator species or injecting leafhoppers with fresh-sap preparations from MLO-infected plants or with spiroplasma cultures and rearing on suitable test plants for symptom development in the greenhouse, are labor intensive and require several months before results are obtained. For species of the *Spiroplasma,* serological criteria are used to differentiate them (Davis *et al.,* 1979; Mouches *et al.,* 1983; Tully *et al.,* 1980). The growth inhibition test (Clyde, 1983) and metabolism inhibition test (Taylor-Robinson, 1983; Tully *et al.,* 1980; Williamson *et al.,* 1978; Williamson, 1983) have been the most frequently used techniques to separate spiroplasma isolates of various origins into serogroups and subgroups (Bove *et al.,* 1983; Davis *et al.,* 1979; Whitcomb *et al.,* 1982a; 1982b; 1983a; 1983b; Williamson *et al.,* 1979) using conventionally produced polyclonal antibodies.

With regard to MLO diseases, it is essential that MLO is correctly identified with high specificity to formulate effective control measures. A highly species-specific test for MLO can also contribute to the delineation of relative relationships among MLO which otherwise is not possible because of the lack of MLO cultures on artificial media.

The improvement in sensitivity and rapidity of plant mycoplasma diagnosis can contribute in a number of important ways to the management of agricultural crops. Rapid, timely diagnosis permits growers not only to eliminate infected plants, which may otherwise serve as reservoirs of plant mycoplasma, but also accurate designation of target plants for tetracycline treatments in such types of perennial crops as fruit trees.

In woody plants, the distribution of mycoplasma is highly irregular (Dijkstra and Hiruki, 1974). High sensitivity of the detection ensures the monitoring of mycoplasma present in minute amounts in different parts of infected plants. To be useful in practical application, any diagnostic procedure must also be highly dependable.

In this chapter, recent progress in four major areas, which offer required high sensitivity, rapidity and dependability, namely enzyme-linked immunosorbent assay, monoclonal antibodies, immunosorbent electron microscopy, and nucleic acid hybridization, are described in relation to mycoplasma diagnosis.

ENZYME-LINKED IMMUNOSORBENT ASSAY (ELISA)

ELISA is simple to perform, gives excellent results and is based on a basic biological principle of the high specificity of antigen-antibody reaction and a chemical reaction leading to the high amplification of chemical reactions involving enzymes. The ELISA procedures in practice, therefore, consist of two steps; the immunological reaction and the enzymatic signal formation to indicate the presence or absence of antigen-antibody reactions. Practical problems concerning specificity may arise, however, when an immunogen which is not sufficiently pure is used for antiserum production.

ELISA, first developed by Nakane and Pierce (1966) for detection of virus antigens in tissue sections, has reached a sensitivity comparable with that of radioimmunoassay (Engvall and Perlmann, 1971, 1972). Further improvement by incorporating the biotin/avidin system has resulted in an extremely high sensitivity of antigen detectability (Hahn *et al.*, 1986; Kendall *et al.*, 1983).

ELISA has many applications where a highly specific antiserum is obtainable and is very useful for screening a large number of samples (Clarke *et al.*, 1978). It has been used to detect *S. citri* in extracts of leaves of sweet orange and periwinkle, in which *Spiroplasma* antigens were detectable at concentrations of as little as 5 to 10 ng (Saillard *et al.*, 1978). ELISA was used to study the distribution of *Acholeplasma* in the tissues of coconut palms affected by lethal yellowing disease (Dollet *et al.*, 1979) and *S. citri* in various tissues of periwinkle during the course of infection by grafting (Archer *et al.*, 1979).

MLO from plants affected by yellows diseases lack a *Spiroplasma*-specific antigen (p55-fibril protein) (Townsend, 1983) and no serological relationship was found between *S. citri* and many isolates of MLO (De Leeuw *et al.*, 1983; Clark *et al.*, 1983). ELISA was effective in detecting aster yellows MLO in the preparation at 4 g/ml (Sinha and Benhamou, 1983) and peach eastern X-disease in purified as well as in partially purified preparations from infected leaves but not in crude plant extracts (Sinha and Chiykowski, 1984). Preparations from other plant species affected by aster yellows, clover phyllody, or clover yellow edge did not give a detectable specific reaction in ELISA with the antiserum against the MLO from celery leaves infected with peach eastern X-disease (Sinha and Chiykowski, 1984). ELISA has been used in field surveys for putative vectors for *S. citri* (Bove *et al.*, 1979; Bove, 1984) and the corn stunt spiroplasma (Gordon *et al.*, 1985). Experimentally, it has been used for testing *S. citri* and the corn stunt spiroplasma in leafhoppers and is sensitive enough to monitor multiplication of the pathogens in the leafhoppers (Archer *et al.*, 1982). However, the sensitivity of the tests has been variable; the method is not capable of detecting spiroplasmas less than about 10 cfu/ml (about 1 ng of protein); the detection threshold for corn stunt spiroplasma antigen was 25 ng/ml and it was about 1-10 ng antigen/ml of extract for other spiroplasmas; in terms of insects, it is capable of detecting one infected insect, even in pooled samples of several insects (Raju and Nyland, 1981; Symons, 1984; Bove, 1984; Archer *et al.*, 1982; Eden-Green, 1982; Gordon, *et al.*, 1985).

The recent progress in the production of monoclonal antibodies against spiroplasma and MLO permits the unlimited production of quality serum and should greatly upgrade dependability of the immunodiagnosis.

MONOCLONAL ANTIBODIES

Monoclonal antibodies prepared by the technique of somatic hybridization of mouse myeloma cells with spleen cells are becoming increasingly important in mycoplasma studies. During the last few years, monoclonal antibodies have been produced against several spiroplasma isolates (Buck *et al.*, 1982; McGarrity *et al.*, 1983; Chen and Chen, 1986; Cheng and Chen, 1986; Jordan *et al.*, 1985; Lei and Chen, 1985; Lin and Chen 1985a, 1985b, 1985c, 1985d, Lin and Chen, 1986) and for non-cultivable MLO (Lin and Chen, 1985d, 1986).

Since the discovery of the corn stunt agent as the first helical wall-less prokaryote (spiroplasma) (Davis *et al.*, 1972), many spiroplasmas have been isolated from diseased plants, flower surfaces, insects and ticks (Davis and Lee, 1982). Thus, the task of identifying spiroplasmas is becoming increasingly difficult and complex. Hybridoma technology, a highly specific technique, is a promising approach to differentiating spiroplasma isolates within the same species or serogroup. Recent results from the comparative studies of 32 spiroplasma isolates belonging to 19 serogroups indicated that the corn stunt spiroplasma-specific monoclonal antibodies, with indirect ELISA, reacted specifically with any of the three corn stunt spiroplasma strains tested, but not with all of the other spiroplasma strains from different groups and subgroups (Lin and Chen, 1985c). In contrast, some non-specific reactions to heterologous spiroplasma strains were obtained when conventional polyclonal antibodies were used (Lin and Chen, 1985c).

In *in vitro* detection by indirect ELISA, the monoclonal antibodies selected for aster yellows MLO reacted specifically with the preparations from the aster yellows-infected plants and were capable of differentiating the aster yellows MLO from other MLOs (Lin and Chen, 1986). Both the plant antigen-absorbed and the non-absorbed polyclonal antibodies, however, cross-reacted with plant antigens. In *in situ* detection by indirect immunofluorescence microscopy, transverse or longitudinal sections of leaf midribs from healthy and aster yellows-infected lettuce were stained with fluorescein isothiocyanate-conjugated antimouse IgG as secondary antibody after application of aster yellows MLO-specific monoclonal antibody. The monoclonal antibody bound specifically to the aster yellows-MLO in the sieve tubes of diseased plants, whereas polyclonal antibody treatment showed fluorescence throughout the sections in both healthy and diseased plants, suggesting non-specific binding to cell walls and membranes (Lin and Chen, 1986).

Using monoclonal antibodies in mycoplasma studies offers many advantages. The most important are their high purity and the certainty that each hybridoma line is a clone producing antibodies of a single type; a single antibody-producing spleen cell fuses with a mouse myeloma cell to form a hybrid producing a single monospecific antibody (Kohler and Milstein, 1975; Galfre and Milstein, 1981). The hybrid cell culture can be propagated continuously, yielding an undisrupted supply of a homogeneous, well-characterized antibody.

The basic technique for producing hybridomas involves fusing myeloma cells with spleen cells immunized with an active antigen (mycoplasma). The spleens of immunized mice providing the highest antibody titer are removed, the spleen cells are fused with myeloma cells in the presence of polyethylene glycol which promotes fusion of cell membranes and the fused cells are plated in selective HAT medium. In 2 to 4 weeks, between 150 and 500 hybrid clonal cells will become visible. Antibody production of each clone is screened by ELISA (Lin and Chen, 1985b; 1985c; 1985d). The positive hybrids selected are cloned. The subclones are again screened for antibody, and each hybridoma is injected into the peritoneal cavity of mice. Ascites fluid, containing large amounts of the monoclonal antibody, accumulated in the cavity can be harvested 10 to 14 days later. The hybridoma cells can be stored frozen until they are needed to generate additional supplies of monoclonal antibody (Lovborg, 1982).

Once a stable hybridoma producing large amounts of specific antibody is selected and stored frozen, the researcher can depend on an indefinite supply of a particular and well-characterized antibody. Coordinated tests between different laboratories become more reliable and accurate. The most suitable hybridoma for a particular purpose can be selected from a collection of hybridoma clones. In addition, a hybridoma with specificity for a target antigen can be selected and its reactivity then amplified (Haber and Hiruki, 1984). This latter feature should make it possible to prepare large quantities of specific antibodies against various epitopes such as those that might be present in a given mycoplasma species. A stable supply of specific antibodies in essentially unlimited quantities also facilitates collaboration in international projects such as ecological studies of mycoplasma distribution in plants and insects, as well as helping to sort out complex problems of mycoplasma serotyping (Lankow *et al.*, 1984).

Since monoclonal antibodies are homogeneous, a selected antibody recognizes only a single antigenic site on complex immunogens. Therefore, if a particular epitope is shared among different immunogens, it does not function as a tool for differentiating the immunogens. Since serological precipitation requires extensive cross-linking to form a lattice, most monoclonal antibodies do not precipitate most proteins (Richman *et al.*, 1984). This means that single monoclonal antibodies may not be useful, in some cases, in routine diagnostic tests such as double immunodiffusion or immunoelectrophoresis. This can be overcome by using a mixture of monoclonal antibodies.

A simple and rapid method of species identification of mycoplasmas can be achieved by immunobinding assay using monoclonal antibodies (Kotani and McGarrity, 1985). It is capable of detecting 9.3 x 10 3 - 7.5 x 10 4 cfu/ml of organisms depending on mycoplasma species. The results with 9 spiroplasma strains of different origins suggested that immunobinding assay on nitrocellulose membrane filters can be useful for serotyping spiroplasmas.

Since monoclonal antibodies are excellent reagents in immunofluorescence tests (Lin and Chen, 1986; Morse *et al.*, 1986), strong possibilities exist for improvement of their sensitivity by the introduction of versatile detection systems. These improvements include incorporation of fluorogenic substrates, chemiluminescent labels, or time resolved fluoroimmunoassay (Soini and Kojola, 1983; Halonen *et al.*, 1983).

IMMUNOSORBENT ELECTRON MICROSCOPY (ISEM)

The application of immunological techniques has greatly enhanced the specificity and sensitivity of electron microscopy as a diagnostic method (Milne and Lesemann, 1984). The visual-ization by EM of the immune complex formed by direct mixing of antiserum and virus was first demonstrated as early as 1941 using tobacco mosaic virus (Anderson and Stanley, 1941). The practical application of ISEM to the detection and identification of plant mycoplasma was introduced in 1976 (Derrick and Brlansky, 1976).

ISEM is an important method for the detection and identification of MLO that are not cultivable and to demonstrate serologically evidence of infection with these microorganisms. MLO antigens were detected by ISEM in preparations partially purified from aster plants experimentally infected with aster yellows agent or from celery leaves but not in crude plant extracts (Sinha and Berhamou, 1983; Sinha and Chiykowski, 1984). Cross reactions between partially purified preparations of peach-X or clover yellow edge MLO and aster yellows MLO antiserum (Sinha and Berhamou, 1983) or between those of aster yellows, clover phyllody or clover yellow edge and peach eastern X MLO antiserum have not been observed (Sinha and Chiykowski, 1984).

The recent development of immuno-gold labelling of sections has also provided a simple, sensitive and versatile modification of the method (Beesley *et al.*, 1984a; 1984b). Labelling antigens with protein A-gold (Pares and Whitecross, 1982) and gammaglobulin (Ig G)-gold (Lin, 1984) complex for positive identification have many advantages such as low non-specific background labelling and easy detection of soluble as well as MLO-associated antigen. Using antiserum produced against whole-cell spiroplasma preparations, the method has proved successful with *S. citri, in situ* (Garnier *et al.*, 1984) as well as *in vitro* (Hiruki, unpublished data). Using specific antigens, a further increase in the sensitivity of the test can be achieved, for example a specific antiserum against spiroplasma fibril protein (Townsend and Plaskitt, 1985) or spiralin, the major membrane-protein (Wroblewski *et al.,* 1977; Mouches *et al.,* 1985).

The identification of fibril protein *in vitro* was used for probing the presence of spiroplasmas in infected plants following extraction of proteins (Townsend *et al.,* 1980). The detection threshold of the test was about 10 6 cfu/g which is similar to the sensitivity of ELISA. However, immunogold labelling offers the possibility of detecting very low levels of MLO infection *in situ,* even a single organism within a tissue sample. Because of the high electron density of the gold marker, labelled bodies can be detected at relatively low magnifications, providing rapid screening of grids for the presence of MLO, an important factor in practical diagnosis. Further increases in sensitivity in the detection of mycoplasmas in both *in vitro* and *in vivo* can be expected by the use of monoclonal antibodies.

Van Lend and Verduin (1985) demonstrated that two viruses of a similar spherical appearance in a mixed infection were distinguishable by using gold particles of different sizes. Such an approach may be useful in studying mixed infections of mycoplasmas and other disease agents.

NUCLEIC ACID HYBRIDIZATION

The techniques of nucleic acid hybridization utilize the homology between the two-hydrogen-bonded strands of nucleic acid either in double-stranded DNA or in hybrid molecules. In practice, a hybrid molecule with homologous sequences is made by using a labeled single-stranded nucleic acid. The double-stranded labeled molecule is then either quantitated, visualized, or further analyzed. The main uses of the nucleic acid hybridization techniques are for detection of specific genomes or portions of them, determination of homology between selected genomes, characterization of DNA structure, and detection and analysis of RNA transcripts (Raab-Traub and Pagano, 1984). Specific DNA probes thus constructed proved very useful for assaying viruses (Symons, 1984) and mycoplasmas (Taylor *et al.*, 1984, 1985). The high sensitivity (< 1 pg of complementary sequence), speed (< 24 h), and simplicity (inexpensive equipment and materials) of the nucleic acid hybridization techniques have made them useful not only for basic research but also for practical disease diagnosis.

Hybridization of total nucleic acid has demonstrated only limited homology, even among some serologically related mycoplasma species (Reich *et al.*, 1966; Somerson *et al.*, 1966; Stanbridge and Reff, 1979). Nevertheless, probes prepared by nick translation of total cell DNA of three spiroplasmas (corn stunt spiroplasma, *S. citri* and BC3) were capable of detecting their respective spiroplasma infections with leafhoppers by the dot hybridization method (Boulton *et al.*, 1984). Corn stunt spiroplasma could be detected in leafhopper *Dalbulus maidis* when titers were as low as 10 2 - 10 3 cfu/insect, which makes the procedure at least 100 times more sensitive than ELISA (Boulton and Markham, 1985). Using two cloned genomic fragments from *Mycoplasma hyorhinis*, species-specific identification of the organism was possible by both Southern hybridization and dot hybridization (Taylor *et al.*, 1985).

This probe selectively recognized *M. hyorhinis* sequences in purified DNA and was proved to be a direct method for differential detection of this species. No background hybridization to the eukaryotic DNA was detected (Taylor *et al.*, 1985). This type of probe can be used directly to screen biological samples, such as infected tissues, for the presence of mycoplasmas, without the need for prior purification of the organism, DNA purification, or restriction endonuclease cleavage analysis.

CONCLUDING REMARKS

In order to respond to various requirements in the diagnosis of plant mycoplasma diseases, needs for further improvements of existing procedures and for developments of new techniques are ever-increasing.

In applying the procedures described in this chapter, it is very important that they are readily usable, yet the results obtained are highly accurate, as the ultimate users include agricultural advisors, seed growers and plant nursery managers in addition to research scientists. In this context, ELISA is the most suitable for the monitoring of a large number of samples for plant mycoplasma infections including both plant and insect specimens. Automated kits for ELISA are commercially available from different sources. Its high sensitivity, rapidity, and simplicity with which results can be obtained make it the method of choice at present.

The use of monoclonal antibodies ensures the supply of an unlimited quantity of standardized reagent with high specificity in detecting both spiroplasmas and MLO. However, its advantage in substantially increasing the sensitivity of ELISA *per se* is yet to be demonstrated. Since ELISA requires a series of washes to remove unbound materials at different steps of the procedure, possible dissociation of antigen-antibody complexes during washing may result in the unintended removal of specifically bound antibody, thus reducing the sensitivity of ELISA (Richman *et al.,* 1984). Furthermore, high titers (1:100,000) of monoclonal antibodies against *Mycoplasma gallisepticum* in ELISA were not related to reactivity in other serological tests such as agglutination, hemagglutination-inhibition and growth-inhibition tests (Morse *et al.,* 1986). Further improvements in the sensitivity of ELISA using monoclonal antibodies will depend on the establishment of high affinity monoclonal antibodies

(Matikainen and Lehtonen, 1984), which may be obtained by selection or molecular engineering.

The ISEM technique has been used successfully for the detection of spiroplasmas and MLO in partially purified preparations. The method is useful for rapid identification of sample materials at the research level but is less suitable for routine assays because of the lengthy purification procedures required and because of the need for electron microscope facilities which may not be available at the field station level. However, immunogold staining offers the possibility of disease diagnosis based on highly specific localization of mycoplasma antigen in a single organism within a tissue sample (Townsend and Plaskitt, 1985).

The use of specific DNA probes has proved highly effective for assaying mycoplasmas (Boulton *et al.*, 1984; Boulton and Markham 1985; Taylor *et al.*, 1985). Considering the rapidity, sensitivity and simplicity of the current nucleic acid hybridization techniques, there is no doubt that this technique will be applied for the detection of uncultivable MLO before too long. Furthermore, detection systems can be improved and the preparation procedures of probes can be simplified. One such direction is the preparation of non-radioactive probes. If successful, it will solve the problems associated with the short shelf life of 32P-labeled probes, the disposal of hazardous radioactive wastes and the health hazards related to radioisotope work. In such an attempt, biotin-labeled nucleotides were incorporated into DNA by nick translation and other enzymatic reactions (Brigati *et al.*, 1983; Langer *et al.*, 1981). The biotinylated probes were hybridized and detected with high sensitivity using biotinylated alkaline phosphatase polymers tagged with avidin (Leary *et al.*, 1983). A rapid and sensitive colorimetric method can be used to visualize biotin-labeled DNA probes hybridized to DNA or RNA immobilized on nitrocellulose. The sensitivity of this improved method is high; 1-10 pg of bound DNA or RNA are detected (Leary *et al.*, 1983; Richman *et al.*, 1984).

The practical problems associated with nucleic acid hybridization are being solved. Thus, it seems likely that the extensive use of this new technology will be realized for the diagnosis of plant mycoplasma diseases.

ACKNOWLEDGMENTS

Studies in the author's laboratory were supported by the Natural Sciences and Engineering Research Council of Canada under Grant No. A3843 and by the Alberta Agricultural Research Trust.

REFERENCES

Anderson, T.F. and Stanley, W.M., 1941. A study by means of the electron microscope of reaction between tobacco mosaic virus and its antiserum. J. Biol. Chem. 139, 339-345.

Archer, D.B. and Best, J. 1980. Serological relatedness of Spiroplasmas estimated by enzyme-linked immunosorbent assay and crossed immunoelectrophoresis. J. Gen. Microbiol. 119, 413-422.

Archer, D.B., Best, J. and Plaskitt, K.A., 1979. The distribution of spiroplasma within infected plants varies during the course of infection. Proc. 4th Meeting Internat. Council Lethal Yellowing. Fort Lauderdale 1979, p. 9, Univ. Florida Pub. FL-80-1.

Archer, D.B., Townsend, R. and Markham, P.G., 1982. Detection of *Spiroplasma citri* in plants and insect hosts by ELISA. Plant Pathol. 31, 299-306.

Beesley, J.E., Day, S.E., Betts, M.P. and

Thorley, C.M., 1984. Immunocytochemical labeling of *Bacteroides nodosus* pili using an immunogold technique. J. Gen. Microbiol. 130, 1481-1487.

Beesley, J.E., Orpin, A. and Adlam, C. 1984b. An evaluation of the conditions necessary for optimal protein A-gold labeling of capsular antigen in ultrathin methacrylate sections of the bacterium. Histochem. J. 16, 151-164.

Boulton, M.I. and Markham, P.G., 1985. The use of dot blotting for the detection of plant pathogens in insect vectors. In: New Developments in Techniques for Virus Detection. Jones, R.A.C. and Torrance, L. (eds.) Assoc. Appl. Biol. Press.

Boulton, M.I. Markham, P.G. and Davies, J. W. 1984. Nucleic acid hybridization techniques for the detection of plant pathogens in insect vectors. Proc. 1984 Brit. Crop Protection Conf. 3, 181-186.

Bové, J.M., 1984. Wall-less prokaryotes in plants. Annu. Rev. Phytopathol. 22, 361-396.

Bové, J.M., Mouches, C., Carle-Junca, P., Degorce-Dumas, J.R., Tully, J.G. and Whitcomb, R.F., 1983. Spiroplasmas of group I: the *Spiroplasma citri* cluster. Yale J. Biol. Med. 56, 573-582.

Bové, J.M., Moutous, G., Saillard, C., Fos, A., Bonfils, J., Vignault, J.C., Nhami, A., Abassi, M., Kabbage, K., Hafidi, B., Mouches, C. and Viennot-Bourgin, G., 1979. Mise en evidence de *Spiroplasma citri*, l'agent causal de la maladie du "stubborn" des argumes dans 7 cicadelles du Maroc. C. R. Hebd. Seanes. Acad. Sci. 288, 335-338.

Brigati, D.J., Myerson, D., Leary, J.J., Spalholz, B., Travis, S.Z., Fong, C.K.Y., Hsiung, G.D and Ward, D.C., 1983. Detection of viral genomes in cultured cells and paraffin-embedded tissue sections using biotin-labeled hybridization probes. Virology 126, 32-50.

Buck, D.W., Kenneth, R.H. and McGarrity, G.J., 1982. Monoclonal antibodies specific for cell culture mycoplasmas. *In vitro* 18, 377-381.

Chen, Z.W., Lei, J.D. and Chen, T.A., 1986. Production of monoclonal antibodies against green leaf bug spiroplasmas. Phytopathology 76, 650-651.

Cheng, W.S. and Chen, T.A., 1986. Production of monoclonal antibodies against honey bee spiroplasma As-576. Phytopathology 76, 658.

Clark, M.F., Barbara, D.J. and Davies, D.L., 1983. Production and characteristics of antisera to *Spiroplasma citri* and clover phyllody-associated antigens derived from plants. Ann. Appl. Biol. 103, 251-259.

Clark, M.F., Flegg, C.L., Bar-Joseph, M. and Rottem, S., 1978. The detection of *Spiroplasma citri* by enzyme-linked immunosorbent assay (ELISA). Phytopathology 92, 332-337.

Clyde, W.A., Jr. 1983. Growth inhibition tests. In: Methods in Mycoplasmology (Razin, S. and Tully, J.G., (eds) Vol. 1, pp. 405-410. Academic Press. New York.

Davis, R.E. and Lee, I.M., 1982. Comparative properties of spiroplasmas and emerging taxonomic concepts: a proposal. Rev.

Infect. Dis. Suppl. 4, S122-S128.

Davis. R.E., Lee, I.M. and Basciano, L.K., 1979. Spiroplasmas: serological grouping of strains associated with plants and insects. Can. J. Microbiol. 25, 861-866.

Davis, R.E., Worley, J.F., Whitcomb, R.F., Ishijima, T. and Steere, R.L., 1972. Helical filaments produced by a mycoplasma-like organism associated with corn stunt disease. Science 176, 521-523.

de Leeuw, G.T.N., Polak-Vogelzang, A.A. and Hagenaars, A.M., 1983. Absence of serological relationship between several mycoplasma-like organism "strains", *Spiroplasma citri*, corn stunt spiroplasma and *Acholeplasma laidlawii*, determined by enzyme-linked immunosorbent assay (ELISA). Phytopathol. Z. 107, 31-36.

Derrick, K.S. and Brlansky, R.H., 1976. Assay for viruses and mycoplasmas using serologically specific electron microscopy. Phytopathology 66, 815-820.

Dijkstra, J. and Hiruki, C., 1974. A histochemical study on sandal *(Santalum album)* affected with spike disease and its diagnostic value. Neth. J. Plant Pathol. 80, 37-47.

Doi, Y., Teranaka, M., Yora, K. and Asuyama, H., 1967. Mycoplasma or PLT group-like microorganisms found in the phloem elements of plants infected with mulberry dwarf, potato witches' broom, aster yellows, or paulownia witches' broom. Ann. Phytopathol. Soc. Jpn. 33, 259-266.

Dollet, M., Saillard, C., Garcia-Jurando, O., Vignault, J.C., Gargani, D., Tully, J.G. and Bové, J.M., 1979. An approach to the serological study of the Mycoplasmas of lethal yellowing in the coconuts in West Africa. Proc. 4th Meeting Internat. Council Lethal Yellowing. Fort Lauderdale 1979, p. 8, Univ. Florida Pupl. FL-80-1.

Eden-Green, S.J., 1982. Detection of corn stunt spiroplasma *in vivo* by ELISA using antisera to extracts from infected corn plants (*Zea mays*) Plant Pathol. 31, 289-297.

Engvall, E. and Perlmann, P. 1971. Enzyme-linked immunosorbent assay, ELISA. III. Quantitation of specific antibodies by enzyme-labelled anti-immunoglobulin in antigen-coated tubes. J. Immunol. 109, 129-135.

Galfré, G. and Milstein, C., 1981. Preparation of monoclonal antibodies: strategies and procedures. Meth. Enzymol. 73, 3-46.

Garnier, M., Clerc, M. and Bove, J.M., 1984. Growth and division of *Spiroplasma citri*: elongation of elementary helices. J. Bacteriol. 158, 23-28.

Gordon, D.T., Nault, L.R., Gordon, N.H. and Heady, S.E., 1985. Serological detection of corn stunt spiroplasma and rayado fino virus in field-collected *Dalbulus spp.* from Mexico. Plant Dis. 59, 108-111.

Haber, S. and Hiruki, C., 1984. Biotechnology and plant virus identification: A prelude to disease control. Symposium on Biotechnology in Plant Science. Agric. Forest. Bull. 7, 41-48.

Hahn, I.F., Bickerhahn, Lenz, W. and Brandis, H., 1986. An avidin-biotin ELISA for the detection of staphylococcal enterotoxins A and B. J. Immunol. Meth. 92, 25-29.

Halonen, P., Meurman, O., Lovgren, T., Hemmila, I., and Soini, E., 1983. Detection of viral antigens by time-resolved fluoroimmunoassay. Curr. Top. Microbiol. Immunol. 133- 146.

Jordan, R., Konai, M., Lee, I.M. and Davis, R. E., 1985. Production and characterization of monoclonal antibodies to *Spiroplasma citri* and corn stunt spiroplasma. Phyto-pathology 75, 1351.

Kendall, C., Ionescu-Matiu, I. and Dreesman, G.R., 1983. Utilization of the biotin/avidin system to amplify the sensitivity of the enzyme-linked immunosorbent assay (ELISA). J. Immunol. Meth. 56, 329-339.

Kohler, G. and Milstein, C., 1975. Continuous culture of fused cells secreting antibody of predefined specificity. Nature 256, 495-497.

Kotani, H. and McGarrity, G.J., 1985. Rapid and simple identification of mycoplasmas by im-munobinding. J. Immunol. Meth. 85, 257-267.

Lankow, R.K., Woodhead, S.H., Patterson, R.J., Massey, R. and Schochetman, G., 1984. Mono-clonal antibody diagnostics in plant dis-ease management. Plant Dis. 68, 1100-1101.

Leary, J.L., Brigati, D.J. and Ward, D.C., 1983. Bio-blots: Rapid and sensitive colori-metric method for visualizing biotin-labelled DNA probes hybridized to DNA or RNA immobilized on nitrocellulose. Proc. Natl. Acad. Sci. USA 80, 4045-4049.

Lei, J.D. and Chen, T.A., 1985. Detection of specific protein of spiroplasmas with mono-clonal antibodies to *Spiroplasma citri* Phytopathology 75, 1351.

Lin, C.P. and Chen, T.A., 1985a. 13 Monoclonal antibodies to specific epitopes on *Spiroplasma citri*. Phytopathology 74, 798.

Lin, C.P. and Chen, T.A., 1985b. Production of monoclonal antibodies against *Spiroplasma citri*. Phytopathology 75, 848-851.

Lin, C.P. and Chen, T.A., 1985c. Monoclonal antibodies against corn stunt spiroplasma. Can. J. Microbiol. 31, 900-904.

Lin, C.P. and Chen T.A., 1985d. Monoclonal antibodies against the aster yellows agent. Science 227, 1233-1235.

Lin, C.P. and Chen, T.A., 1986. Comparison of monoclonal antibodies and polyclonal antibodies in detection of the aster yellows mycoplasma-like organism. Phytopathology 76, 45-50.

Lin, N.S., 1984. Gold-IgG complexes improve the detection and identification of viruses in leaf dip preparations. J. Virol. Meth. 8, 181-190.

Lovborg, U., 1982. Monoclonal antibodies. Production and maintenance. pp. 66. William Heinemann Medical Books, London.

Matikaine, M.T. and Lehtonen, O.P., 1984. Relation between avidity and specificity of monoclonal anti-chlamydial antibodies in culture supernatants and ascitic fluids determined by enzyme immunoassay. J. Immunol. Meth. 72, 341-347.

McGarrity, G.J., Kenneth, R., Megraud, F. and Buck, D., 1983. Preparation of monoclonal antibodies against mycoplasmas, acholeplasmas and spiroplasmas. Yale J. Biol. Med. 56, 860-861.

Milne, R.G. and Lesemann, D.E., 1984. Immunosorbent electron microscopy in plant virus
studies. In: Methods in Virology, Maramorosch, K. and Koprowski, H., (eds.) Vol.
8, 85-101. Academic Press, New York.

Morse, J.W., Boothby, J.T. and Yamamoto, R.,
1986. Detection of *Mycoplasma gallisepticum* by direct immunofluorescence using a
species-specific monoclonal antibody.
Avian Dis. 30, 204-206.

Mouches, C., Candresse, T., Barroso, G., Saillard, C., Wroblewski, H. and Bove, J.M.,
1985. Gene for spiralin, the major membrane protein of the helical mollicute
Spiroplasma citri: cloning and expression in *Escherichia coli.* J.
Bacteriol. 164, 1094-1099.

Mouches, C., Candresse, T., McGarrity, G.J. and
Bove, J.M., 1983. Analysis of spiroplasma
proteins: Contribution to the taxonomy of
group IV spiroplasma and the characterization of spiroplasma protein antigen. Yale
J. Biol. Med. 56, 451-457.

Nakane, P.K. and Pierce, G.G., 1966. Enzyme-
labelled antibodies: Preparation and application for the localization of antigens. J.
Histochem. Cytochem. 14, 929-931.

Pares, R.D. and Whitecross, W.I., 1982. Gold-
labelled antibody decoration (GLAD) in the
diagnosis of plant viruses by immunoelec-
tron microscopy. J. Immunol. Meth. 51,
23-28.

Raab-Traub, N. and Pagano, J.S., 1984. Hybrid-
ization of viral nucleic acids: Newer meth-
ods on solid media and in solution. In:
Methods in Virology (Maramorosch, K. and
Koprowski, H. (eds.) Vol. 8, pp. 1-39.
Academic Press, New York.

Raju, B.C. and Nyland, G., 1981. Enzyme-linked immunosorbent assay for the detection of corn stunt spiroplasma in plant and insect tissues. Curr. Microbiol. 5, 101-104.

Reich, P.R., Somerson, N.L., Hybner, C.J., Chanock, R.M. and Weissman, S.M., 1966. Genetic differentiation by nucleic acid homology. I. Relationships among *Mycoplasma* species of man. J. Bacteriol. 92, 302-310.

Richman, D.D., Cleveland, P.H., Redfield, D.C., Oxman, M.N. and Wahl, G.M., 1984. Rapid viral diagnosis. J. Infect. Dis. 149, 298-310.

Saillard, C. and Bove, J.M., 1983. Application of ELISA to spiroplasma detection and classification. In: Methods in Mycoplasmology. Razin, S. and Tully, J.G., (eds.) Vol. 1, pp. 471-476. Academic Press, New York.

Saillard, C., Dunez, J., Garcia-Jurado, O., Nhami, A. and Bove, J.M., 1978. Detection de *Spiroplasma citri* dans les agrumes et les pervenches par la technique immuno-enzymatique "ELISA". C.R. Hebd. Seanes Acad. Sci. 286, 1245-1248.

Sinha, R.C. and Benhamou, N., 1983. Detection of mycoplasma-like organism antigens from aster yellows-diseased plants by two serological procedures. Phytopathology 73, 1199-1202.

Sinha, R.C. and Chiykowski, L.N., 1984. Purification and serological detection of mycoplasma-like organisms from plants affected by peach eastern X-disease. Can. J. Plant Pathol. 6, 200-205.

Soini, E. and Kojola, H., 1983. Time-resolved fluorometer for lanthanide chelates--a new generation of non-isotopic immunoassays. Clin. Chem. 29, 65-68.

Somerson, N.L., Reich, P.R., Walls, B.E., Chanock, R.M. and Weissman, S.M., 1966. Genetic differentiation by nucleic acid homology. II. Genotypic variations within two *Mycoplasma* species. J. Bacteriol. 92, 311-317.

Stanbridge, E.J. and Reff, M.E., 1979. The molecular biology of mycoplasma. pps. 157-187. In: The Mycoplasmas, Barile, M.F. and Razin, S. (eds.) Vol. 1, Academic Press, New York.

Symons, R.H., 1984. Diagnostic approaches for the rapid and specific detection of plant viruses and viroids. In: Plant Microbe Interactions; 1 Molecular and Genetic Perspectives. Kosuge, T. and Nester, E.W., (eds.) pp. 93-124. MacMillan, New York.

Taylor, M.A., Wise, K.S. and McIntosh, M.A., 1984. Species-specific detection of *Mycoplasma hyorhinis* using DNA probes. Isr. J. Med. Sci. 20, 778-780.

Taylor, M.A., Wise, K.S. and McIntosh, M.A., 1985. Selective detection of *Mycoplasma hyorhinis* using clones genomic DNA fragments. Infect. Immun. 47, 827-830.

Taylor-Robinson, D., 1983. Metabolic inhibition tests. In: Methods in Mycoplasmology. Razin, S. and Tully, J.G., (eds.) Vol. 1, pp. 411-418. Academic Press, New York.

Townsend, R., 1983. Mycoplasma-like organisms from plants with 'yellows' disease lack a spiroplasma-specific antigen. J. Gen. Microbiol. 129, 1959-1964.

Townsend, R., Burgess, J. and Plaskitt, K.A., 1980. Morphology and ultrastructure of helical and non-helical strains of *Spiroplasma citri*. J. Bacteriol. 142, 973-981.

Townsend, R. and Plaskitt, K.A., 1985. Immunogold localization of p55-fibril protein and p25 spiralin in spiroplasma cells. J. Gen. Microbiol. 131, 983-992.

Tully, J.G., Rose, D.L., Garcia-Jurado, O., Vignault, J.C., Saillard, C., Bove, J.M., McCoy, R.E. and Williamson, D.L., 1980. Serological analysis of a new group of spiroplasmas. Curr. Microbiol. 3, 369-372.

Van Lent, J.W.M. and Verduin, B.J.M., 1985. Specific gold labeling of antibodies bound to plant viruses in mixed suspensions. Neth. J. Plant Pathol. 91, 205-213.

Whitcomb, R.F., 1980. The genus of spiroplasmas. Annu. Rev. Microbiol. 34, 677-709.

Whitcomb, R.F., Tully, J.G., Clark, T.B., Williamson, D.L. and Bove, J.M., 1982a. Revised serological classification of spiroplasmas, new provisional groups and recommendations for serotyping of isolates. Curr. Microbiol. 7, 291-296.

Whitcomb, R.F., Tully, J.G., McCawley, P. and Rose, D.L., 1982b. Application of the growth inhibition test to *Spiroplasma taxonomy*. Int. J. Syst. Bacteriol. 32, 387-394.

Whitcomb, R.F., Tully, J.G. and Wroblewski, H., 1983b. Spiralin: major membrane protein specific for subgroup I-1 spiroplasmas. Curr. Microbiol. 9, 7-12.

Whitcomb, R.F., J.G. Tully, J.M. Bové, and P. Saglio. 1973. Spiroplasmas and Acheloplasmas: multiplication in insects. Science 182: 1251-1252.

5

PROSPECTS FOR RAPID IDENTIFICATION

OF SPIROPLASMAS IN

PLANTS AND ANIMALS

by

Alan Liss

Department of Biological Sciences

University Center at Binghamton

Binghamton, New York 13901, USA

INTRODUCTION

Mycoplasmas are cell wall-less prokaryotic organisms. Placed in the class Mollicutes (soft skinned), mycoplasmas represent a diverse collection of microbes (Tully and Razin, 1977). These organisms are unique in that they each are surrounded solely by a single biological membrane. A complete taxonomic scheme exists for mycoplasmas (Ghosh and Raychaudhuri, 1972). The plant pathogenic mycoplasmas primarily belong to the taxonomic family, *Spiroplasmataceae* (commonly called spiroplasmas) (Maramorosch, 1980). Members of the family *Mycoplasmataceae* are generally known for being etiologic agents of animal disease (e.g. *Mycoplasma pneumoniae* causes primary atypical pneumonia in man). Members of the family *Acholeplasmataceae* are generally considered saprophytes (e.g. *Acholeplasma laidlawii*). However, as shall be noted later, all 3 taxonomic families may have members capable of causing plant disease.

Most mycoplasmas are relatively unknown organisms. Despite years of promulgation of seemingly mystical attributes, (.e.g. the replication of mycoplasmas via elementary bodies), what little we have learned suggests that mycoplasmas share most (if not all) properties with classical eubacteria. This commonality with eubacteria is, in fact, a promising situation, because it suggests that classical microbiological experiments can be used (perhaps in a modified form) to generate data concerning mycoplasmas.

Characterization of the etiology of a particular disease is an age-old problem. Of the many factors involved, one must consider the problems of recognizing the presence of a potential pathogen, as contrasted with being able to grow this organism. Certainly fulfillment of the terms of Koch's postulates (to identify an organism as an etiologic agent) requires one to be able to grow and purify a presumptive etiologic agent. However, often practical problems (e.g. economic or health requirements) demand immediate pathogen identification. For spiroplasmas, the term "immediate" almost precludes analyses of growth studies. Early and accurate disease diagnosis is a priority for limiting and eradicating plant, animal or human infections. The advent of immunodiagnostic procedures (e.g. immunofluorescence, radioimmunoassay, enzyme linked immunosorbent assay) signaled a time of rapid identification of a variety of disease causing agents. When used to diagnose viral diseases, an immunodiagnostic process taking hours replaced a growth procedure that may have taken weeks.

Attempts to grow spiroplasmas are often made difficult because; 1) not all spiroplasmas can be propagated on a single growth medium and 2) not all apparently spiroplasma-like organisms can be grown at all. In fact, there exists a long list of plant diseases whose cause appears related to a cell wall-less agent which can be demonstrated in electron micrographs but not cultivated (e.g. Anantha Padmanabha *et al,* 1973; Caudwell and Kuszak, 1984; Esau *et al,* 1976). Collectively called mycoplasma-like organisms (MLO), these agents become recognized as spiroplasma, acholeplasma or mycoplasma isolates only when grown in cell-free medium.

Several immunodiagnostic procedures have been used to identify MLOs in plants. Fluorescent antibody procedures have proven moderately successful (Kumar *et al,* 1975). However, an antiserum is only as specific as the antigen used to generate it. Also, as shown in *in vitro* studies, sero-diagnostic procedures for spiroplasmas must also reconcile the problem of the great diversity present within the genus *Spiroplasma* (Mouches *et al,* 1982). Additionally, the realm of the MLOs is not exclusively that of spiroplasmas yet to be grown. I have isolated a number of *Acholeplasma sp.* and *Mycoplasma sp.* from diseased, MLO containing plant material. (Liss, unpublished data). The eventual fulfillment of Koch's postulates with these isolates would have turned MLO's into non-spiroplasma plant pathogens, adding even more antigenic diversity to the immunodiagnostic picture. To try to remedy this complex problem I have attempted some non-cultivation, non-immunodiagnostic detection procedures.

Viruses have been found associated with *Mycoplasma sp.,Acholeplasma sp.* and *Spiroplasma sp.* (Maniloff *et al.,* 1982). Acholeplasmaviruses have been the most studied. A variety of viral isolates can be placed into one of three groups (based on morphologic and serologic criteria) designated group 1 (naked bullet-shaped particles), group 2 (enveloped particles with no obvious internal capsid) and group 3 (short-tailed polyhedrons) (Maniloff *et al.*, 1982).

Viruses infecting spiroplasmas are also placed into three groups: spiroplasma virus (SpV) group 1 (naked rods), SpV group 2 (long-tailed polyhedrons), SpV group 3 (short-tail polyhedrons) (Maniloff *et al.*, 1982). Over 90% of all spiroplasmas examined (n=100) by electron microscope techniques have revealed presence of one or more of these viruses. However, only members of SpV group 1 and SpV group 3 can be propagated *in vitro*.

Propagation of acholeplasma-, spiroplasma-, or mycoplasmaviruses requires specific indicator hosts. So far, with few repeatable exceptions, there has been no virus capable of crossing genus lines. At best, then, we would need three separate indicators to identify all viral isolates of members of the Mollicutes.

Using a single spiroplasma isolate (Christiansen *et al.*, 1980), (isolated from *Opuntia tuna monstrosa*) I attempted to screen diseased plant tissue for presence of spiroplasmaviruses. In a limited survey, I detected spiroplasmaviruses (as plaque forming units) in 10 of 20 diseased samples and 0 to 5 controls. The virus assay took 24 hrs. as compared to days or weeks to cultivate the presumptive spiroplasmas. Before extending this study, I concluded that a "viral" diagnostic procedure for identifying spiroplasmas was not a foolproof procedure for a number of reasons, including: 1) not all isolates of a single SpV group will plaque on a single indicator host; and 2) not all plant pathogenic Mollicutes are spiroplasmas. However, there could be some improvement in this approach in regard to MLOs. I have been able to identify plaque forming viruses from the spiroplasma of Drosophila sex-ratio disorder before it could be cultivated in cell culture. More research must be undertaken before effectively using these virsues for diagnostic purposes.

A more successful procedure may involve the sophisticated and sensitive procedure involving

DNA probes. A standard detection test using a DNA probe would be as follows:

1. a single strand sequence complementary to a specific target sequence is made;
2. a test sample is prepared so that it's DNA is available for the probe;
3. the probe is added and an assay to detect double stranded hybrids is performed;
4. presence of hybrid probe: target sequences denote a positive DNA probe assay.

The ideal probe molecule is one which can hybridize only if the complementary target sequence is present. When incubated alone or in a sample where no target sequence is present, the single stranded probe molecule cannot form a double stranded molecular configuration. The specificity of a nucleic acid hybridization assay lies in the probe molecule and the conditions under which the reaction takes place. The more unique the probe, the more specific it can be in any identification protocol.

The strategy I am suggesting involves making a DNA probe complementary to the ribosomal RNA (R-RNA) of spiroplasmas. This probe can then be used to detect spiroplasmas, by demonstrating presence of R-RNA, in almost any test situation (e.g. plant tissue, vector material).

In co-operation with Dr. David Kohne (Gen-Probe, San Diego, DA), preliminary experiments have been performed involving the production of spiroplasma DNA-probes. A probe has been made that does not hybridize to eucaryotic organellular R-RNA but does hybridize to:

1. *Mycoplasma* species
2. *Acholeplasma* species
3. *Spiroplasma* species

Each of the above hybridization occurs at levels of 55% or greater and takes place within one hour. Currently, hybridization procedures are being developed to increase the specificity of the probe to hybridize, uniquely, with spiro plasma species. Sensitivity appears high. More testing must be and is being done to improve the DNA probe procedure.

One important aspect of DNA probes must be emphasized here. You need not grow the test organism--either to generate the DNA probe or to test for the target organism. This point should be especially emphasized for studies involving MLOs- non-cultivatable mycoplasma-like organisms. Assuming that MLOs are pro-karyotic cells, one can estimate that about 90% of their total RNA is R-RNA (one estimate would be that a typical procaryotic cell has 2.5 x 10 gms of R-RNA). Therefore, if one could par-tially purify an MLO from plant tissue, a probe could be made. Currently the hybridization pro-tocols used by Kohne need about 5 x 10 micro-grams of probe per reaction (test screening for MLO). This amount of R-RNA would come from about 2 x 10 MLO "bodies".

Obviously no technique will be universally accepted or universally successful. DNA probes offer a viable alternative to growth--requiring identification procedures because of the time involved and the absence of needing a catholic growth milieu. One does not need clean anti-gens for preparation of antisera as is needed for the immunodiagnostic procedures. Instead, nucleic acid is used to generate useful probe sequences. I have suggested the use of R-RNA,, however, other spiroplasma specific DNA se-quences should prove successful. DNA probes will not truly identify pathogens--any organism carrying the target sequence will be found. How-ever, for many of us, the completion of Koch's postulates is second to the search for possible etiologic agents leading to possible strategies for disease control.

In summary, *Mollicutes* are certainly im-portant members of plant flora--either as eti-ologic agents or innocent by-standers. More basic research on these organisms is necessary before we can properly assess their role in plant diseases. While waiting for this informa tion, we must continue to develop rapid diagnostic procedures for acholeplasmas, mycoplasmas, spiroplasmas, and MLOs. It will prove better to have attacked this problem indirectly than to wait for the final words on the disease causing role of these diverse cell wall-less prokaryotes.

ABSTRACT

Cell wall-less procaryotes are placed in the taxonomic class, Mollicutes. Often called "mycoplasmas", these microbes have diverse characteristics yet all share being surrounded only by a single biological membrane. Though they share size and filterability properties with the larger viruses, mycoplasmas can generally be cultivated in artificial, cell-free media. The spiroplasmas are motile, spiral-shaped mycoplasmas responsible for causing a number of plant diseases. When observed during electron microscope surveys of diseased tissues, they share the apparent pleomorphic appearance of the group of putative microbes called "mycoplasma-like organisms" or MLO's. Unlike MLO's, spiroplasmas are often cultivatable *in vitro*. However, no one growth medium is useful for all spiroplasmas and, at best, the spiroplasma isolation procedure is laborious. New growth procedures are necessary to improve the isolation success rates for known spiroplasmas as well as to create new spiroplasmas from the current ranks of the MLO's.

We have been trying to develop novel, non-growth requiring methods for identifying the presence of spiroplasmas in plants and insects. While many have been concentrating on spiroplasma surfaces (e.g. with mono-clonal antibodies), we have been looking at intracellular characteristics. Many spiroplasmas have viruses that infect their cells. Screening test material for the presence of these viruses is rapid (overnight) and has shown some limited success. The main problems with this approach are: 1) not all spiroplasmas carry virus; and

2) not all viruses grow on the same spiroplasma indicator strain. A second approach involves what have been called "gene probes". We currently are utilizing spiroplasma ribosomes as a source of cell specific segments of spiroplasma nucleic acid. From these ribosomes, the ribosomal RNA is purified and then used as a template for production of a complementary DNA strand (cDNA). This cDNA is used to probe test samples for presence of its complementary stretch of DNA--the genetic information encoding for spiroplasmal ribosomal RNA. In preliminary tests, a cDNA probe has been constructed that can detect low levels of spiroplasmas. It does not cross-react with plant or animal RNA populations, although some cross-reaction with other mycoplasmal RNA is seen. More work must be done, but this does appear to be a promising alternative to current methodologies. Also, we should be able to use this technique to detect non-cultivable organisms, such as MLO's, as long as we can isolate enough "bodies" from which we can purify ribosomal RNA.

ACKNOWLEDGEMENTS

A special thanks to Dr. David Kohne for his contributions to the gene probe experiments. Thanks also go to B.E. Ritter for careful reading of the manuscript and to Anna Sefcovic for final manuscript preparation.

REFERENCES

Anantha Padmanabha, H.S., S.P. Bisen, and R. Nayar. 1973. Microplasma-Like Organisms in Histological Sections of Infected Sandal Spike (*Santalum album* L.). Experientia 29: 1571.

Caudwell, A. and C. Kuszak. 1984. Bioassay Techniques to Assess Plant Pathogenicity of Mycoplasma-like Organisms (MLO). Ann. Microbiol. (Inst. Pasteur) 135A: 255-261.

Christiansen, C., E.A. Freundt, and K. Maramorosch. 1980. Identity of Cactus and Lettuce Spiroplasmas with *Spiroplasma citri* as Determined by DNA-DNA Hybridization. Current Microbiol. 4: 353-356.

Esau, K., A.C. Magyarosy, and V. Breazeacle. 1976. Studies of the Mycoplasma-Like Organism (MLO) in Spinach Leaves Affected by the Aster Yellows Disease. Protoplasma. 90: 189-203.

Ghosh, S.K., and S.P. Raychandhuri, 1972. Mycoplasmas: The New Chapter in Plant Pathology. Curr. Sci. 42: 235-241.

Kumar, D., S.K. Ghosh, S.P. Raychandhuri, T.K. Nariani and A. Varma. 1975. Fluorescent Antibody Test for Detection of Citrus Greening Mycoplasma. Curr. Sci. 44: 170-171.

Maniloff, J., K. Haberer, R.N. Gourlay, J. Das and R. Cole. 1982. Mycoplasma Viruses. Intervirology 18 177-188.

Maramorosch, K. 1980. Spiroplasmas of Plants and Animals: An Overview. National Aca. of Sciences, India Golden Jubilee Comm. Vo: 43-62.

Mouches, C., A. Menara, J.G. Tully, and J.-M. Bove. 1982. Polyacrylamide Gel Analysis of Spiroplasmal Proteins and its Contribution to the Taxonomy of Spiroplasma. Rev. Infect. Dis. 45: 5141-5147,

Tully, J.G., and S. Razin. 1977. The Mollicutes. In: CRC Handbook of Microbiology, A.J. Laskin and H. Lechevalier, eds. 2nd Vol I, CRC Press, Boca Raton, Fl 405-459.

6

TRENDS IN RESEARCH ON PLANT MYCOPLASMAS

by

R.E. Davis and I.-M. Lee

Microbiology and Plant Pathology Laboratory

ARS-USDA, Beltsville, MD and

Department of Botany, University of Maryland

College Park, Maryland

INTRODUCTION

Nearly two decades have passed since the discovery of cell wall-less prokaryotes in diseased plants (13,16,19). In that period, numerous diseases have been attributed to wall-less prokaryotes of two principal types; the (helical) spiroplasmas and the (non-helical) mycoplasma-like organisms (MLO's). The discovery and culture *in vitro* of spiroplasmas has made possible rapid progress in understanding these prokaryotes (14,17,45-47), but, inability thus far to culture the MLO's has greatly hindered research on these non-helical agents (24). Now, developments in molecular biology and hybridoma technology have rekindled investigations on the MLO's and have also opened new and exciting opportunities for deeper understanding of the spiroplasmas and the diseases they cause. Already, important progress in the culture *in vitro* of *Spiroplasma,* in the development of new approaches to pathogen detection, and in the molecular biology of wall-less prokaryotes is setting a trend for future research.

CULTIVATION *IN VITRO*

Thus far, of the cell wall-less prokaryotes in plants, only spiroplasmas have been cultured *in vitro*. Attempts to cultivate plant-pathogenic wall-less prokaryotes other than those known with certainty to be spiroplasmas have failed. In the few cases where non-helical mycoplasmas have been isolated in culture from plants, those studied in detail have proven to be non-phytopathogenic *Acholeplasma spp.* and *Mycoplasma spp.*. The failure to culture plant-pathogenic, non-helical MLOs has left research on diseases, presumably caused by these agents, in a comparatively quiescent phase. Although much more effort ought to be devoted to the study of the non-helical MLOs, few laboratories continue to struggle with the highly challenging problem of their culture *in vitro*.

Serum-free Media

Several media have been developed for the primary isolation and subcultivation of spiroplasmas (1,10,17,24,27,28,48). These media have permitted the isolation of spiroplasmas from a wide variety of habitats including the interior of diseased plants, the bodies of insects and ticks, and the surfaces of flowers (1,6,11,24,48). Nearly all the media designed for isolation and culture of plant-pathogenic spiroplasmas are, like media for culture of animal mycoplasmas, enriched with several complex ingredients, such as pleuropneumonia-like organism (PPLO) broth base, animal serum, and yeast extract or yeastolate. The major difference between most spiroplasma culture media and conventional mycoplasma culture media is that media for culture of plant-pathogenic spiroplasmas have high osmolarities (600-700 mOsm).

Serum, one of the major components in most media for culture of spiroplasmas and animal mycoplasmas, contains a variety of lipids and other ingredients, and its exact composition varies from batch to batch. Some lipids (e.g. cholesterol, fatty acids, and/or phospholipids) present in serum are essential for growth of spiroplasmas and mycoplasmas; while certain lipids or other ingredients may be toxic and inhibitory to their growth. Recently, several chemically defined and semi-defined media were developed for *in vitro* culture of plant-pathogenic and non-phytopathogenic spiroplasmas (2-5,25,28). Preliminary studies of nutritional requirements of *S. citri* in a chemically defined medium revealed that the biosynthetic capacity of this organism may be far more limited than previously thought (24,25).

Additional information on lipid requirements of two plant-pathogenic spiroplasmas was recently obtained by using a semi-defined medium in which defined lipids and albumin were added to replace serum. Using this serum-free medium and its modifications to study the lipid metabolism of *S. citri* and the corn stunt (CS) spiroplasma, we realized that the use of lipids differs markedly between these spiroplasmas and non-phytopathogenic spiroplasmas. Indeed, modifications of the lipid content in the serum- free medium proved to be of major importance in culture of certain fastidious strains of the corn stunt spiroplasma, which grew poorly, if at all, in serum-containing media (23,24,26).

Culture of Corn Stunt Spiroplasma

Isolation of several strains of *S. kunkelii* (Whitcomb *et al.*, 1986) (the corn stunt spiroplasma) defective in motility and helicity was made possible recently through the use of media lacking serum (23,26). In these media, serum was replaced by a mixture of lipids and bovine serum albumin. Morphology of colonies formed by the aberrant strains was related to cell shape and capacity for translational motility, with non-helical, non-motile strains producing minute (<0.1 mm diam) colonies and helical, motile strains forming large (up to 2 mm diam) colonies on agar-solidified serum-free media. However, many of these strains did not develop colonies on solid media containing serum, nor could they be continuously subcultured in serum-containing broth media. For more than a decade, helical cell shape has been used as the major criterion to differentiate spiroplasmas from mycoplasma-like organisms in routine diagnostic practice (7,8). This approach is based largely on the assumption that "non-helical" mutants of spiroplasmas are rare. The isolation and culture of numerous CS spiroplasma strains defective in helicity indicates that incidence of non-helical strains of spiroplasma in nature may be much more common than previously assumed (23). This notion encourages wider application of other criteria, in addition to cell shape, to determine the possible spiroplasma nature of cell wall-free prokaryotes.

Advantages and Applications of Serum-free Media

We found serum-free media to yield higher titers of fastidious strains of the corn stunt spiroplasma in broth and to permit colony formation by these strains on agar versions. These advantages have prompted us to employ serum-free media routinely for work in our laboratory. For example, we have used, successfully, serum-free media for primary isolation and sub-cultivation of *S. citri* as well as for plague assay of spiroplasma viruses on sensitive indicator lawns (9,10,12). Efficiency of primary isolations, titers and growth rates attained in broth media, and efficiency of plaque formation on agar were comparable in serum-free and serum-containing media in the case of *S. citri*. Our findings, thus far, have indicated that a serum-free medium (containing a chemically defined serum substitute in place of serum) can be as useful as a serum-containing medium for primary isolation and culture of spiroplasmas from plants and insects, and may be more useful than some serum-containing media for primary isolation and culture of fastidious strains. Use of serum-free media obviates certain disadvantages associated with use of serum containing media for culture of fastidious microbes. For example, serum can be variable in growth supporting ability; serum may be contaminated with mycoplasmas; serum can contribute a major cost to a medium; high quality serum can be difficult to obtain or unavailable in some parts of the world; and animal sera may contain components that inhibit growth of spiroplasmas or other fastidious microbes that do not naturally occur in vertebrate hosts.

We expect that increasing attention will be devoted to attempts to culture the MLO's *in vitro* during the next several years. Use of serum-free media may be important in these endeavors. Alternative approaches to culturing the MLO's will undoubtedly include the use of live insects (36) as intermediate hosts to develop high titer inocula, and the use of phloem sap (37) as well as of insect cell (18) and of plant cell cultures as means to adapt microbes to, or to select strains for, *in vitro* growth. Preparation of inocula by new methods including sieve cell isolation from diseased plants may also contribute to eventual MLO culture.

Chemically Defined Media

The development of chemically defined media for spiroplasma culture has overcome obstacles to investigations of spiroplasma nutrition and metabolism (2-5,24,25). Moreover, these media have made it possible to select mutant strains for later research in molecular genetics (S. McCammon and R.E. Davis, unpubl.). Not only are chemically defined media expected to be utilized increasingly in fundamental studies of *Spiroplasma*, but such media will probably be investigated in attempts to culture the plant MLO's. Just as some components in serum may be detrimental to the growth of fastidious microbes like spiroplasmas (30,31), components of media basal ingredients (e.g., PPLO broth base) may inhibit MLO growth. Use of chemically defined basal ingredients may aid eventual *in vitro* culture of MLO's.

DETECTION AND DIAGNOSIS

Correct diagnosis is the necessary prelude to any successful disease control. However, diagnosis of plant mollicute diseases has often been one of the more difficult aspects in the study of these diseases. This is due, principally to the general lack of methodologies for detection (whether by culture or other means) of the pathogens in the field or in quarantine material, where the natural disease is encountered and where control measures must be applied. In addition, the relatedness among various MLO's, and the epidemiologies of many of the MLO diseases, could not be efficiently approached for study until recently. Fortunately, development and exploitation of new approaches to pathogen detection now promise to overcome many of these problems that hinder correct diagnosis and control.

Monoclonal Antibodies

Hybridoma technology provides the potential to produce cell lines generating a continuous supply of high quality antibodies with known specificity. Thus, monoclonal antibodies (MAb's) have recently been produced against several *Spiroplasma* species and against a few MLO's (20,22,32,33). Such antibodies have been demonstrated to be capable of varying degrees of specificity, depending on the individual antibody, in the detection of *Spiroplasma* (12,32) or MLO's (33) in diseased hosts. Specific serodiagnostic tests with either MAb's or polyclonal antisera (for example, see 42) permit detection and identification of a pathogen without the necessity of its isolation in culture, a special advantage in the case of uncultured MLO's.

Nucleic Acid Probes

The use of cloned genomic cDNA has for some time been employed in the detection of specific viruses in plant and insect hosts. The use of cloned probes promises also to become a widely used practice in the study and control of MLO and *Spiroplasma* diseases of plants. Cloned random fragments of DNA and known nucleotide sequences from pure-cultured *Spiroplasma* have potential for detection of these pathogens. Conceivably, for example, selected nucleotide sequences could be employed to detect members of the genus *Spiroplasma*, whereas, other less conserved sequences of species level specificity could be selected and employed for detection and identification of species.

Although the plant MLO's have not yet been pure cultured *in vitro*, it should be possible to employ recombinant DNA for their detection and identification as well. It should be possible, for example, to employ methods for cloning of nucleotide sequences in low abundance to achieve cloned MLO DNA fragments from mixtures of host and MLO DNA. Enrichment of preparations of MLO DNA by partial purification of MLO's was employed in the development of cloned sequences capable of detecting a pathogen in plants with western X disease MLO (21).

We envision that cloned nucleotide sequences will become increasingly employed for the detection/identification of both spiroplasmas and MLO's; as well as, for the study of interrelatedness, plant and insect host range, and epidemiology of the pathogens; for the study of gene organization and expression in these agents; and for the further development of serodiagnostic reagents (viz. specific antibodies) through exploitation of cloned sequences. This list mentions but a few of the potential applications of this powerful approach in detection/identification/diagnosis of plant diseases attributed to cell wall-less prokaryotes.

MOLECULAR BIOLOGY

The molecular biology of spiroplasmas is a subject of some intense research at present (1, 15,29,34,35,39-41,43,44). Soon, the same will be said of MLO's. Characterization of the size, base composition, and organization of the genome, understanding the genetics of phages and phage infections, development of gene transfer systems, and understanding genetic determinants of host range and pathogenicity are objectives of on-going studies in molecular biology of the spiroplasmas and MLO's. Genome sizes, base compositions, and sequence homologies of *Spiroplasma*, and tRNA sequences have been investigated, as have restriction enzyme patterns of DNA from some strains and the occurrence of extrachromosomal DNA. Cloning of DNA sequences from *Spiroplasma spp.* has been performed, and sequences common to plasmids have been found in several *Spiroplasma* strains in chromosome, plasmid, and phage DNA. However, it is not known what functions may be encoded by these plasmids; no *Spiroplasma* strain has been reported cured of a plasmid; and transfer of a plasmid from one spiroplasma strain to another has not been achieved.

Some laboratories have recently shown interest in research directed toward the development of a recombinant DNA and gene transfer system for *Spiroplasma*. For example,

Pascarel-Devilder, *et al.*, (41) have reported isolation, restriction mapping, and cloning of DNA of the replicative form (RF) from a naked isometric spiroplasma virus containing circular single-stranded DNA. McCammon, *et al.*, (34-35) have characterized a new rod-shaped spiroplasma virus isolated from *S. melliferum*. This latter virus also contains single-stranded circular DNA. Restriction mapping of the RF of this virus revealed single cleavage sites for the restriction enzymes *Eco* R1 and Msp 1 and multiple sites for *Acc* 1, *Hin* f1, and *TAQ* 1. Both research groups (34,35,41) found single-stranded viral DNA, as well as the respective double-stranded replicative forms, to be capable of transfecting sensitive spiroplasma hosts. Pascarel-Devilder, *et al.*, (41) found that the RF of the virus they studied could transfect spiroplasma after cloning in *Escherichia coli*. Further study of systems such as these could lead to construction of a desired molecular gene vector. Development of a genetic system involving recombinant DNA for *Spiroplasma* will provide important means needed to study many properties of this microorganism, including mechanisms regulating host range and pathogenicity, and how these mechanisms eventually might be controlled to reduce disease.

CONCLUDING REMARKS

The foregoing discussion sketches only some of the important trends in research on plant mycoplasmas. There are others. For example, the use of DNA-staining compounds has increased and may continue to be refined as a general method for agent detection. This methodology has potential particularly in studies of distribution and movement of known pathogens in hosts, particularly plants. Serological procedures are similarly experiencing wider application. This is in part due to the development of improved serological reagents including monoclonal antibodies. It is also due to the increasing use of efficient ELISA and dot immunobinding procedures. Moreover, new methods, such as immunofluorescence of *Spiroplasma*, while retaining helical cell shape, have potential in enhancing pathogen detection (22).

As noted, major progress is anticipated from future use of hybridoma technology, of recombinant DNA methodologies, and of new methods for pathogen cultivation *in vitro*. These and other approaches now beginning to receive emphasis in phytopathology have broad and far reaching implications for plant quarantine, for understanding disease epidemiology and pathogen relatedness, for our ability to culture additional phytopathogenic wall-less prokaryotes *in vitro*, and for the development of fundamental understanding of pathogen genetics and the disease process.

REFERENCES

1. Bove, J.M., 1984. Wall-less prokaryotes of plants. Ann. Rev. Phytopathol. 22: 361-396.

2. Chang, C.J., 1985. Lipid utilization of two flower spiroplasmas and honey bee spiroplasma. Can. J. Microbiol. 31: 173-176.

3. Chang, C.J. and Chen, T.A., 1979. A PPLO broth base-free medium for the cultivation of *Spiroplasma citri*. Abstr. 71st Ann. Meet. Am. Phytopthol. Soc. St. Paul: Am. Phytopathol. Soc. p. 647.

4. Chang, C.J. and Chen, T.A., 1982. Spiroplasmas: Cultivation in chemically defined medium. Science 215: 1121-22.

5. Chang, C.J. and Chen, T.A., 1983. Nutritional requirements of two flower spiroplasmas and honey bee spiroplasma. J. Bacteriol. 153: 452-57.

6. Davis, R.E., 1979. Spiroplasmas: Newly recognized arthropod-borne pathogens. In: Leafhopper Vectors and Plant Disease Agents. Maramorosch K. and Harris, K.F., (Eds.)

7. Davis, R.E., 1976. Spiroplasma: Role in the diagnosis of corn stunt disease. Proc. Intnl. Maize Virus Dis. Colloq. Workshop, August 16-19, 1976. Ohio Agr. Res. and Dev. Center, Wooster, pp. 92-96.

8. Davis, R.E., 1974. New Approaches to diagnosis and control of plant yellows disease. 3rd Intern. Virus Diseases Ornamental Plants 11-15 Sept. 1974, pp. 289-302. College Park, MD, Tech. Commun. Intern. Soc. Hort. Sci., Acta Horticulturae, No. 36.

9. Davis, R.E. and Lee, I.-M., 1985. Plaque assay of spiroplasma virus on sensitive indicator lawns in serum-free medium. Phytopathology 75: 1380.

10. Davis, R.E. and Lee, I.-M., 1985. Primary isolation of *Spiroplasma citri* and corn stunt spiroplasma in serum-free medium. Phytopathology 75: 1380.

11. Davis, R.E. and Lee, I.-M., 1982. Pathogenicity of spiroplasmas, mycoplasma-like organisms, and vascular-limited fastidious walled bacteria. (Chapter 21, pages 491-513. In: Phytopathogenic Prokaryotes. (Mount, M.S., and Lacy, G., (Eds.)

12. Davis, R.E. Lee, I.-M. Jordan, R. and Konai, Meghnad., 1987. Detection of corn stunt spiroplasma in plants by ELISA employing monoclonal antibodies and by isolation of the pathogen and its phage in serum-free medium. In: Plant pathogenic bacteria, Proc. VI Intl. Conf. Pl. Path.

Bacteria, June 2-7, 1985. Martinus Nyhoff (E.L. Civerolo, A. Collmer, R.E. Davis, A.G. Gillaspie, (Eds.) pp. 306-312.

13. Davis, R.E. and Worley, J.F., 1973. Spiroplasma: Motile, helical microorganism associated with corn stunt disease. Phytopathology 65, 403-408.

14. Davis, R.E., Worley, J.F., Whitcomb, R.F., Ishijima, T. and Steere, R.L., 1972. Helical filaments produced by a mycoplasma-like organism associated with corn stunt disease. Science 176 (4034), 521-523.

15. Dickinson, M.J. and Townsend, R., 1984. Characterization of the genome of a rod-shaped virus infecting 5Spiroplasma citri. Gen. Virology 65: 1607-1610.

16. Doi, Y., Teranaka, M., Yora, K. and Asuyama, H., 1967. Mycoplasma- or PLT-like microorganisms found in the phloem elements of plants infected with mulberry dwarf, potato witches' broom, aster yellows or Paulownia witches' broom. Ann. Phytopathol. Soc. Jpn. 33: 259-66. 1969. Rev. Plant Prot. Res. 2: 84-88 (English trans.)

17. Fudl-Allah, A.E.-S.A., Calavan, E.C. and Igwegbe, E.C.K., 1972. Culture of a mycoplasma-like organism associated with stubborn disease of citrus. Phytopathology 62, 729-731.

18. Hackett, K.J. and Lynn, D.E., 1985. Cell assisted growth of a fastidious spiroplasma. Science 230: 825.

19. Ishiie, T., Doi, Y., Yora, K. and Asuyama, H., 1967. Suppressive effect of antibiotics of the tetracycline group on symptom development of mulberry dwarf disease. Ann. Phytopathol. Soc. Jpn. 33: 267-75.

1969. Rev. Plant Prot. Res. 2: 91-5 (English Trans.)

20. Jordan, R., Konai, M., Lee, I.-M., Hsu, H.T. and Davis, R.E., 1987. Development of monoclonal antibodies to spiroplasma utilizing a mixture of *Spiroplasma citri* and corn stunt spiroplasmas as immunogen. In: Plant pathogenic bacteria, Proc. VI Intol. Conf. Pl. Path. Bacteria, June 2-7, 1985. Martinus Myhoff Publ. E.L. Civerolo, A. Collmer, R.E. Davis, A.G. Gillaspie, (Eds.) pp. 340-346.

21. Kirkpatrick, B.C., Stenger, D.C., Morris, T.J. and Purcell, A.H., 1985. Detection of X-disease mycoplasma-like organisms in plant and insect hosts using cloned, disease specific DNA. Phytopathology 75: 1351.

22. Konai, M., Davis, R.E. and Jordan, R., 1986. Immunofluorescence microscopy using monoclonal antibodies to study *Spiroplasma* with preserved helical cell shape. Phytopathology. 76.

23. Lee, I.-M and Davis, R.E., 1987. Strains of corn stunt spiroplasma defective in helicity and motility. In: Plant pathogenic bacteria, Proc. VI Intl. Conf. Pl. Path. Bacteria, June 2-7, 1985 Nartinus Nyhoff Publ. E.L. Civerolo, A. Collmer, R.E. Davis, A.G. Gillaspie, (Eds.), pp. 364-369.

24. Lee, I.-M. and Davis, R.E., 1986. Prospects for *in vitro* culture of plant-pathogenic mycoplasma-like organisms. Ann. Rev. Phytopathol. 24: 339-354.

25. Lee, I.-M. and Davis, R.E., 1983. Chemically defined medium for cultivation of several epiphytic and phytopathogenic

spiroplasmas. Appl. Environ. Microbiol.
46(6) : 1247-51.

26. Lee, I.-M. and Davis, R.E., 1985. Strains
of corn stunt spiroplasma defective in
morphology and motility. Phytopathology
75: 1380 (Abstr.).

27. Lee, I.-M. and Davis, R.E., 1984. New media
for rapid growth of *Spiroplasma citri*
and corn stunt spiroplasma. Phytopathology
74: 84-89.

28. Lee, I.-M. and Davis, R.E., 1982. Serum-
free medium for cultivation of phytopathogenic and epiphytic spiroplasmas. Phytopathology 72: 1004.

29. Lee, I.-M. and Davis, R.E., 1980. DNA
homology among diverse spiroplasma strains
representing several serological groups.
Can. J. Microbiol., 26: 1356-1363.

30. Liao, C.H., Chang, C.J. and Chen, T.-A.,
1979. Spiroplasmastatic action of plant
tissue extracts. Republic of China-U.S.
Cooperative Sci. Program, Joint Sem. on
Mycoplasma Dis. Plants, Taipei, 27-31 March
1978, NSC Symposium Series No. 1, pp.
99-103 National Science Council, Taipei,
Taiwan.

31. Liao, C.H. and Chen, T.A., 1980. Presence
of spiroplasma-inhibitory substances in
plant tissue extracts. Can. J. Microbiol.
26: 2807-11.

32. Lin, C.P. and Chen, T.A., 1985. Production
of monoclonal antibodies against *Spiroplasma citri*. Phytopathology 75: 848-
851.

33. Lin, C.P. and Chen, T.A., 1985. Monoclonal
antibodies against the aster yellows agent.

Science 227: 1233-1235.

34. McCammon, S.L. and Davis, R.E., 1987. Spiroplasma virus SVTS2: partial characterization and transfection. (In Press).

35. McCammon, S. and Davis, R.E., 1987. Trans-
fection of *Spiroplasma citri* with DNA
of a new rod-shaped spiroplasma virus. In:
Plant pathogenic bacteria, Proc. VI Intl.
Conf. Pl. Path. Bacteria, June 2-7, 1985.
Martinus Nyhoff, Publ. E.L. Civerolo, A.
Collmer, R.E. Davis, A.G. Gillaspie, (Eds.)
pp. 458-464.

36. McCoy, R.E., Davis, M.J. and Dowell, R.V.,
1981. *In vivo* cultivation of
spiroplasmas in larvae of greater wax
moth. Phytopathology 71: 408-11.

37. McCoy, R.E., 1976. Plant phloem sap: a
potential mycoplasma growth medium. Proc.
Soc. Gen. Microbiol. 33:155 (Abstr.).

38. McCoy, R.E., 1977. Growth of mycoplasmas
in phloem sap from lethal yellowing resistant Malayan dwarf coconut palm. Proc. Am.
Phytopathol. Soc. 4:108 (Abstr.)

39. Mouches, C., Barroso, G. and Bove, J.M.,
1983. Characterization and molecular
cloning in *Escherichia coli* of a
plasmid from the mollicute *Spiroplasma
citri*. J. Bacteriol. 156: 952-955.

40. Mouches, C., Candresse, T., Barroso, G.,
Saillard, C., Worblewski, H. and Bove, J.
M., 1985. Gene for spiralin, the major
membrane protein of the helical mollicute
Spiroplasma citri: cloning and expression in *Escherichia coli*. J. Bacteriol.
164: 1094-1099.

41. Pascarel-Devilder, M.-C., Renaudin, J. and

Bove, J.M., 1986. The spiroplasma virus 4 replicative form cloned in *Escherichia coli* transfects spiroplasmas. Virology 151: 390-393.

42. Raju, B.C., and Nyland, G., 1981. Enzyme-linked immunosorbent assay for detection of corn stunt spiroplasma in plant and insect tissues. Current Microbiol. 5:101-104.

43. Razin, S. 1984. Molecular and biological features of Mollicutes. Ann. Microbiol. (Inst. Pasteur) 135A: 9-15.

44. Rogers, M.J., Steinmetz, A.A. and Walker, R.T., 1986. The nucleotide sequence of a tRNA gene cluster from *Spiroplasma melliferum*. Nucleic Acids Res. 14: 3145.

45. Saglio, P., D. Lafleche, D., L'Hospital, M., Dupont, G. and Bove, J.M., 1972. Isolation and growth of citrus mycoplasmas. In: Pathogenic Mycoplasmas, CIBA Foundation Symposium, London, England: 25-27 January 1972 (J. Birch, Ed.) pp. 187-198. Elsevier and North Holland, Amsterdam.

46. Saglio, P., D. Lafleche, D., L'Hospital, M., Dupont, G., Bove, J.M., Tully, J.G. and Freundt, E.A. *Spiroplasma citri* gen. and sp. n: a mycoplasma-like organism associated with "stubborn" disease of citrus. Int. J. Syst. Bacteriol. 23, 191-204.

47. Saglio, M.P., Lafleche, D., Bonissol, C. and Bove, J.M., 1971. Isolement et culture *in vitro* des mycoplasmes associes au stubborn des agrumes et leur observation au microscope electronique. C.R. Acad. Sci. (Paris) Ser. D272, 1387-90.

48. Whitcomb, R.F., 1981. The biology of spiroplasmas. Ann. Rev. Entomol. 15: 405-464.

Table 1. Composition of serum-containing media LD8A and LD8A3 and serum-free
media LD57,LD58,LD59[a].

	LD8A	LD8A3	LD57	LD58	LD59
Amino acids (g/l)					
L-Arginine	0.6	0.6	0.6	---	0.6
L-Asparagine	0.6	0.6	0.6	---	0.6
L-Cystein HCl	0.4	---	0.4	---	---
L-Methionine	0.4	0.4	0.4	---	0.4
Organic acids (g/l)					
a-Keteglutaric acid	0.4	0.4	0.4	0.4	0.4
Pyruvic acid	0.4	0.4	0.4	0.4	0.4
Carbohydrates (g/l)					
Fructose	4.0	1.0	4.0	1.0	1.0
Sucrose	90.0	120.0	90.0	120.0	120.0
Other Components (g/l)					
PPLO broth	12.0	15.0	12.0	15.0	15.0
Yeastolate	1.5	---	1.5	---	---
Lactalbumin hydrolysate	---	2.0	---	2.0	2.0
HEPES (buffer)[b]	20.0	7.0	20.0	7.0	7.0
Fetal bovine serum (ml)	100.0	100.0	---	---	---
Lipids-BSA solution (ml)	---	---	100.0	100.0	100.0
Deionized wat (ml)	750.0	720.0	750.0	720.0	720.0

[a] Final pHs of media were 7.4 - 7.5

[b] HEPES: N-2-hydroxyethylpiperazine-N-2-ethane-selfonic acid.

7

NEW DEVELOPMENTS IN THE CULTURE

OF *SPIROPLASMA KUNKELII,*

THE CORN STUNT SPIROPLASMA

by

I.-M. Lee and R.E. Davis

Department of Botany

University of Maryland, College Park and

Microbiology and Plant Pathology Laboratory

ARS-USDA

Beltsville, MD

INTRODUCTION

Spiroplasma kunkelii, the corn stunt spiroplasma (CSS), is one of several *Spiroplasma* species that have been isolated from the plant habitat. Of those proven or suspected to be plant pathogenic, *S. kunkelii* is the most fastidious in its requirements for growth *in vitro.* *S. kunkelii* was first isolated in culture in 1975 (2,21), about four years after the isolation of *S. citri* (7,8,20) and the discovery of CSS (6). Initially, the techniques and media utilized for culture of *S. citri* were employed in attempts to isolate the CSS. However, CSS proved to be rather sensitive to inhibitory substances derived from host tissues (16,18) and medium components. Even with the development of media capable of supporting growth of the CSS, the strains initially isolated often failed to develop colonies on solid agar versions of the media, and for quantitation of the CSS researchers frequently resorted to cell counts or "color changing units" in broth medium (9).

Our recent attention was drawn to problems in the isolation and culture of CSS particularly by outbreaks of CSS disease in California and Florida, and by experience, in laboratories including our own, with difficulty in isolating CSS from plants in these outbreaks. We had also had similar experience with some CSS-infected plants from Central and South America, as well as from plants from Mississippi. Strains of CSS from some plants collected in these locations proved to be much more fastidious than the CSS strains first isolated in culture in 1975. This circumstance underscored the necessity to investigate means to improve the frequency of CSS isolation. Several laboratories undertook to modify the original formulations of media to improve CSS isolation and culture. The result of this renewed effort in the culture of CSS has been improvements in conventional serum-containing media, the development of new media lacking serum, and the isolation of highly fastidious and even aberrant strains of the CSS.

CONVENTIONAL SERUM-CONTAINING MEDIA

Since 1975, many artificial media have been formulated for the primary isolation and routine subculture of the CSS. The first media for cultivation of the CSS were C-3G (1,17,19) and M1 (21), both developed in the initial isolation of the CSS in 1975. Broth versions of these media are capable of supporting growth of strains including the type strain I747, but most CSS strains form colonies with difficulty, if at all, on agar versions of the media. Subsequently, medium LD8A was formulated in our laboratory for the primary isolation and subculture of the CSS (Table 1). In this medium, strain 1747 reached titers of over 10^9 cells/ml in broth, with a doubling time of 11 to 12 h at 31 C. In C-3G broth, CSS strain 1747 grew with a doubling time of 20 h. The relatively high titers reached in LD8A made it feasible to estimate the growth (titer) of strain 1747 and certain other strains by measuring turbidity. Moreover, these strains formed colonies readily on agar-solidified versions of LD8A, and it was possible to assess the titer of viable spiroplasma on the basis of colony forming units (CFU) (11,13).

In 1981-82, new outbreaks of corn stunting diseases were reported in California and Florida (3,4,5,10). Spiroplasmas were isolated and cultured *in vitro* from diseased plants in these outbreaks and from insect vectors collected from diseased plants. In studies to verify the identities of the spiroplasma strains associated with these outbreaks, we employed medium LD8A for primary isolations and subsequent subcultures (4). The spiroplasmas involved in the outbreaks all proved to be strains of the CSS (3,4,10), however, we found that, in particular, some strains from Florida grew poorly in medium LD8A broth and failed to form colonies on agar-solidified LD8A. In his investigations of the Florida CS disease outbreak, M. J. Davis had similar experience with medium LD8A and formulated a new medium, L20, which he found more useful under anaerobic conditions for the culture of the Florida CSS (3). Anaerobic conditions also favored the growth of Florida CSS in medium C-3G. Other studies (I.-M, Lee and R.E. Davis, unpubl.) have indicated that anaerobic conditions of incubation favor the growth of more fastidious strains of the CSS by retarding the oxidation of serum components and the consequent formation of compounds toxic to CSS (see Serum-Free Media, below). Because of the experiences with medium LD8A in the culture of Florida CSS, we undertook the reformulation of LD8A. This approach resulted in medium LD8A3 (Table 1).

Medium LD8A3 was found to support more rapid growth of Florida CSS, as well as of a Mississippi CSS, under air or in candle jar than did either C-3G or LD8A. Although most strains of CSS grew relatively well in broth LD8A3, C-3G,and other serum-containing media, the more highly fastidious strains (including some Florida and Mississippi strains) formed colonies with low efficiency, or not at all, on agar versions of the media even under anaerobic conditions of incubation.

SERUM-FREE MEDIA

In 1982, we formulated an initial serum-free medium, LD57, by replacing serum in LD8A with a mixture of lipids and bovine serum albumin for nutritional investigations of *Spiroplasma* including the plant pathogens *S. citri* and *S. kunkelii* (Table 1) (12). In the course of this work, we found that growth of *S. kunkelii* strains I747 and PU8-17 in LD57 was comparable to that in medium LD8A. When we applied the serum-free medium to the culture of more fastidious strains of the CSS, we found it to yield higher cell titers in broth, more consistent colony formation on agar versions, and higher plating efficiencies than media containing serum. Primary isolations of CSS were also achieved with greater efficiencies in serum-free medium than in serum-containing medium.

Further improvements in the serum-free medium resulted in LD58 and LD59 (Table 1). These media employed, as their base, the basal (non-serum) components of the serum-containing medium LD8A3. These improved serum-free media have subsequently been employed for primary isolation and routine subculture of CSS from Florida, California, Mississippi, Nicaragua, and Mexico, as well as of highly fastidious aberrant strains of the CSS (14,15). Some of the CSS strains cultured in these serum-free media could not be continuously cultured in any of the serum-containing media we tested.

The superiority of serum-free media for the culture of fastidious strains of the CS spiroplasma in our experiments may be attributable to a number of factors. For example, BSA-lipids complex in serum-free media may provide lipids, which are required for growth and are in a form more readily incorporated by these strains. Also, we found that the serum-free media maintained their growth-supporting capacity much longer than did serum-containing media tested. Long lasting growth-supporting capacity of a culture medium is a property that is indispensable for the culture of slow-growing organisms. This is especially true for certain unusual fastidious strains of CS spiroplasma (doubling time over 20 h) where one desires colony formation on agar plates. For example, strain I-2S of CS spiroplasma required more than 20 days of incubation to form visible colonies on serum-free agar medium LD58. Our data indicated that the growth-supporting efficacies of serum-containing media dropped rapidly after being stored for seven days.

The inhibitory effects of aging during long periods of incubation thus may be important factors that can seriously limit growth supporting capacity of serum-containing broth or agar media. Anaerobic incubation, which improved the growth of fastidious CS spiroplasma strains in serum-containing medium, may, in fact, have simply facilitated preservation of the growth-supporting capacity of the medium by slowing deterioration of its serum components. An anaerobic environment is not required for good growth of these fastidious strains in freshly prepared serum-free medium LD58.

CULTURE OF ABERRANT STRAINS

In the course of primary isolation of CSS from diseased plants from Mississippi, we isolated a number of strains with varying degrees of defects in helicity and motility. The spiroplasma strains produced three general types of colony on agar-solidified LD59. One consisted of large uniformly diffuse colonies (up to 2 mm in diam.); one consisted of very minute (<0.1mm) well-defined "fried-egg" shaped colonies; and the third consisted of intermediate-sized, poorly defined "fried-egg" shaped colonies. surrounded by granular satellite colonies. Colony type was related to cell shape and motility. For example, diffuse colonies were formed by helical and highly motile strains, whereas, minute "fried-egg" colonies were formed by strains that lacked helical cell shape and therefore also lacked translational motility. These latter strains were the first non-helical strains reported for CSS. The existence of these non-helical spiroplasmas cautions consideration of morphology as the major criterion for distinguishing spiroplasmas from mycoplasma-like organisms (MLO's) in plants.

CONCLUDING REMARKS

It is now evident that there exists in nature a wide variety of CSS strains differing markedly in their ease of culture *in vitro*. The isolation of fastidious CSS strains in serum-free media, and the failure of some such strains to grow thus far in serum-containing media, indicate that the *Spiroplasma* strains isolated previously from diseased hosts provide a selected representation of less fastidious portions of *Spiroplasma* populations. The isolation of strains representing the population as a whole could be the key to eventual understanding of the naturally occurring disease, of epidemiology, and how to achieve effective disease control.

REFERENCES

1. Chen, T.A. and Davis, R.E., 1979. Cultivation of spiroplasmas. In: The Mycoplasmas, Whitcomb, R.F. and Tully, J.G., 3: 65-82.

New York/San Francisco/London: Academic.
Press. pp. 351.

2.   Chen,   T.A.   and   Liao,   C.H.,   1975.   Corn
      stunt   spiroplasma: Isolation, cultivation,
      and   proof   of   pathogenicity.   Science 188:
      1015-17.

3.   Davis,   M.J.,   Tsai,   J.H.   and McCoy, R.E.,
      1984.   Isolation of corn stunt spiroplasma
      from   maize   in   Florida.   Plant Dis. 68:
      600-4.

4.   Davis,   R.E.,   and Lee. I.M., 1984.   Identi-
      ties   of   new spiroplasmas reported in maize
      in   the   United States.   In: Proc. 2nd Int.
      Maize Virus Dis. Colloq. Workshop 1982,
      Gordon,   D.T.,   Knoke, J.K., Nault, L.R. and
      Ritter,   R.M.,   (eds.) pp. 51-55. Wooster:
      Ohio State Univ. Press.   pp. 261.

5.   Davis.   R.E.   and   Lee, I.M., 1985.   Primary
      isolation   of   *Spiroplasma citri* and   corn
      stunt spiroplasma in serum-free medium. Phy-
      topathology 75: 1380   (Abstr.)

6.   Davis,   R.E.,   Worley, J.F., Whitcomb, R.F.,
      Ishijima,   T. and Steere, R.L., 1972.   Heli-
      cal   filaments produced by a mycoplasma-like
      organism associated with corn stunt dis-
      ease.   Science 176: 521-23.

7.   Fudl-Allah,   A.E.A.,   Calavan,   E.C. and
      Igwegbe,   E.C.K.,   1971.   Culture of a myco-
      plasma-like organism associated with stub-
      born disease of citrus. Phytopathology 61:
      1321.   (Abstr.)

8.   Fudl-Allah,   A.E.S.,   Calavan,   E.C. and
      Igwegbe,   E.C.K.,   1972.   Culture of a
      mycoplasma-like organism associated with
      stubborn disease.   Phytopathology 62:
      729-31.   (Abstr.)

9. Jones, A.L., Whitcomb, R.F., Williamson, D.L. and Coan, M.E., 1977. Comparative growth and primary isolation of spiroplasmas in media based on insect tissue culture formulation. Phytopathology 76: 738-46.

10. Kloepper, J.W., Garrott, D.G., Kirkpatrick, B.C. and McCutcheon, O.D., 1982. Association of spiroplasmas with a new disease of corn in California. Calif. Plant Pathol. 56: 5-6.

11. Lee, I.M. and Davis., R.E., 1978. Identification of some growth-promoting components in an enriched cell-free medium for cultivation of *Spiroplasma citri*. Phytopathol. News 12(9): 215. (Abstr.).

12. Lee, I.M. and Davis., R.E., 1982. Serum-free medium for *in vitro* cultivation of phytopathogenic and epiphytic spiroplasmas. Phytopathology 72: 1004. (Abstr.).

13. Lee, I.M. and Davis., R.E., 1984. New media for rapid growth of *Spiroplasma citri* and corn stunt spiroplasma. Phytopathology 74. 84-89.

14. Lee, I.M. and Davis., R.E., 1985. Strains of corn stunt spiroplasma defective in morphology and motility. Phytopathology 75: 1380. (Abstr.).

15. Lee, I.M. and Davis., R.E., 1987. Strains of corn stunt spiroplasma defective in helicity and motility. In: Plant Pathogenic Bacteria, Proc. VI Conf. Intl. Plant Pathogenic Bacteria. Martin Nyhoff Publ. (E.L. Civerolo, A. Collmer, R.E. Davis, A.G., Gillaspie, Eds.) pp. 364-369.

16. Liao, C.H. and Chen, T.A., 1975. Inhibitory effect of corn stem extract on the

growth of corn stunt spiroplasma. Proc. Am. Phytopathol. Soc. 2: 53. (Abstr.).

17. Liao, C.H. and Chen, T.A., 1977. Culture of corn stunt spiroplasma in a simple medium. Phytopathology 67: 802-807.

18. Liao, C.H. and Chen, T.A., 1977. Presence of spiroplasma-inhibitory substances in plant tissue extracts. Can. J. Microbiol. 26: 807-11.

19. Liao, C.H. and Chen, T.A., 1982. Media and methods for culture of spiroplasmas. In: Plant and Insect Mycoplasma Techniques. Daniels, M.J. and Marhan, P.G., (eds.) pp. 174-200. London: Croom Helm. pp. 369.

20. Saglio, P., Lafleche, D., Bonissol, C. and Bove, J.M., 1971. Isolement, culture et observation au microscope electronique des structures de type mycoplasme associees a la maladie due stubborn des agrumes et leur comparaison avec les structures observees dans le cas de la maladie du greening des agrumes. Physiol. Veg. 9: 569-82.

21. Williamson, D.C. and Whitcomb, R.F., 1975. Plant mycoplasmas: A cultivable spiroplasma causes corn stunt disease. Science 188: 1018-20.

8

COMPARATIVE MORPHOLOGY OF

MYCOPLASMA-LIKE ORGANISMS

by

Bijan K. Ghosh, Robert Wood Johnson Medical School, UMDNJ, Piscataway, NJ 08854.

S. Misra, Department of Botany, University of Rajasthan, Jaipur, India.

V. Muniyappa, Department of Plant Pathology, University of Agricultural Sciences, Hebbal, Bangalore 560024, India.

INTRODUCTION

Mycoplasma-like organisms (MLO) are pathogens affecting a large variety of plants. The MLOs infect mono- and dicotylidenous plants, ranging from large trees to cultivated crops and weeds. Mycoplasmas are wall-less pleomorphic, prokaryotic organisms extensively studied from mammalian sources. They are parasitic to animals and can be grown in pure culture, while MLOs from plant and insect sources have not yet been cultivated outside the hosts. Hence, no information on the physiology and genetics of MLOs are available. Some recent reports including a few papers presented in this volume indicate that the cultivation of MLOs might be routinely possible in the near future. When such material will become available, similarities and differences between plant MLOs and animal mycoplasmas will be studied in detail. The name MLO has been coined to indicate that not enough information is available to date to establish identity or differences between them and mammal-infecting mycoplasmas. The term "plant mycoplasmas" is often used to describe both MLOs and the culturable spiroplasmas. Besides examination of the pathological symptoms of MLO infected plants, electron microscopy has been the major tool for the detection of MLO infection in plants or for the determination of their presence in the insect vectors. (Daniels and Markham, 1982) The refinement of existing techniques and the application of new technology can provide means for early diagnosis of MLO induced diseases and document the progressive development of the disease.

The present study was initiated in an effort to refine existing techniques of fixation of field samples for electron microscopic examination of MLO infected plants.

MATERIALS AND METHODS

Plant samples were collected during a planned field trip in India. Two widely varying types of material were collected: i. *Santalum album* in a reserve forest of Bangalore; ii. weeds growing in a vegetable field, which were suspected to harbor mycoplasma. The weeds used in study were *Kirganellia reticulata, Tecoma stans* and *Averrhoa carambola*. In addition, we used *Nicotiana rustica* which was experimentally infected with aster yellows MLO.

Different parts of the plants were collected and cut immediately into small pieces by hand dissection on the site; these small pieces were washed with drops of 0.85% saline and transferred into vials of fixative. The fixative was prepared by mixing freshly depolymerized paraformaldehyde and glutaraldehyde to a final concentration of 1% each in phosphate buffer, pH 7.4 containing 0.85% saline. The fixative, prepared in the United States, was sealed in screw capped vials under nitrogen. These vials were opened on site and were again sealed just after putting the tissue material into the vials. The material in these vials was further processed in the laboratory in the United States. The fixative from the vials was drained out and the tissue samples were treated with a mixture of glutaraldehyde and tannic acid (1.0% each) for 1 hr. Following this the tissues were washed repeatedly with cacodylate buffer pH 6.5 and further treated with 1.0% osmium tetroxide for 10-12 hr (overnight) in the dark. Next morning, the tissues were washed free of the osmium tetroxide and were treated with 0.1% uranyl acetate for 90 min. After this treatment, the samples were processed by usual methods of dehydration in graded ethanol and embedded in Spurr medium (Ghosh, 1977). Sections from the samples were cut in a Sorvall MT2B microtome using a diamond knife. The resulting thin sections were stained with Reynolds lead citrate (Reynolds, 1963) and examined in a JEM 100C electron microscope.

RESULTS

The aim of the study was to examine the differences in the morphology of MLO in various tissues of the same plant. In addition, these differences were compared in different plants. The following is a brief presentation of the data obtained from the study.

Santalum album

Petiole:
The segments of tissues were collected from the base of small leaves with characteristic symptoms of the spike disease (Varmah, 1981). It is known that the sieve elements of the petiole contain MLO. In thin sections, sieve elements were thoroughly examined and heavily infected regions could be detected (Fig. 1). It was seen in a low magnification micrograph that

Young Bark

The MLOs in the young bark were highly pleomorphic, only a few dark staining spherical bodies were present. The bilayered membranes were intact around some of the organisms. In contrast to the MLOs present in the petiole and the shoot,the highly rarefied appearance of the cytoplasmic material of the MLOs in the bark suggested that these were degenerated bodies (Fig. 9).

Old Bark

In the old bark, a large number of both spherical (Fig. 10 a,b) and pleomorphic MLOs (Fig. 10 C) were detected. Although most of these bodies were empty, a few containing cytoplasmic material, allowed their identification as MLOs. The MLOs at this location were extensively degenerated.

K. reticulata and *A. carambola*

In both these weed plants a thorough search was made in the petiole and in the main leaf vein for MLOs. In some degenerated sieve elements highly pleomorphic membrane bound bodies were boserved (Fig. 11 a,b). These bodies rarely contained any ribosome or fibrillar nuclear material (Fig. 11 b). However, this material was abundant and frequent, suggesting that the MLOs were present, but these might have degenerated after an initial proliferation. Again, one may question if these bodies were a product of poor fixation. However, the presence of highly organized Golgi bodies and rough endoplasmic reticulum in the adjacent healthy portion of the tissue argued against this possibility (Fig. 12). Examination of *A. carambola* also revealed the presence of similar degenerated MLOs.

the space inside the sieve elements showed necrosis while others showed healthy areas. Under
a slightly higher magnification (Fig. 2) the
packing of the MLOs was tight, the distribution
was uniform and the pleomorphism was low. Mostly, the MLOs were spherical (Fig. 3 a,b,c); however, some elongated and tortuous bodies could
be observed (Fig. 2 and 3). This observation
suggested that organisms of comparable sizes
aggregated in one location. Bilayered plasma
membrane, ribosomal material and fibrilar nuclear material were seen (Fig. 3). The quality
of fixation could only be tested if different
subcellular bodies were preserved. This was
evident from the cluster of Golgi bodies, present in the uninfected portion of the tissue,
and exhibiting their characteristic organization (Fig. 4). The chloroplast, plasmodesmata
and cellulose fibrils of the wall were also distinct (Fig. 5).

Terminal Shoot:

The sieve elements of the terminal shoot
also contained large numbers of MLOs. These
were in the form of spherical bodies, greatly
varying in size. In the terminal shoot pleomorphism was rare; besides a few elongated
bodies there was remarkable uniformity of
shape. It seems that the MLOs were densely
distributed in association with the plasma
membrane. Unlike the petiole, the sizes of
these MLOs varied widely in the terminal shoot
(Fig. 6). In addition, at some location, ellip-
tical bodies were also seen; however, cytoplasmic material in these bodies was disorganized
suggesting their degenerated nature (Fig. 7)
(Hirumi and Maramorosch, 1972). In Fig. 8,
mitochondria with characteristic cristae and a
MLO were seen. Close association between the
membranes of these two bodies clearly indicated
the difference in the structure of the
membrane.

N. rustica:

Presence of MLO has been extensively studied in this plant. Laboratory infected plants were used in the present study. The leaf vein of such an infected plant was examined for the presence of MLO. The sieve element of the leaf vein contained a large number of MLOs. Although they were mostly spherical, there were a significant number of pleomorphic bodies and all of these showed a wide variation of sizes (Fig. 13 a,b). The bilayered membrane was distinct and frequently dividing MLOs were found (Fig. 14 a, b). The presence of comparable dividing bodies has been reported earlier (Hirumi and Maramorosch, 1973). The MLOs comparable to those present in the leaf vein were also observed in the petiole and young shoot. In addition, it was interesting to detect pleomorphic MLOs in the root. These generally accumulated in the space between the plasma membrane and the wall; this distribution is unique for the root (Fig. 15).

Many chloroplasts in the leaf vein showed varying degrees of degeneration which is seen by comparing Fig. 16 a and b. The chloroplasts showed various degrees of swelling and pleomorphic MLOs were present inside these degenerating chloroplasts. Some of these degenerated chloroplasts disintegrated and released MLOs. These abnormal chloroplasts were difficult to identify (Fig. 17). This intraorganellar development of the MLOs had a significant toxic effect on the plant.

DISCUSSION

Demonstration of MLO in a plant is not a novel observation. This is an established plant pathogen carried by insect vectors. The electron microscopy has been the tool for the demonstration of MLOs. It is well known that specimen preparation technique for electron microscopy, if improperly applied, may generate artifacts. This is particularly true for plant samples for MLO examination because these are collected in the field where adequate facilities for good fixation are unavailable. On the other hand, due to high resolution, electron microscopy of carefully prepared specimens may provide information on the growth, distribution and the characteristics of polymorphism of the MLOs. The present study has been initiated in an attempt to correlate field and laboratory experiments. Our results emanate from three dissimilar specimens, a large tree *(S. album)* from a forest reserve in India, weeds from vegetable crop fields in India and an aster yellow-diseased plant experimentally infected in the United States. The results presented here, although preliminary, are significant and show the future possibility of studies on the basic understanding of pathogenicity of MLO.

The present data show that pleomorphism and distribution are interrelated. At the site of fast growth i.e., sieve element of petiole and the main vein of the leaf, MLOs have uniformly spherical shape. These MLOs elongate while they divide, by constriction. A small variation in the sizes of profiles in random sections suggest that the sizes of MLOs in sieve elements do not vary widely. The concentration of MLOs appears to be greater in the vicinity of the plasma membrane. The size variation appears to be snall in *S. album* and large in *N. rustica*. Furthermore, the organism seems to form clusters in the sieve element of *S. album*. In the young shoot, the organism shows differences in size and shape. This becomes more pronounced in the young bark. Intracellular material of the MLOs in these locations become rarefied and frequently vacuolated. In the old bark, MLOs contain very little cytoplasmic material and show wide variations in sizes and shapes. In the root, the MLOs are pleomorphic, presenting a mixture of rarefied and dense bodies. It seems from these observations that MLOs, after infection,may grow very rapidly in sieve elements but later become widely distributed in different tissues. In these tissues, the MLOs possibly grow poorly and gradually degenerate.

Another interesting observation is the possible intraorganellar growth of MLOs particularly inside chloroplasts. It is possible that after entering the plant body, especially leaves, the MLOs start to grow inside chloroplasts and then these are released into sieve elements. In this site due to the availability of adequate nutrients, fast growth of MLOs could occur. Due to the rapid flow of fluid through sieve elements, the MLOs become widely distributed in the plant body. Although this concept of infection is hypothetical and based only on preliminary observation, a study should be conducted in this direction to gain a clear understanding on the pathogenicity of MLOs.

Observation of the weeds in the vegetable crop field gives strong indication for the presence of MLOs. However, it is possible that these plants are naturally resistant to MLO infection. Hence, MLOs entering into these plants through the feeding of insect vectors grow very poorly and then degenerate. However, storage of MLOs over a period of time, may provide an intermediate state in the spread of infection.

REFERENCES

Daniels, M.J., and Markham, P.G. (eds.) 1982. Plant and Insect Mycoplasma Techniques. John Wiley and Sons, pp. 369.

Ghosh, B.K., 1977. Techniques to study the ultrastructure of bacteria. Handbook of Microbiology, Vol. 1, H.A. Lechevalier and A.I. Laskin, (Eds.). CRC Press, pp. 31-38.

Hirumi, H., and Maramorosch, K. 1972. Natural degeneration of mycoplasma-like bodies in an aster yellows infected host. Phytopathol. Z. 75: 9-26.

Hirumi, H., and Maramorosch, K. 1973. Ultra-structure of the aster yellows agents: Myco-plasma-like bodies in sieve tube elements of *Nicotiana rustica*. Ann. N.Y. Acad. Sci. 225: 522-530.

Reynolds, E.S., 1963. The use of lead citrate at high pH as an electron opaque stain in electron microscopy. J. Cell Biol. 17. p. 208.

Varmah, J.C. 1981. Sandal *(Santalum album)* spike disease. In: Mycoplasma Diseases of Trees and Shrubs. K. Maramorosch and S.P. Raychaudhuri, eds. Academic Press: 253-258.

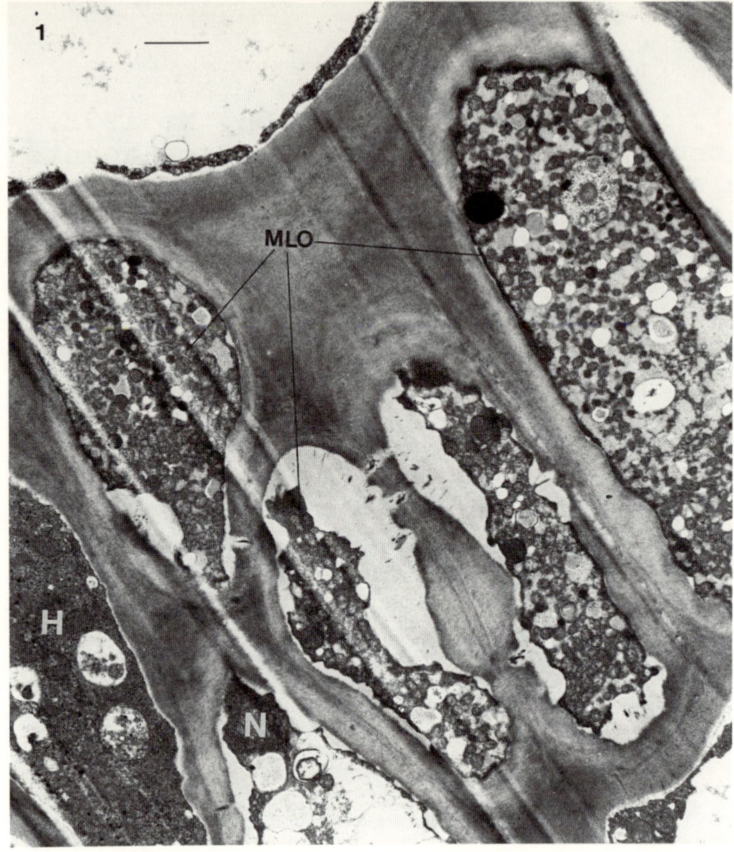

Figure 1. A low magnification micrograph of a thin section of the sieve element of the petiole showing a large number of MLOs; M, MLO; N, necrotic tissue; H, healthy tissue, Marker = 1.0 μ in all micrographs except Figure 14.

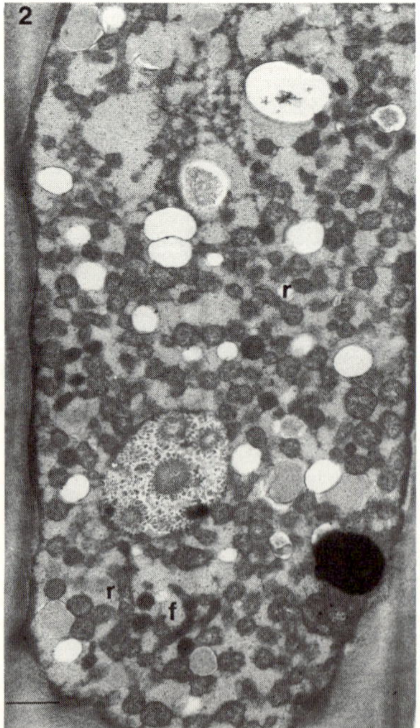

Figure 2. A slightly higher magnification micrograph showing uniform packing of the above sieve elements with MLOs; note one filamentous body (1) and a few rod-shaped structures (r).

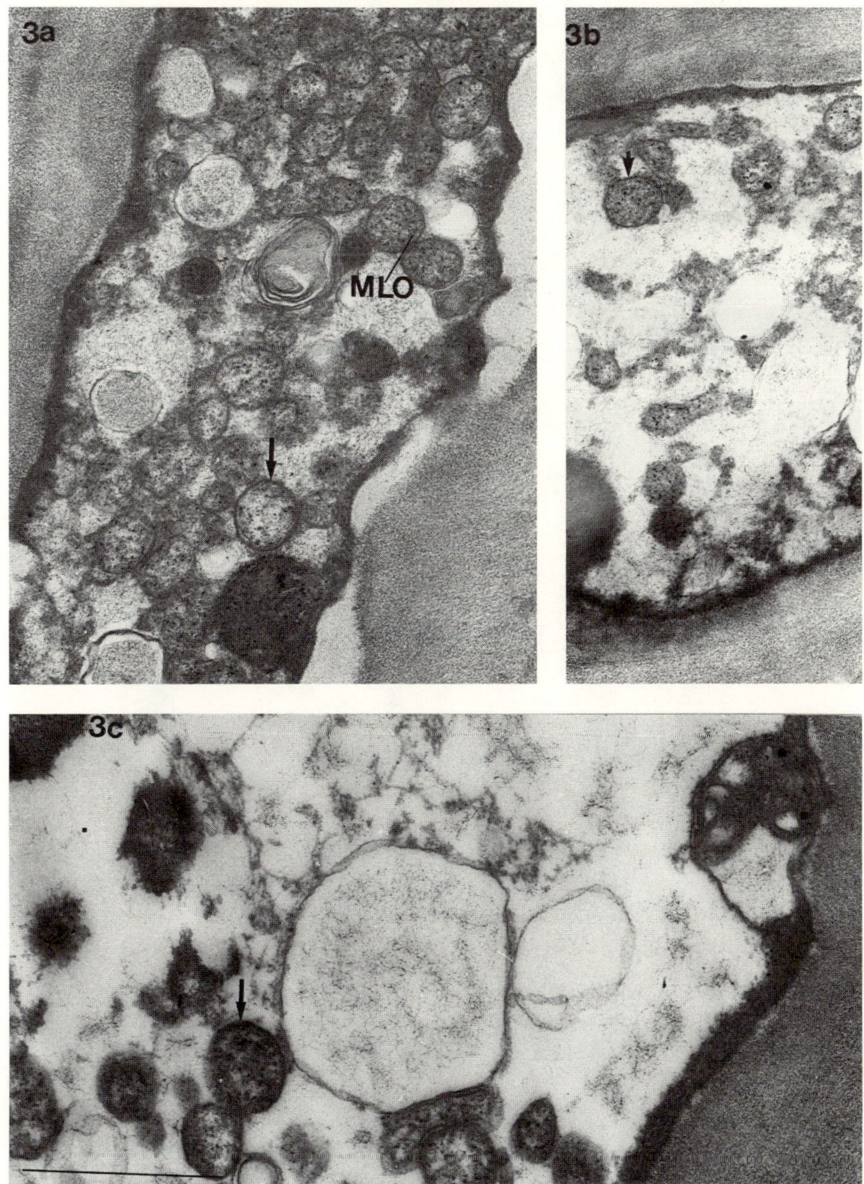

Figure 3. a,b,c. Closely packed spherical MLOs (a) and bilayered membranes (a,b,c, arrows).

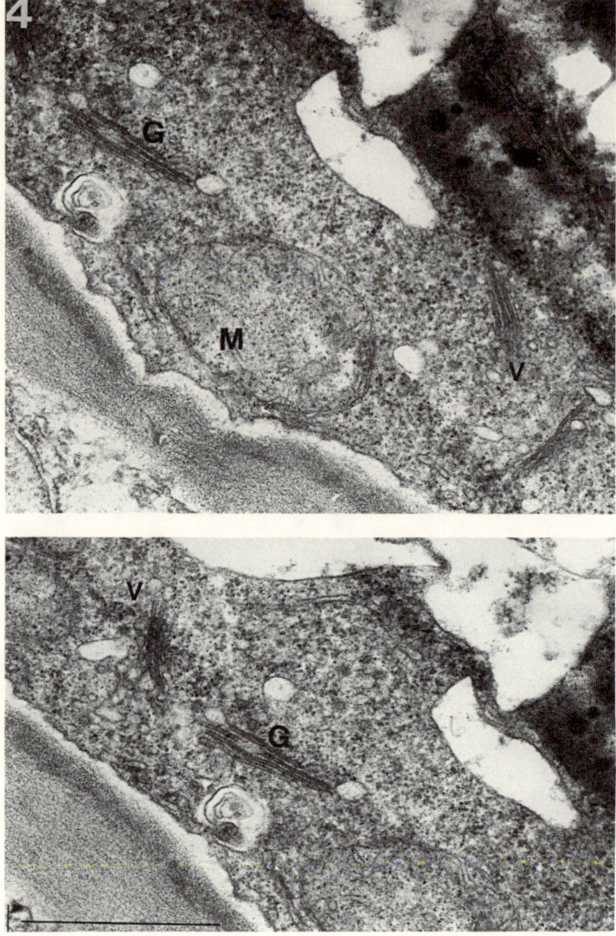

Figure 4. Uninfected portion of the plant. Highly characteristic multiple layers of Golgi bodies (G) and many vesicles (v) around these Golgi bodies are also seen. Mitochondria (m) show superior quality of fixation.

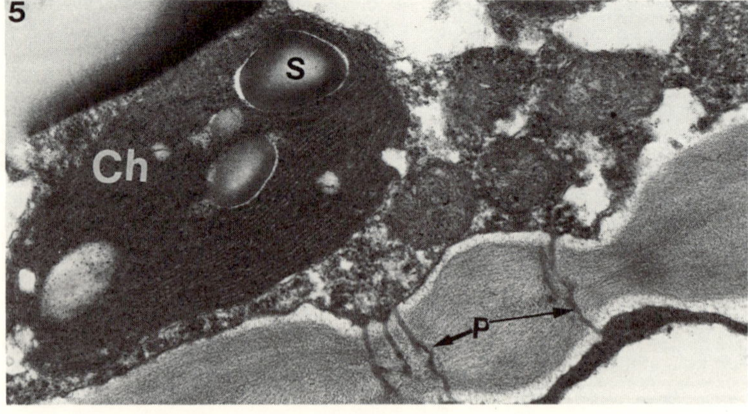

Figure 5. Plasmodesmata (P). Chloroplast (ch) with starch granule (s) is also seen.

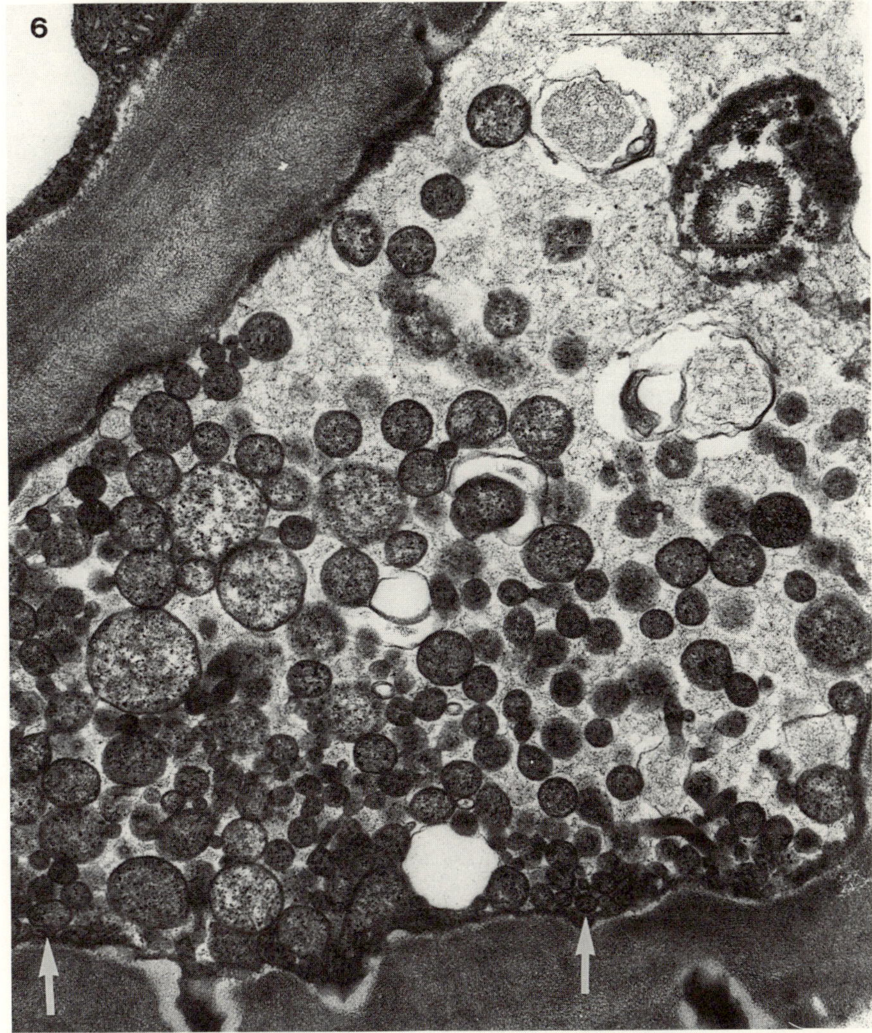

Figure 6. MLOs in the sieve element of the terminal shoot. Note the large variation in sizes. Although these are mostly spherical, rare filamentous bodies are seen (f). MLOs are densely packed in association with the plasma membrane (arrows).

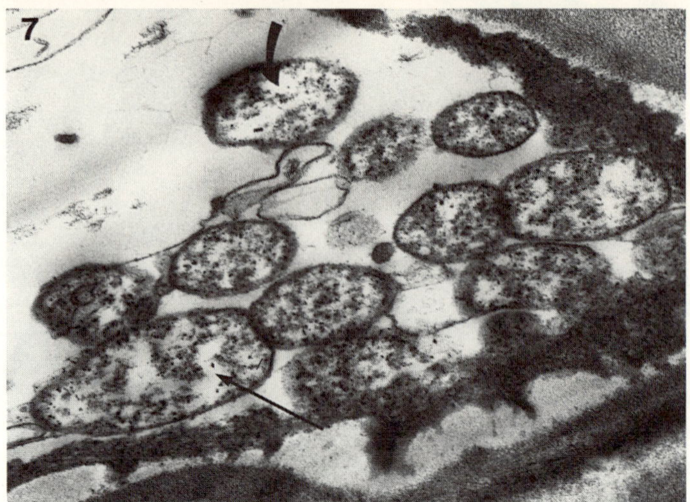

Figure 7. Some MLOs in the shoot appear oblong. Although intact, the cytoplasmic material is very rarefied and vacuolated (arrows).

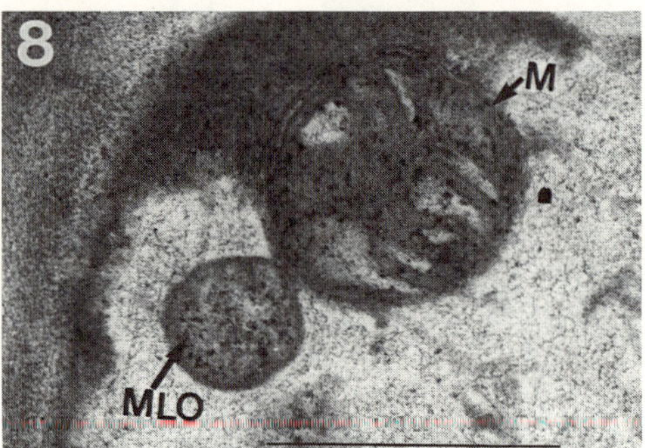

Figure 8. Closely associated MLO and a mito-chondria (M). Note the difference in the thick-ness of the membrane.

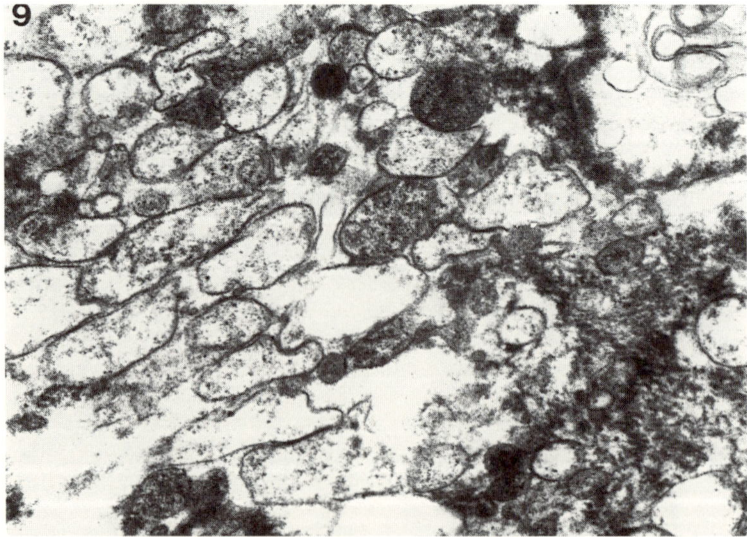

Figure 9. In young bark pleomorph, MLOs are evident. Besides a few dark-staining bodies, most of these appear to be degenerated.

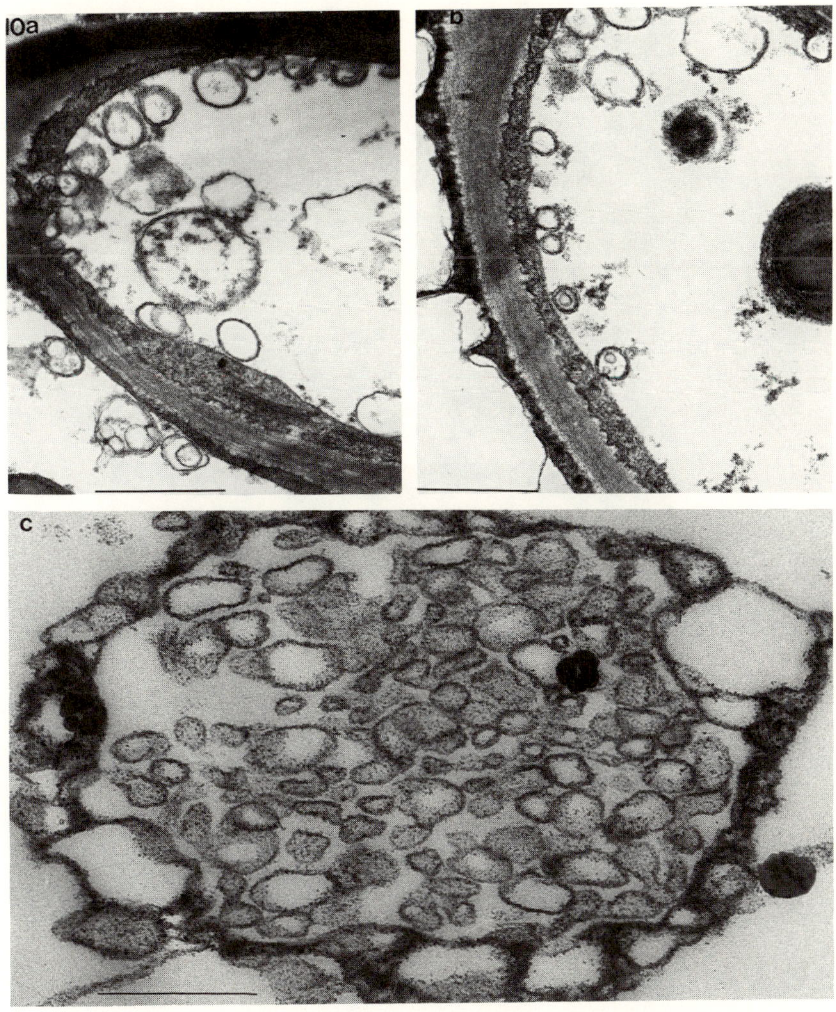

Figure 10. a,b,c. In the old bark, MLOs are pleomorphic and highly vacuolated without any cytoplasmic material.

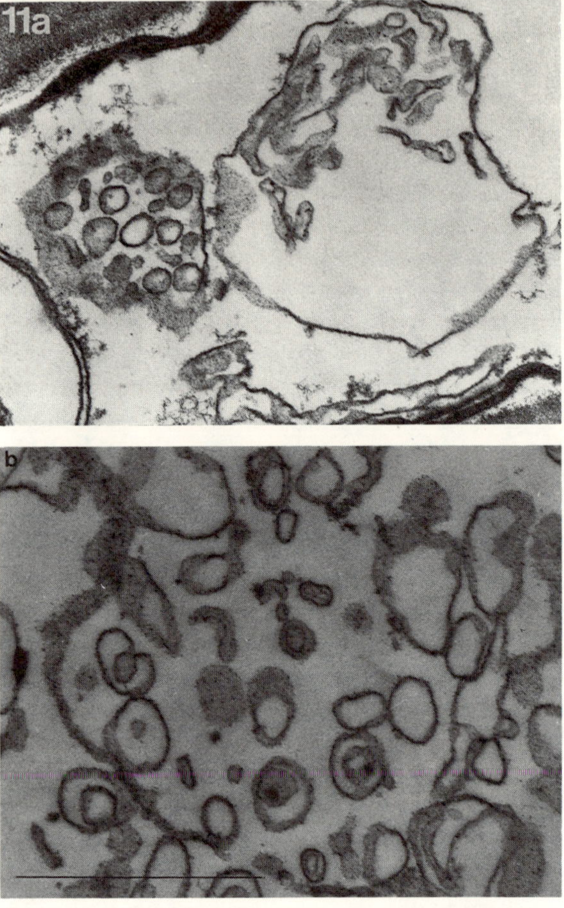

Figure 11. a.b. In *K. reticulata*, MLOs are extremely pleomorphic and mostly lack cytoplasmic material.

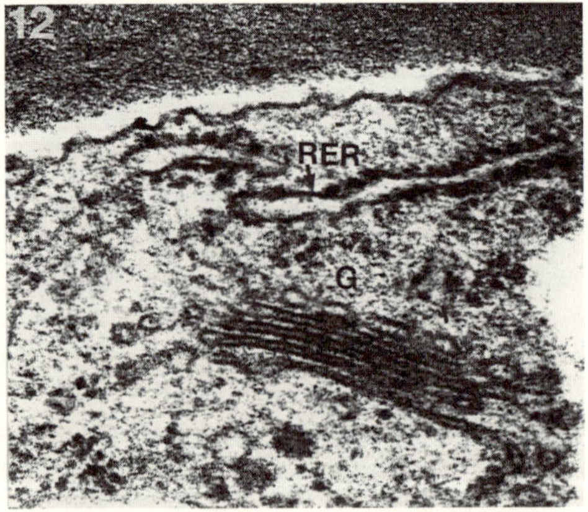

Figure 12. In *K. reticulata*, the adequate fixation condition is assured by intact Golgi bodies (G) and rough endoplasmic reticulum (RER).

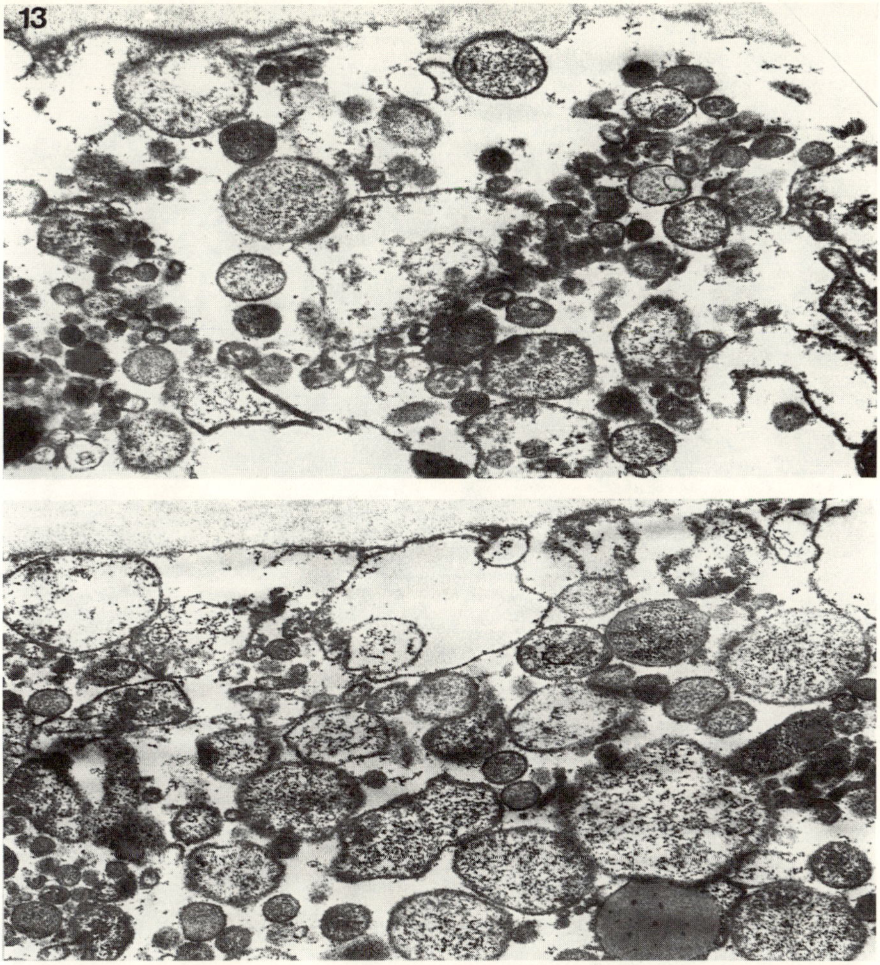

Figure 13. In infected *N. Rustica* leaf vein, the MLOs are plentiful. They are largely spherical bodies, but show a wide variation of sizes. There are also some dividing bodies.

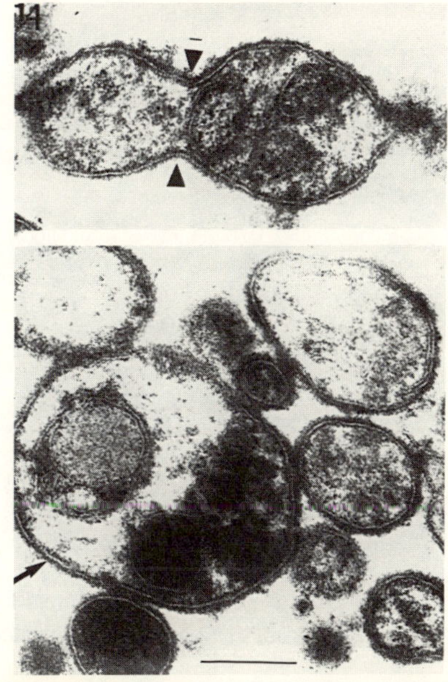

Figure 14. In *N. Rustica*, MLOs show a distinct bilayered membrane (arrows) and clear constriction of dividing body (arrowheads). Marker = 0.1 μ.

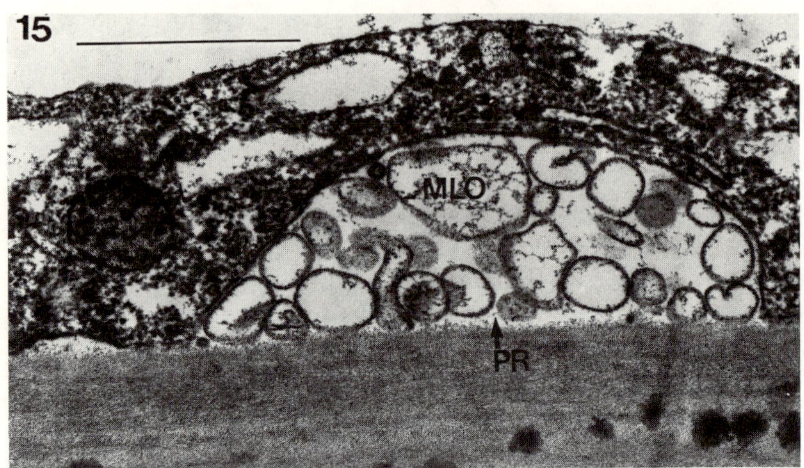

Figure 15. In *N. Rustica*, pleomorphic MLOs are seen in the root; these are located in the periplasmic region (PR).

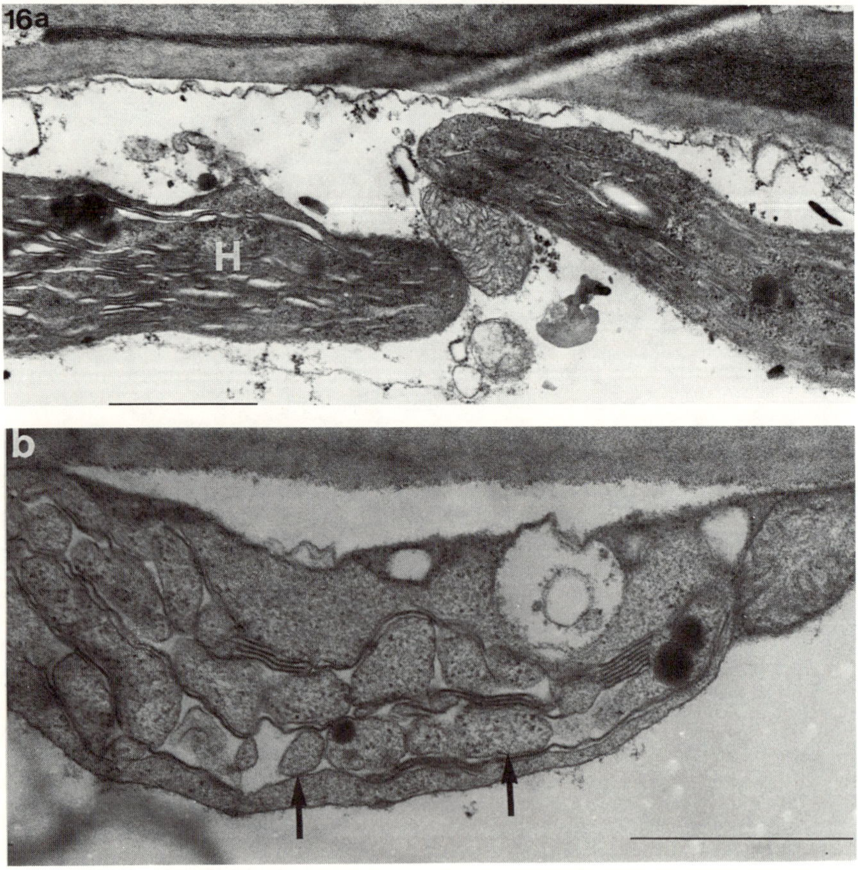

Figure 16. a,b. Comparison of healthy (H) and MLO-infected (arrow) chloroplasts. Note the distinct ribosomes, membrane, and nuclear material.

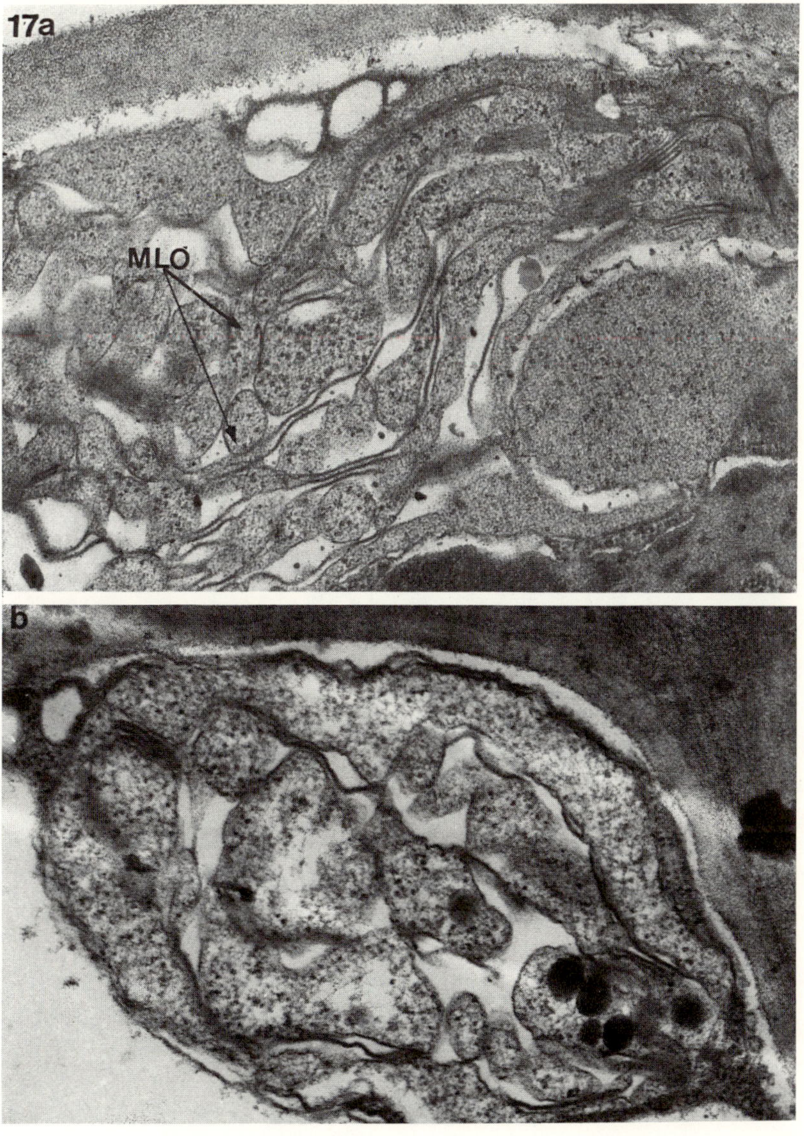

Figure 17. a,b. In *N. rustica*, MLOs are released from infected chloroplasts. The MLOs are pleomorphic at this stage.

9

THREE DIMENSIONAL MORPHOLOGY OF MLO AND SPIROPLASMAS STUDIED BY SEM

by

G.T.N. de Leeuw, P.A.M. van Vught,

A.A. Polak-Vogelzang, and R.A. Samson

Willie Commelin Scholten Phytopathological Laboratory, Baarn, National Institute of Public Health and Environmental Hygiene, Bilthoven, Centraal Bureau voor Schimmelcultures, Baarn,

The Netherlands.

INTRODUCTION

Mycoplasmas grown in liquid and on semi-solid media, and mycoplasma-like organisms (MLO) in the sieve tubes of affected plants, have manifested their pleomorphism when studied with light microscopes equipped with darkfield- or phase contrast optics, or with transmission-(TEM) and scanning electron microscopes (SEM). This pleomorphism is mainly attributed to their lack of a rigid wall. The absence of rigid walls greatly contributes to their facility to change form rapidly and reversibly in liquid media (Bredt *et al.*, 1973), and also makes them very sensitive to osmotic shocks, even during fixation (Lemcke, 1972; de Leeuw *et al.*, 1985).

The identification of MLO in sieve tubes is based on the presence of polymorphic bodies such as: filaments, branching forms, dumbbels, budding spherical bodies and spherical bodies connected in long chains by a continuous membrane structure (Haggis and Sinha, 1978). "Octopus-like structures consisting of numerous filaments attached to a central body were common in declining plants (Hearon *et al.*, 1976). Waters and Hunt (1980) showed that many previous interpretations of MLO morphology may have been too simple. With serial ultrathin sections they showed that the MLO associated with lethal yellowing disease of coconut exhibited a range of different morphologies, and they described the various forms as filiform, moniliform, cylindrical, erythrocyte-like or saccate. The circular profiles 80 to 120 μ m in diameter appeared to be sections of filiform cells, but they could not find elementary bodies in the diseased plants. *Spiroplasma citri* may show characteristic helical filaments in the sieve tubes of affected plants and in liquid culture (Bové and Saillard, 1979). Garnier *et al.*, (1981) studied the morphology and size distribution of *S. citri* in liquid culture and indicated the influence of culture age, pH, and nutritional requirements. Under optimal conditions, the organisms divide rapidly and large proportions of small helices with no more than two turns are formed. Under unfavorable conditions the proportion of long helices with less pronounced helicity increases, and asteroid, coccoid, and swollen structures may appear. Numerous round bodies were observed in older broth cultures, but helices were never seen to arise from them. *S. citri* cultures often contain "medusa-like" aggregates in which helices radiate from a central mass of poorly defined bodies (Townsend *et al.*, 1980).

Scanning Electron Microscopy

Polak-Vogelzang *et al.*, (1979) reviewed previous SEM studies on *Mycoplasma* and

Acholeplasma spp. and described the formation of sheets by the irregular shaped cells of
young *A. laidlawii* PG8A colonies and the
occurrence of sharply boundered peripheral
zones formed by irregular shaped cells around
Acholeplasma sp. FC 097-2 colonies. The
formation of these peripheral zones may result
from local disturbances in water balance or may
be induced by nutritional discrepancies.

It is almost impossible to extract MLO from
the sieve tubes of diseased plants for SEM without inducing artefacts either on extraction or
during observation. Distortion of MLO cells may
be avoided by fixation of colonies *in situ*
Petzold *et al.,* (1977) studied by SEM the
morphology and the distribution pattern of
different MLO in the sieve tubes of prefixed
plants after critical point drying of deparaf-
finized sections. In most cases, masses of
round bodies aligning the sieve tube walls were
found. In some cases, flat multilobed and
filiform cells were detected. Thin threads
interconnecting separate MLO possibly are rem-
nants of a reticulum or a slime layer, functioning as a sticking material for the mycoplasma cells. However, such threads may also
be considered to be phloem protein (P-protein)
or just an artefact. The results presented by
Petzold *et al.,* (1977) are remarkable, but
it must be kept in mind that their method is
very time consuming and that possibly the large
amount of solvents may have induced some artefacts.

Haggis and Sinha (1978) avoided the paraf-
fin used by Petzold *et al.,* (1977) by
freeze-fracturing roots and petioles of MLO-infected China aster *(Callistephus chinensis)*
plants. Longitudinal fractioned sieve tubes
clearly showed spherical bodies and branching
polymorphous forms, which were seen in infected, but not in healthy plants. Tiny threads
adhearing MLO to the sieve tube wall can
clearly be distinguished. The detected MLO
forms and adhearing tiny threads are comparable

to those shown by Petzold *et al.,* (1977) in the paraffin-treated material.

The possible misinterpretation of spherical non-mycoplasmal bodies present in the sieve tubes was avoided by Maramorosch and Pillips (1981) by studying with SEM the morphology of *S. citri* cells in colonies grown on a semi-solid agar medium. Regular cells, irregular bodies, twisted spiral forms and chains of beaded cells were detected. Comparative SEM work on *Spiroplasma, Acholeplasma* and *Mycoplasma* colonies grown on agar facilitated the study of the influence of the fixation vehicle osmolality on Mycoplasmatales morphology (de Leeuw *et al.,* 1985). Depending on colony age, the spiroplasma colonies showed elongated mycelium-like threads (Figs. 1,2,3, 8), which were sometimes branched (Fig. 3), or showed a helical appearance (Figs. 4,8), beaded chains (Fig. 6), "octopus-like" structures (Fig. 5), and round bodies (Fig. 6). The fixation vehicle osmolality appeared to influence the morphology of the spiroplasma cells, whereas *A. laidlawii* and *M. hyorhinis* cells did not react to the different fixation mixtures used. M 199 medium, commonly used for Vero (African green monkey kidney) cell culture, appeared to aid the preservation of the spiroplasma cell morphology when used as a vehicle of the glutaraldehyde fixation mixture.

Considerations and Future Perspectives

The great advantages of SEM for the detection of MLO in plants are: first, the relative ease in which large areas, such as whole cross- and longitudnal sections of petioles and stems can be examined at high magnifications and with great depths of field and second, the great amount of information that can be gathered about the spatial distribution of MLO and their tissue preference. In our SEM studies with corn stunt spiroplasma colonies grown on agar, we obtained some typical, tiny, coral-like outgrowths after fixation in 5% glutaraldehyde in M 199 cell culture medium with 0.25 M sucrose (Fig. 7). However, despite their intriguing outlook and unknown origin, these structures probably are artefacts.

Electron microscopists always have to overcome the possibility of introducing artefacts during fixation. The desiccating environment of the biological specimen is one of the major problems. Repeated sample handling involves great risks of specimen disturbance and damage. Therefore, more than one technique has to be employed to determine the actual morphology of the mycoplasma and MLO cells *in situ*, and only a very cautious interpretation of cell morphology may be given. The morphological characterization of different cell types in presumed reproduction cycles and in taxonomical aspects can only hold firm when the fixation methods have been improved and standardized. Therefore, it seems worthwhile to study the morphology of mycoplasmas and MLO also with a SEM equipped with a cryo-system.

The origin and function of the fibrillar network occasionally found to connect separate mycoplasma cells to each other or to a substrate, is open to discussion. Also bacteria (Tyson *et al.*, 1985) and fungi (Cooper and Wood, 1974), growing inside tracheary elements, sometimes are surrounded by a kind of reticulum, when studied by SEM. Mucilaginous layers and glycoprotein films are known to be disrupted during tissue preparation or investigation of the treated material in the electron beam of the SEM. Possibly, such layers could better be studied by making use of low temperature SEM (Kaneko *et al.*, 1985; Rhodes *et al.*, 1985), thus avoiding the chemical fixation and additional procedures utilized in ordinary SEM.

REFERENCES

Bové, J.M. and Saillard, C., 1979. Cell biology of spiroplasmas. In: R.F. Whitcomb and J. G. Tully (Eds.), The Mycoplasmas, Volume III. Plant and Insect Mycoplasmas, 85-153.

Bredt, W., Heunert, H.H., Höfling, K.H., and Milthaler, B., 1973. Microcinematographic studies of *Mycoplasma hominis* cells. J. Bacteriol. 113: 1223-1227.

Cooper, R.M. and Wood, R.K.S., 1974. Scanning electron microscopy of *Verticillium albo-atrum* in xylem vessels of tomato plants. Physiol. Pl. Pathol. 4: 443-446.

Garnier, M., Clerc, M., and Bové, J.M., 1981. Growth and division of spiroplasmas. Morphology of *Spiroplasma citri* during growth in liquid medium. J. Bacteriol. 147: 642-652.

Haggis, G.H. and Sinha, R.C., 1978. Scanning electron microscopy of mycoplasma-like organisms after freeze fracture of plant tissues affected with clover phyllody and aster yellows. Phytopathology 68: 677-680.

Hearon, S.S., Lawson, R.H., Smith, F.F., McKenzie, J.T., and Rosen, J., 1976. Morphology of filamentous forms of a mycoplasma-like organism associated with Hydrangea virescence. Phytopathology 66: 608-616.

Kaneko, Y., Matsushima, K., Wada, M., and Yamada, M., 1985. A study of living plant specimens by low-temperature scanning electron microscopy. J. Electron. Microsc. Techn. 2: 1-6.

Leeuw, G.T.N. de, Vught, P.A.M. van, Samson, R.A., and Polak-Vogelzang, A.A., 1985.

Scanning electron microscopy of the influence of fixation vehicle osmolality on cell morphology of *Spiroplasma, Acholeplasma* and *Mycoplasma spp.* grown on agar. Phytopath. Z. 114: 149-159.

Lemcke, R.M., 1972. Osmolar concentration and fixation of mycoplasmas. J. Bacteriol. 110: 1154-1162.

Maramorosch, K. and Phillips, D.M., 1981. Scanning electron microscopy of aster yellows and cactus spiroplasmas. Phytopath. Z. 102: 195-200.

Petzold, H., Marwitz, R., Özel, M. und Goszdziewski, P., 1977. Versuche zum rasterelektronenmikroscopischen Nachweis von mykoplasmaähnlichen Organismen. Phytopath. Z. 89: 237-248.

Polak-Vogelzang, A.A., Samson, R.A. and Leeuw, G.T.N. de, 1979. Scanning electron microscopy of *Acholeplasma* colonies on agar. Can. J. Microbiol. 25: 1373-1380.

Rhodes, M.J.C., Robins, R.J., Turner, R.J., and Smith, J.I., 1985. Mucilaginous film production by plant cells immobilized in a polyurethane or nylon matrix. Can. J. Bot. 63: 2357-2363.

Townsend, R., Burgess, J., and Plaskitt, K.A., 1980. Morphology and ultrastructure of helical and non-helical strains of *Spiroplasma citri*. J. Bacteriol. 142: 973-981.

Tyson, G.E., Stojanovic, B.J., Kuklinski, R.F., DiVittorio, T.J., and Sullivan, M.L., 1985. Scanning electron microscopy of Pierce's disease bacterium in petiolar xylem of grape leaves. Phytopathology 75: 264-269.

Waters, H. and Hunt, P., 1980. The *in vivo* three-dimensional form of a plant mycoplasma-like organism by the analysis of serial ultrathin sections. J. Gen. Microbiol. 116: 111-113.

EXPLANATION OF FIGURES

Fig. 1 Corn stunt spiroplasma colony, four days old. 7300x

Fig. 2 Elongated mycelium-like threads in a four-days old *S. citri* colony. 15000x

Fig. 3 Branched and elongated mycelium-like BC3 spiroplasma cells in a seven days-old colony. 28000x

Fig. 4 Helical *S. citri* cells in a four days-old colony. 28000x

Fig. 5 "Octopus-like" structure in a four days-old *S. citri* colony. 33000x

Fig. 6 Beaded chains and round bodies in an 11 days-old *S. citri* colony. 30000x

Fig. 7 Typical tiny "coral-like" outgrowths of a five days-old Corn stunt spiroplasma colony. 7100x

Fig. 8 Elongated and helical *S. citri* cells in a five days-old colony. 30000x

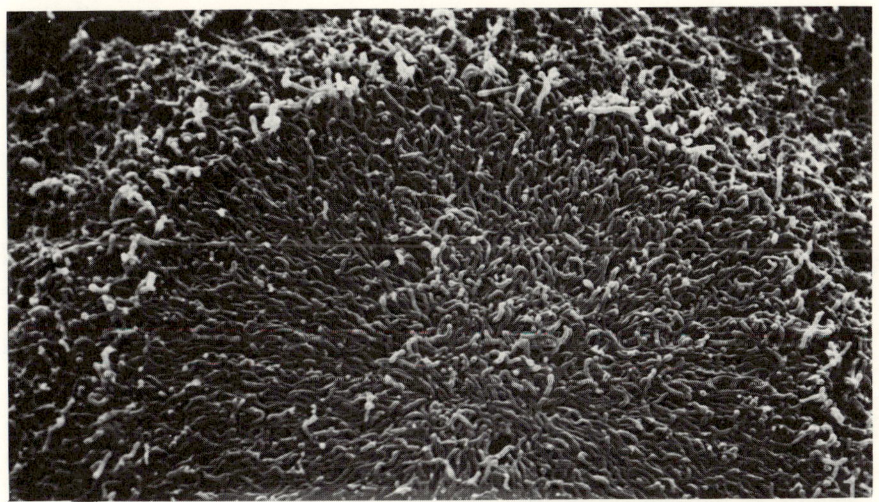

Figure 1.

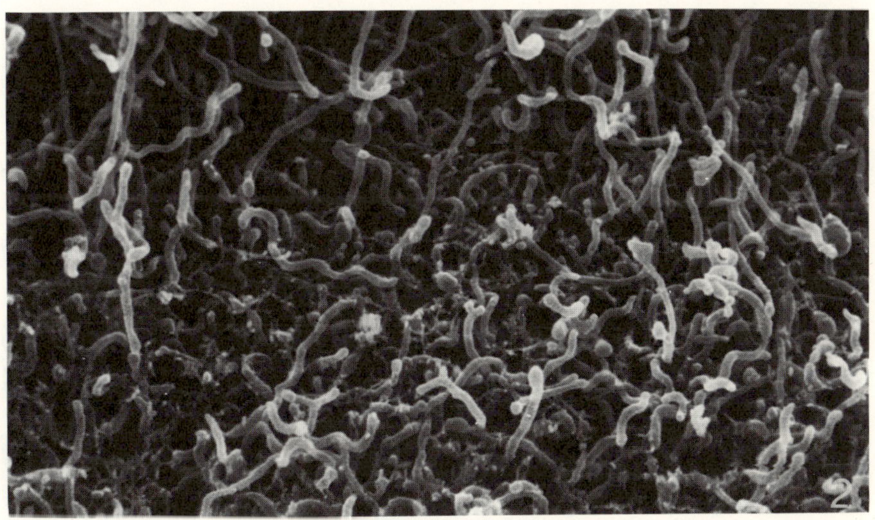

Figure 2.

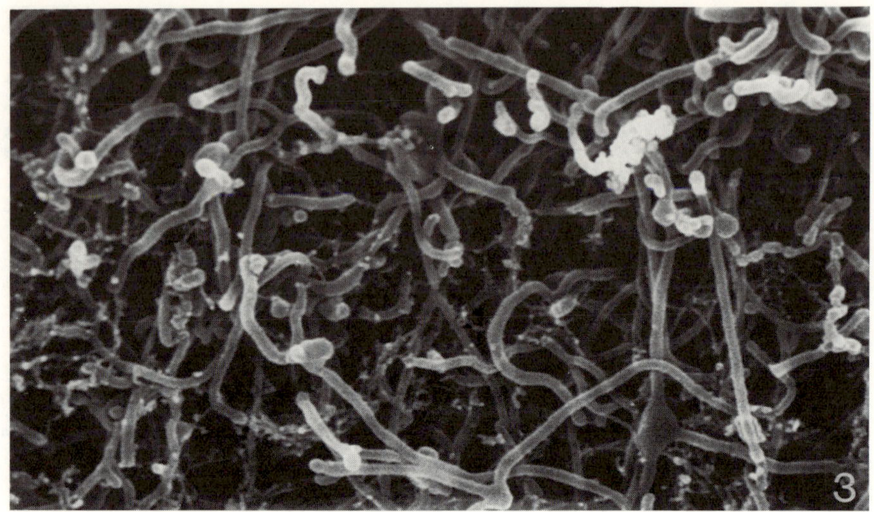

Figure 3.

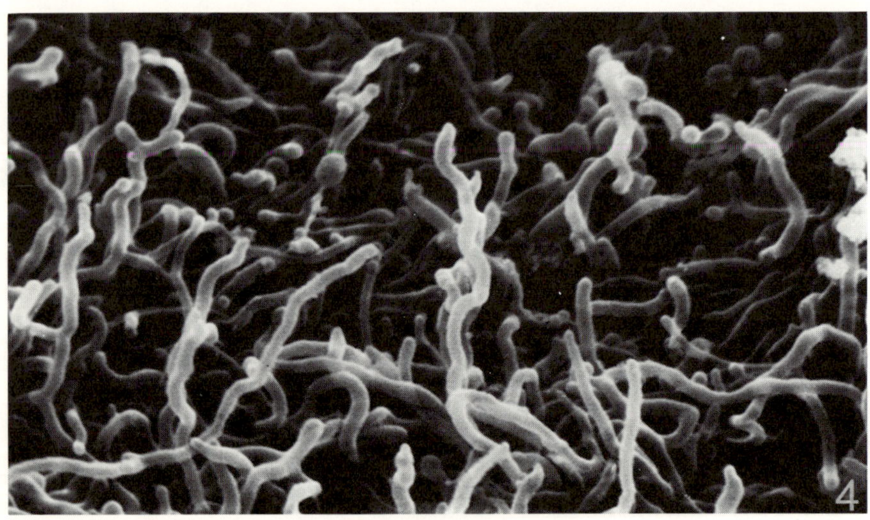

Figure 4.

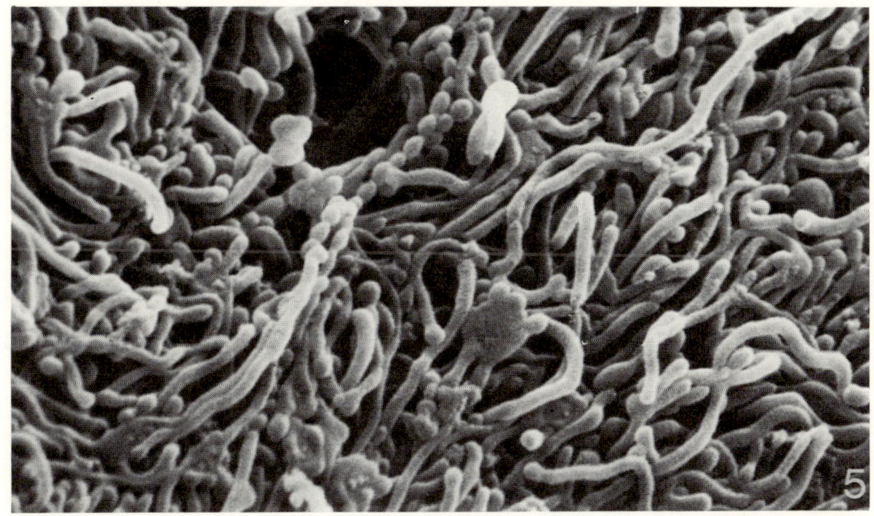

Figure 5.

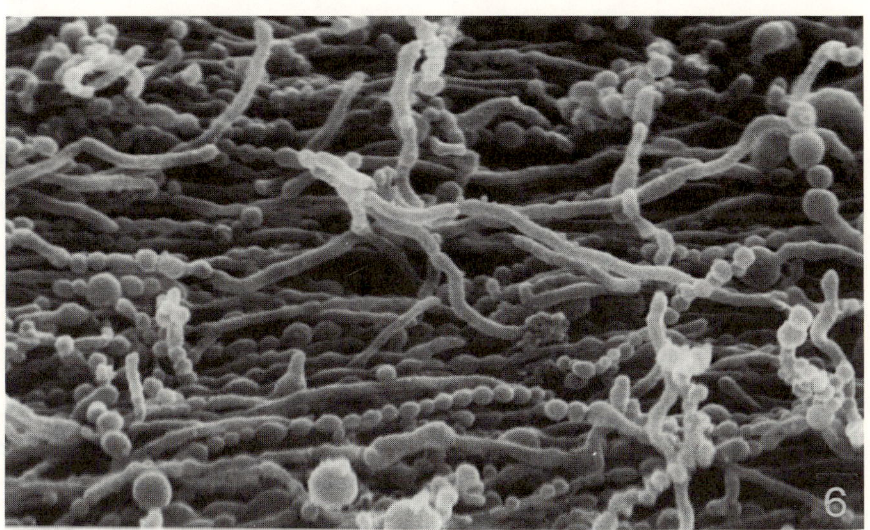

Figure 6.

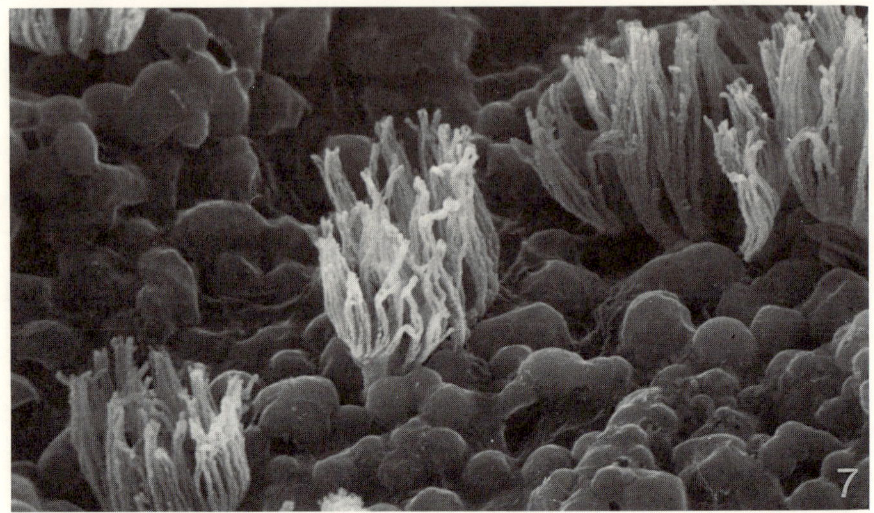

Figure 7.

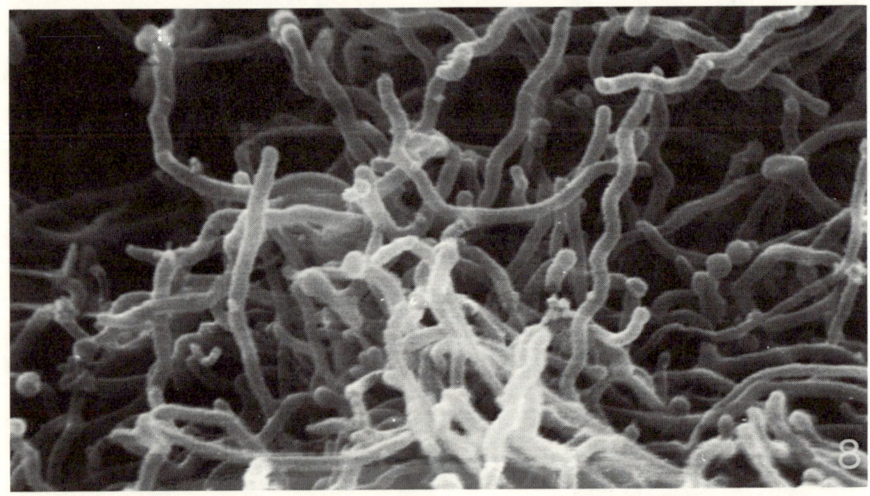

Figure 8.

Part II: Interactions with Plants, Insects, and Viruses

10

ECOLOGICAL ASSOCIATIONS OF

SPIROPLASMA CITRI

WITH INSECTS, PLANTS, AND OTHER

PLANT MYCOPLASMAS IN THE

WESTERN UNITED STATES

by

George N. Oldfield

USDA, ARS, Boyden Fruit and Vegetable

Insects Laboratory, University of California

Riverside, CA 92521

INTRODUCTION

The discovery of the helical, plant pathogenic mollicute, *Spiroplasma citri*, has its
roots in the recognition of a disease condition
of cultivated citrus first observed in California in 1915 and referred to as "stubborn" according to Fawcett *et al.* (1944). A similar condition, called locally "little leaf",
was recognized from citrus in Palestine as
early as 1928 (Reichert and Perlberger, 1931).
Demonstrated to be graft-transmissible several
decades ago, the nature of the causative agent
of the condition called stubborn in several
countries and little leaf in the eastern Mediterranean eluded discovery for several decades.
Then, after Doi *et al.*, (1967) found mycoplasma-like bodies in phloem elements of yel-

lows diseased plants, and Ishiie *et al.*, (1967) reported remission of symptoms of mulberry dwarf disease after treatment with tetracycline, Igwegbe and Calavan (1970) reported similar findings with citrus plants graft-inoculated with stubborn. Almost immediately, investigators in France (Saglio *et al.*, 1971) and California (Fudl-Allah *et al.*, 1972) reported *in vitro* cultivation of a mycoplasma-like organism from stubborn-diseased citrus. Formally described and named *Spiroplasma citri* by Saglio *et al.*, (1973), Markham *et al.*, (1974) succeeded in proving its role as the causative agent of little leaf (= stubborn) by injecting the European leafhopper, *Euscelis plebejus* with *S. citri* and demonstrating its transmission to citrus. This completion of Koch's postulates proved particularly noteworthy since *E. plebejus* neither reproduces on citrus nor survives long enough on stubborn-infected citrus to become inoculative. Thus, after *E. plebejus* had been injected with *S. citri* cultures, leafhoppers were allowed to feed for about three weeks on clover plants, then allowed to feed on citrus. As a result, typical symptoms of stubborn disease developed on citrus, and the clover plants developed chlorosis, wilted and died prematurely. This single experiment demonstrated pathogenicity of *S. citri* to plants, its role as the specific causative agent of citrus stubborn disease, and its ability to infect and cause disease in non-citrus plants. The report by Markham *et al.*, and the isolation of spiroplasmas from the beet leafhopper [*Circulifer tenellus* (Baker)] collected from stubborn-diseased citrus and associated weeds (Lee *et al.*, 1973) in California constituted a substantial base from which the knowledge of the field ecology of *S. citri* in the western United States has been significantly expanded in the last decade. The report that follows outlines the present status of knowledge of the association of

S. citri with vector and nonvector leaf-hoppers, wild and cultivated plants, and other plant mycoplasmas.

Leafhopper Vectors.

Currently, three species of Cicadellidae, *C. tenellus*, *Scaphytopius nitridus* (De Long), and *Scaphytopius acutus delongi* (Young), are recognized as natural vectors of *S.citri* in the western United States. The work of Rana *et al.*, (1975) offered the first evidence that *C. tenellus* and *S. nitridus* might be capable of natural acquisition and transmission of *S. citri*. They demonstrated that these two leafhoppers were capable of acquiring *S. citri* from an *in vitro* suspension by feeding through a membrane, retaining *S. citri* in their bodies, and eventually transmitting it to citrus. Reports of plant-to-plant transmission of *S. citri* by *S. nitridus* (Kaloostian *et al.*, 1975; Oldfield *et al.*, 1977), by *C. tenellus* (Oldfield *et al.*, 1976) and *S. acutus delongi* (Kaloostian *et al.*, 1979) soon followed.

Recently, O'Hayer *et al.*, (1983) reported a low rate of transmission of an isolate of *S. citri* from horseradish in Illinois by the aster leafhopper [*Macrosteles fascifrons* (Stål)]; however, despite continued searches, no additional natural vectors have been reported from *S. citri* in the western United States. In our tests, *M. fascifrons* failed to transmit *S. citri* isolated from stubborn-diseased citrus in California, and known to be transmitted by the three demonstrated California vectors (Oldfield *et al.*, 1984). We further found that several leaf-hopper species commonly encountered in California citrus plantings acquired and retained *S. citri* when given access to infected plants, but failed to transmit the pathogen. Although we attempted to demonstrate natural association of *S. citri* with several leaf-hopper species frequently collected with *S. citri*-inoculative *C. tenellus*, these attempts failed except that spiroplasmas, easily cultivated in *S. citri* media and presumed to be *S. citri*, were cultured from the bodies of the leafhopper, *Ollarianus strictus* (Ball) on several occasions (Oldfield and Kaloostian, 1979). Still, field-collected and laboratory-reared *O. strictus* repeatedly failed to transmit *S. citri* even though they could acquire it from plants in the laboratory and retain it. On the basis of our findings, even if under closer scrutiny, spiroplasmas isolated from field-collected *O. strictus* proved to be *S. citri*, their presence in this species would ostensibly be of no epidemiological importance with respect to further spread of *S. citri* to plants. On the other hand, the question of pathogenicity to *O. strictus* is perhaps worthy of investigation. The early report of pathogenicity of *S. citri* to *M. fascifrons* (Whitcomb et al., 1973) and the demonstration of Liu et al., (1983) that *S. citri* adversely affects *C. tenellus* suggests that other insects that acquire *S. citri* in nature may suffer deleterious effects. To date, no one has investigated the effects of *S. citri* on its other reported natural vectors.

Between 1974 and 1982 we collected leaf-hoppers in and near citrus plantings in California and assayed them for *S. citri* by either feeding them on test plants or attempting to culture the organism from their bodies (Oldfield *et al.*, 1984). Of the leafhoppers we tested, only *C. tenellus* commonly harbored and transmitted *S. citri*. Field-collected specimens of this species transmitted *S. citri* to 117 of 1464 test plants (mean of 10/plant) and we cultured *S. citri* from the bodies of 88 of 358 other groups of *C. tenellus* (mean 13/group). By contrast, field-collected specimens of only one other species

S. nitridus transmitted *S. citri* directly to test plants. However, this species transmitted *S. citri* to only one of 136 test plants (mean of 25/plant) and we were able to culture *S. citri* from the bodies of only one of 76 groups (mean of 14 insects/ group) of the fieldcollected specimens of this species. Tests with *S. acutus delongi*, the other reported vector, failed to demonstrate the presence of *S. citri* in the bodies of any of 21 groups of field-collected specimens (mean of 14/ group), and none of 25 test plants exposed to field-collected *S. acutus delongi* (mean of 20/plant) became infected. Other frequently collected leafhoppers for which we were unable to demonstrate association with *S. citri* by culturing or feeding on test plants included *Aceratagallia* spp. (3,596 insects), *Graminella sonora* (Ball), a vector of the corn stunt spiroplasma (401 insects), and *Empoasca* spp. (9,462 insects). None of 8,778 field-collected *O. strictus* fed on 126 test plants transmitted *S. citri*; however, spiroplasma were cultured from 6 of 162 groups (mean of 13/group) from the field. Notably, all the *O. strictus* from which we cultured spiroplasmas were collected during the autumn of one year from a weedy truck crop field where we collected many *S. citri*-inoculative *C. tenellus*. Most of the other *O. strictus* we tested were collected from asparagus, a reproductive host of this species but not a known host of *S. citri*.

Some of the same species or species groups for which we were unable to demonstrate an association with *S. citri* in the field, nevertheless acquired *S. citri* from infected plants in the laboratory and retained it for 12 days. Thus, we demonstrated acquisition and retention by laboratory-reared *Aceratagallia curvata* Oman, (9 of 12 groups of 15 insects each), a species prominently represented among species of *Aceratagallia* collected in the field, and *G. sonora* (8 of 48 groups of 12 insects each). Interestingly, El-Bolok (1984) reported isolation of a leafhopper transmissible strain of *S. citri* from the bodies of field-collected specimens of a species of *Aceratagallia* from California. M. fascifrons was seldom represented in our collections of leafhoppers and we found no *S. citri* naturally associated with this species. In the laboratory, we were able to culture *S. citri* from only one of 85 groups of *M. fascifrons* two weeks after they were fed on infected plants. Although 490 *A. curvata*, 1,060 *G. sonora*, and 2,731 *M. fascifrons* were used in laboratory vector tests, none transmitted *S. citri*.

The demonstrated ability of *A. curvata*, *G. sonora*, and *M. fascifrons* to acquire and retain, but not transmit *S. citri*, and the failure to detect *S. citri* naturally associated with these species suggests that these species seldom encounter *S. citri* in areas of California where we know it occurs. The frequency with which we collected *G. sonora* and *M. fascifrons* on monocotyledonous plants none of which have been reported as natural hosts of *S. citri* offers a possible explanation for our failure to find *S. citri* associated with these two leafhoppers. Apparently opportunities for encountering *S. citri* are less for these species than for species that often feed on known hosts of *S. citri*.

Each of the three identified vectors are known to reproduce on at least one recognized host of *S. citri*. *S. nitridus*, a species reported to reproduce on citrus in Southern California by Kaloostian and Pierce (1972) commonly encounters *S. citri*-infected citrus, particularly in the inland citrus-growing areas. However, as suggested by the relatively few field-collected specimens tested by us for *S. citri* since 1974, populations in citrus are usually very low. Similarly low populations of

S. acutus delongi are usually the rule in the San Joaquin Valley of California where this species reproduces on citrus. *C. tenellus*, the only leafhopper which commonly harbors *S. citri*, does not reproduce on citrus. *C. tenellus* reproduces on members of several plant families including Brassicaceae, Chenopodiaceae, Amaranthaceae, and Plantaginaceae. Our investigations point to frequent infection of many brassicaceous and plantaginaceous plants in the western United States by *S. citri*.

None of the vectors of western American isolates if *S. citri* are likely to have evolved with *S. citri* if citrus is its original host since citrus is from Southeastern Asia, *C. tenellus* is originally from the Mediterranean area and *Scaphytopius* is a new world genus. A case might be made that citrus is the original host since citrus alone of the known plant hosts of *S. citri* is never killed by it. Is this "equilibrium" evidence of a long-standing ecological relationship or is citrus a recently encountered host of *S. citri*? The prospect that *S. citri* and *C. tenellus* may have coexisted in the Mediterranean area for eons is suggested by a similar host range of the two organisms.

Host Plants

Since the mid-1970's, *S. citri* has been found to cause experimental disease in plants representing 19 dicotyledonous families and one monocotyledonous family (Oldfield and Calavan, 1982). However, many experimental hosts have not been shown to be infected with *S. citri* in nature. Experimental hosts (almost all of which are annuals) usually die prematurely when infected with *S. citri*. General symptoms of *S. citri* infection include chlorosis, shortened internodes, wilting and death.

Plants from which naturally occurring *S. citri* has been cultured include several wild plants and several cultivated plants in California and Arizona, and cultivated turnip in Washington. Allen and Donndelinger (1982) reported *in vitro* cultivation of spiroplasmas of the *S. citri* serogroup from several plants in Arizona including zinnia, Aztec marigold, viola, common foxglove (all cultivated plants), and two weed species, wild turnip (*Brassica tournefortii*) and London rocket (*Sisymbrium irio*).

The geographical range of *S. citri* in plants in the western United States may correspond closely to the wide geographical range of *C. tenellus*. *C. tenellus* is commonly encountered in semi-arid areas of most states west of the Rocky Mountains and in some states of the central United States. The present author has collected *C. tenellus* in extreme southern Canada, and it is known to occur in Mexico and in the central states of the United States. We have cultured spiroplasmas from several turnip plants from Washington exhibiting disease symptoms indistinguishable from those exhibited by turnips experimentally infected with *S. citri* cultured from stubborn-diseased citrus in California. Spiroplasmas cultured from turnips from Washington were transmissible by *C. tenellus* and *S. nitridus* and caused stubborn disease in citrus.

Presently, the known geographical range of *S. citri* in the western United States very much reflects the amount of effort expended in searching for naturally infected plants in various areas. Hence, in Southern California and Arizona where stubborn is an important disease of citrus, other natural hosts have been identified. Because most efforts to identify noncitrus natural hosts have occurred in southern California, the greatest numbers of natural hosts are known from that area. Oldfield and Calavan (1981) listed non-rutaceous plants naturally infected with spiroplasmas presumed or confirmed to be *S. citri* in southwestern United States. These dnncluded *Catharanthus roseus, Sedum praeltum,* and *Viola* sp. (all introduced ornamental plants found infected in Southern California) and many species or cultivars of Brassicaceae. Listed as natural hosts were eight brassicaceous crop plants: broccoli, brussels sprout, cabbage, Chinese cabbage, pak-choi, radish, rutabaga and turnip. Wild brassicaceous hosts included *Brassica geniculata, B. nigra, B. tournefortii, Raphanus raphanistrum, Sisymbrium irio* and *S. orientale.* Another wild plant, *Plantago ovata,* was reported as a host in southeastern California. The many brassicaceous plants and *P. ovata* are reproductive hosts of *C. tenellus* but not of *S. nitridus* and *S. acutus delongi.*

Although not all spiroplasmas cultured from *Plantago ovata* and various wild and cultivated brassicaceous plants in the Western United States have been serologically tested, those found by us at Riverside were cultured from plants that exhibited symptoms typical of those exhibited by the same species inoculated by *S. citri* in the laboratory. Further evidence that many isolates from naturally infected plants were *S. citri* was obtained by demonstrating their transmissibility by both *C. tenellus* and *S. nitridus* to citrus and other known hosts of *S. citri* and the development of disease symptoms indistinguishable from those caused by known strains of *S. citri.*

Perhaps of significance with respect to the lack of evidence for a natural association of *S. citri* with several grass-feeding leafhoppers, only two monocotyledonous species, onion and leek (both *Allium* sp.) have been experimentally infected with *S. citri* (Oldfield *et al.,* 1978; Oldfield and Calavan, 1982); attempts to infect other monocots such as corn and wheat have failed (Oldfield, unpublished). Also, no monocots have been found to be naturally infected with *S. citri.*

So far, in California, most of the species that regularly exhibit symptoms of *S. citri* infection belong to the Brassicaceae. Much evidence points to brassicaceous plants as important links in the epidemiological cycle of *S. citri*. By comparison with the cultivated Brassicaceae, the wild hosts, by virtue of their ubiquitous occurrence, are probably more important sources from which *S. citri* is spread by *C. tenellus*.

Spiroplasmas possibly assignable to *S. citri* have been reported from some other perennial plants in California. Raju *et al.*, (1983) reported occasional isolation of spiroplasmas, serologically indistinguishable from *S. citri*, from pear decline-diseased pear trees, but their inability to transmit such spiroplasmas by injection of pear psylla (the vector of the pear decline agent) indicated that such spiroplasmas are not the causal agent of pear decline disease. Kloepper and Garrott (1983) reported that cherry trees with and without symptoms of X-disease in California (caused by another leafhopper-transmitted agent) are occasionally infected with *S. citri*. All wild brassicaceous hosts are annuals or biennials which under temperatures of about 30 C in the greenhouse are killed prematurely by *S. citri*. Our observations that some infected annual hosts persist longer at lower temperatures suggest that wild hosts may harbor *S. citri* for comparatively long periods under lower ambient temperatures. Hence, the importance of an infected plant as an inoculum source may vary with temperature, as well as other factors. To my knowledge no studies have been conducted comparing the efficiency in which *S. citri* is acquired by vectors from plants at different temperatures.

Our studies indicate that *S. citri* is cultivable from wild and cultivated brassicaceous plants during each season of the year. i.e., at times in which daytime temperatures varied from about 40 C to about 15 C. If wild hosts survive longer at lower temperatures, *B. geniculata* in particular may thus be capable of surviving for many months should inoculation occur at the start of the cooler part of the year.

Other Plant Mycoplasmas

When we first demonstrated transmission of *S. citri* to Madagascar periwinkle plants by naturally inoculative field-collected *C. tenellus* (Oldfield *et al.*, 1976) we also observed that other Madagascar periwinkle plants upon which field-collected *C. tenellus* were fed developed floral virescence (Oldfield *et al.*, 1977). Further investigations (Oldfield *et al.*, 1980) revealed the presence of mycoplasma-like organisms in Madagascar periwinkle plants experimentally inoculated by *C. tenellus* and revealed that this virescence agent naturally infects several natural hosts of *S. citri*. These included *B. geniculata, R. sativus, S. irio* and broccoli (all Brassicaceae), an ornamental *Phlox* species, and Madagascar periwinkle. In southern California, plants of each of these species have been found to be infected with both the virescence agent and *S. citri*. Madagascar periwinkle plants inoculated by field-collected *C. tenellus* with only the viresence agent or with both agents survived indefinitely while those inoculated with only *S. citri* died prematurely. Madagascar periwinkle plants inoculated with *S. citri* by *S. nitridus* that fed previously on doubly infected plants developed stubborn disease and died prematurely. Kloepper *et al.*, (1982) found that field-grown broccoli plants infected with both virescence and *S. citri* had fewer *S. citri* organisms than those infected with only *S. citri*.

S. citri also coinfects plants with the aster yellows mycoplasma-like organism in California. In the San Joaquin Valley, diakon plants infected with *S. citri*, the aster yellows agent, or both disease agents have been found frequently during the last five years. We have repeatedly been able to transmit *S. citri* (with *C. tenellus*) and the aster yellows MLO (with *M. fascifrons*) from doubly infected plants exhibiting symptoms of aster yellows. Double infection by *S. citri* and the aster yellows mycoplasma results in early death of experimentally inoculated diakon and Madagascar periwinkle plants.

SUMMARY

A little more than a decade ago, *Spiroplasma citri* was discovered and identified as the causual agent of citrus stubborn disease. Since then, the known host range of *S. citri* in nature has been expanded from citrus to members of several plant families. Under experimental conditions, *S. citri* infects members of 20 plant families, most of which are annual dicotyledonous plants. Three natural vectors are recognized in the western United States. One, *C. tenellus*, reproduces on brassicaceous and plantaginaceous plants that are natural hosts of *S. citri*. The other two, *S. nitridus* and *S. actus delongi*, reproduce on citrus in southern and central California, respectively. *C. tenellus* is the only vector frequently found to be naturally inoculative. *S. citri* cultured from field-collected *C. tenellus* and from many brassicaceous plants causes stubborn disease in citrus. The etiological role in plant disease (if any) of isolates from cherry and pear trees is currently unknown.

S. citri occurs together with at least two other plant pathogenic mycoplasmas in certain brassicaceous plants in California; plants that are infected with both *S. citri* and a *C. tenellus*-transmitted virescence agent are not killed by *S. citri*.

REFERENCES

Doi, Y., M. Teranaka, K. Yora, and H. Asuyama. 1967. Mycoplasma- or PLT group-like microorganisms found in the phloem elements of plants infected with mulberry dwarf, potato witches' broom, aster yellows, or Paulownia witches' broom. Ann. Phytopathol. Soc. Japan 33: 259-266.

El-Bolok, M.M., 1981. Specific and nonspecific transmission of spiroplasmas and mycoplasmalike organisms by leafhoppers (Cicadellidae: Homoptera) with implication for etiology of aster yellows disease. Ph.D. Dissertation in Economic Entomology, Cairo University. Giza, Egypt.

Fawcett, H.S., J.C. Perry, and J.C. Johnston. 1944. The stubborn disease of citrus. Calif. Citrograph 29: 146-147.

Fudl-Allah, A.E.-S.A., E.C. Calavan, and E.C.K. Igwegbe. 1972. Culture of a mycoplasmalike organism associated with stubborn disease of citrus. Phytopathology 62: 729-731.

Igwegbe, E.C.K., and E.C. Calavan. 1970. Occurrence of mycoplasma-like bodies in phloem of stubborn-infected citrus seedlings. Phytopathology 60: 1525-1526.

Igwegbe, E.C.K., and E.C. Calavan. 1973. Effect of tetracycline antibiotics on symptom development of stubborn disease and infectious variegation of citrus seedlings. Phytopathology 63: 1044-1048.

Ishiie, T., Y. Doi, K. Yora, and H. Asuyama. 1967. Suppressive effects of antibiotics of tetracycline group on symptom development of mulberry dwarf disease. Ann. Phytopathol. Soc. Japan 33: 267-275.

Kaloostian, G. H., and H.D. Pierce. 1972. Notes on *Scaphytopius nitridus* in California. J. Econ. Entomol. 65: 880.

Kaloostian, G. H., G.N. Oldfield, H.D. Pierce. E.C. Calavan, A.L. Granett, G.L. Rana, and D.J. Gumpf. 1975. Leafhopper-Natural vector of citrus stubborn disease? California Agriculture 29 (2): 14-15.

Kaloostian, G. H., G.N. Oldfield, H.D. Pierce. E.C. Calavan. 1979. *Spiroplasma citri* and its transmission to citrus and other plants by leafhoppers. In: K. Maramorosch and K.F. Harris (eds.) Leafhopper Vectors of Plant Disease Agents, pp. 447-450. Academic Press, New York.

Kloepper, J.W. and D.G. Garrott. 1983. Evidence for a mixed infection of spiroplasmas and non-helical mycoplasma-like organisms in cherry with X-disease. Phytopathology 73: 357-360.

Kloepper, J.W. and D.G. Garrott, and G.N. Oldfield. 1982. Quantification of plant pathogenic spiroplasmas in infected plants. Phytopathology 72: 577-581.

Lee, I.-M. G. Cartia, E.C. Calavan, and G.H. Kaloostian. 1973. Citrus stubborn disease organism cultured from beet leafhopper. California Agriculture 27 (11): 14-15.

Liu, H-Y., D.J. Gumpf, G.N. Oldfield, and E.C. Calavan. 1983. The relationship of *Spiroplasma citri* and *Circulifer tenellus*. Phytopathology 73: 585-590.

Markham. P.G., R. Townsend, M. Bar-Joseph, M.J. Daniels, A. Plaskitt, and B.M. Meddins. 1974. Spiroplasmas are the causal agents of citrus little-leaf disease. Ann. Appl. Biol. 78: 49-57.

O'Hayer, K.W., G. A. Schultz, C.E. Eastman, J. Fletcher, and R.M. Goodman. 1983. Transmission of *Spiroplasma citri* by the aster leafhopper *Macrosteles fascifrons* (Homoptera: Cicadellidae) Ann. Appl. Biol. 102: 311-318.

Oldfield, G.N. 1980. A virescence agent transmitted by *Circulifer tenellus* (Baker); aspects of its plant host range and association with *Spiroplasma citri*. Abstracts 3rd Conf. Int. Org. for Mycoplasmology. Custer, S.D. p. 46.

Oldfield, G.N., and E.C. Calavan. 1982. Stubborn disease in non-rutaceous plants. In J. Bové and R. Vogel (eds.), Description and Illustration of Virus and Virus-like Diseases of Citrus: A Collection of Color Slides (2nd ed.) SETCO-IRFA, Paris.

Oldfield, G.N., and G.H. Kaloostian. 1979. Vectors and host range of the citrus stubborn disease pathogen, *Spiroplasma citri*. Proc. R.O.C. - United States Cooperative Science Seminar on Mycoplasma Diseases of Plants. NSC Symposium Series No. 1. 119-125.

Oldfield, G.N., G.H. Kaloostian, H.D. Pierce, E.C. Calavan, A.L. Granett, and R.L. Blue. 1976. Beet leafhopper transmits citrus stubborn disease. California Agriculture 30 (6): 15.

Oldfield, G.N., G.H. Kaloostian, H.D. Pierce, E.C. Calavan, A.L. Granett, R.L. Blue, G.L. Rana, and D.J. Gumpf. 1977. Transmission of *Spiroplasma citri* from citrus to citrus by *Scaphytopius nitridus*. Phytopathology 67: 763-765.

Oldfield, G.N., G.H. Kaloostian, H.D. Pierce, A.L. Granett, and E.C. Calavan. 1977.

Beet leafhopper transmits virescence of periwinkle. California Agriculture 31 (6): 14-15.

Oldfield, G.N., G.H. Kaloostian, D.A. Sullivan, E.C. Calavan, and R.L. Blue. 1978. Transmission of the citrus stubborn disease pathogen to a monocotyledonous plant. Pl. Dis. Reptr. 62 (9): 758-760.

Oldfield, G.N., D.A. Sullivan, and E.C. Calavan. 1984. Inoculativity of leafhopper vectors of stubborn disease in California. Proc. 9th Conf. Int. Org. Cit. Virol. 125-130.

Rana, G.L., G.H. Kaloostian, G.N. Oldfield, H.D. Pierce, E.C. Calavan, A.L. Granett, R.L. Blue and D.J. Gumpf. 1975. Acquisition of *Spiroplasma citri* through membranes of homopterous insects. Phytopathology 65: 1143-1145.

Raju, B.C., A.H. Purcell, and G. Nyland. 1983. Current status of the etiology of pear decline. Phytopathology 73: 350-353.

Reichert, I., and J. Perlberger. 1931. Littleleaf disease of citrus and its cause. Hadar 4: 193-194.

Saglio, P., D. Laflèche, C. Bonissol, and J.M. Bové. 1971. Culture in vitro des mycoplasmaes associes au "Stubborn" des agrumes et leur observation au microscope electronique. C.R. Acad. Sci. (Paris) Ser. D. 272: 1387-1390.

Saglio, P., M. L'Hospital, D. Laflèche, G. Dupont, J.M. Bové, J.G. Tully, and E.A. Freundt. 1973. *Spiroplasma citri* gen. and sp. n.: A Mycoplasma-like organism associated with "stubborn" disease of citrus. Int. J. Syst. Bacteriol. 23: 191-204.

Whitcomb, R.F., J.G. Tully, J.M. Bove, and P. Saglio. 1973. Spiroplasmas and Acheloplasmas: multiplication in insects. Science 182: 1251-1252.

11

THE OCCURRENCE AND INTERACTION OF

PLANT VIRUSES AND MOLLICUTES IN

PLANTS AND INSECT VECTORS

by

Ernest E. Banttari

Department of Plant Pathology

University of Minnesota

St. Paul, MN 55108

INTRODUCTION

The literature is extensive concerning dual
infections of plants with viruses and other
pathogens causing interactions ranging from interference to synergism. A few recent examples
indicate increased virulence of one of the pathogens (26,29,39,40,43,44,47,48); others show
evidence for interference or reduced virulence
of one of the components (18,34,42,50). The references for virus-Mollicute interactions in
plants are more recent and limited (Table 1).
There are few reports that suggest interactions
of these pathogens in dual infections of plants
and/or vectors (6,7,9,10,13,21,33). Previous
reviews (4,49) have summarized the literature
regarding this topic and this report will incorporate and emphasize more recent pertinent
findings.

There are, obviously, abundant opportunities for combined infection of plants with viruses and Mollicutes since there are broad overlapping host ranges of these pathogens. Furthermore, many leafhoppers and planthoppers are vectors for both viruses and Mollicutes and some plants that are susceptible to both pathogens are also preferred hosts of the vectors. Many species of wild and cultivated plants may be chronically infected with latent viruses, which increases the probability for dual virus-Mollicute infections. Many of these examples have not been examined because there are no apparent interactions or any observed interactions are weak.

The examples in which plants exhibit different, diminished or more severe symptoms than those occurring in plants infected with either pathogen alone have naturally attracted attention and, in a few cases, prompted more thorough investigation. The oat blue dwarf virus-aster yellows disease complex in flax, *Linum usitatissimum L.* is one example (6). Both the virus and the mycoplasma-like organism (MLO) have a common leafhopper vector, both pathogens infect and replicate in phloem and the infection results in some symptoms not observed in independent infections with individual pathogens. Examples such as this suggest interesting cytological and histological abnormalities and possible effects on one or both pathogens as well. Likewise, there could be changes in transmissibility of the pathogens by a common vector, there might be alterations in fecundity and longevity of the vector and there might be changes in the epidemiology of the disease.

In Table 1, dual infections of plants with viruses and Mollicutes are listed. Those under A) have been investigated sufficiently so that both pathogens have been identified and, in some cases, an interaction has been characterized. In the second group a virus (or virus-like particles (VLP) and a Mollicute have been observed in sectioned material, however, the identity of the virus or VLP was not accomplished. In the third group viruses or VLP that are suspected pathogens of Mollicutes are listed.

Since an extensive review of this topic was provided in an earlier publication (4) this discussion will serve only to update information on previously reported examples or to point out dual infections reported since the review.

Oat Blue Dwarf Virus and Aster Yellows Mycoplasma-Like Organism

This dual infection of plants was first investigated prior to the true identity of aster yellows (21). The disease in flax, *Linum usitatissimum L.* resulted in symptoms (Fig. 1) that were different and more severe than that caused by either pathogen (6). Since both pathogens are phloem-limited, and both pathogens are transmitted by the aster leafhopper, *Macrosteles fascifrons* Stål, investigations of this dual infection in plants, as well as in the vector produced some interesting findings. In brief, the histological examination of this dual infection in flax supported the hypothesis that the virus replicates in very young phloem tissue, probably immature phloem elements. On the other hand, aster yellows mycoplasma-like organism (AY-MLO) probably utilizes metabolites occurring in cells of a wider maturity range in its replicative cycle. The deleterious effects of OBDV to the phloem are compounded by the effects of AY-MLO to the same tissue. When young flax plants are dually infected with both pathogens the plants are severely stunted, there is swelling, chlorosis and deformation of the stem apex. Enations develop on leaf veins and, usually, the flowers are sterile and the plants die prematurely. Histological studies revealed extensive hyperplasia of phloem elements and hyperplasia and hypertrophy of fibers and cortical parenchyma in stems. The phloem was disoriented, and occluded and probably non-functional. Although both pathogens were also occasionally found in close proximity in some phloem elements, there appeared to be no visible effect of one on the other.

Studies on the effects of dual transmission of AY-MLO and OBDV on the aster leafhopper suggested that there was interference between these pathogens in the vector (33). Means of 76% and 34% of leafhoppers given one week of acquisition feeding on either AY- or OBDV- infected asters and oats, respectively, transmitted the individual pathogens. However, only 42 and 21% transmitted AY-MLO and OBDV respectively, that had acquisition-access on AY-infected asters for one week, then OBDV-infected oats for the 2nd week. Means of 30 and 31% of leafhoppers transmitted AY-MLO and OBDV, respectively, that were given acquisition-access feeding for one week each on the reverse sequence of hosts. Only a few leafhoppers (mean of 5.5%) that had either sequence of acquisition access feeding on these infected plants transmitted both pathogens to susceptible plants. Hypotheses to explain interference between strains of viruses or MLO in plants or in leafhopper or planthopper vectors are speculative. However, interference between certain viruses and MLO in vertebrate tissue cultures has been explained as a result of competition for replicative sites or substrates (46). Another hypothesis is that insects have primitive mechanisms to resist invading microorganisms (53) and these might be activated by OBDV and, to a lesser extent, by AY-MLO. Recent research (Azar and Banttari, unpublished) shows that as much as 96.7% of individual leafhoppers carry OBDV at levels detectable by ELISA, transmission assay or by inoculation assay. However, only 53.4% transmitted virus during three weeks of inoculation access feeding on oat seedlings. These experiments also suggested that aged insects were less efficient vectors of OBDV than young adult leafhoppers. Further experiments were made to examine if those pathogens effected interference when leafhoppers were inoculated. Inoculation of aster leafhoppers that had been caged continuously on OBDV-infected oats with clarified triturates of AY-infected leafhoppers or inoculation of leafhoppers that had been caged continuously on AY-infected asters with triturates of OBDV-infective leafhoppers also suggested that there is interference between these pathogens in the vector (Banttari, unpublished data). Age-synchronized healthy adult leafhoppers were anaesthetized with CO_2 gas and injected abdominally with clarified preparations of either OBDV or AY-MLO. The inoculated insects were assayed on oats and aster seedlings. Means of 58% and 72% of these insects transmitted OBDV and AY-MLO, respectively. Only 35% of individual leafhoppers transmitted AY-MLO to asters when they were raised continuously on OBVD-infected oats prior to the injection of AY-MLO, 51% transmitted OBDV to oat seedlings and none transmitted both pathogens. When the insects were hatched and reared continuously on AY-infected asters and then injected with clarified triturates of OBD-viruliferous leafhoppers, 36% transmitted OBDV, 81% transmitted AY-MLO and 5% transmitted both pathogens.

The availability of antisera to both AY-MLO (36,45) and OBDV (3) and sensitive ELISA techniques may enable experiments to examine the titer of both OBDV and AY-MLO in individual leafhoppers that have had acquisition access feeding in either sequence of AY-MLO and OBDV-infected plants. Such experiments may show if pathogen titers are changed due to interference and which could affect transmissibility. The data suggested that transmission of either pathogen acquired by feeding was not affected markedly by subsequent injections of the other pathogen. These experiments also demonstrated that only occasional leafhoppers that had acquired one of the pathogens by natural feeding and the other by injection simultaneously transmitted both agents.

Both pathogens replicate in the aster leaf-hopper (5,37). Paracrystalline aggregates or inclusions of OBDV have been observed in fat bodies, salivary glands and in cells of the neural lamella surrounding the supraesophageal ganglia of OBD-viruliferous leafhoppers. Hirumi and Maramorosch (32) found AY-MLO in the sali-vary glands of aster leafhoppers carrying the aster yellows agent. Hemmati (28) found MLO's of dwarf aster yellows mycoplasma-like agent in mid-guts of the aster leafhopper. Since both OBDV and AY-MLO occur in the salivary gland it would be a likely organ to investigate by elec-tron microscopy as well as by serology for the occurrence and possible interfering effects of these pathogens.

Greening-MLO and Tristeza Virus in Kagzi Lime, *Citrus aurantifolia*

Bhagabati and Nariani (9) reported that there was a synergistic effect of these pathogens in the host resulting in a 52.6% reduction in growth. The dual infection induced symptoms that were characteristic of both pathogens.

Mosaic Viruses and *Spiroplasma citri* in Horseradish

Fletcher *et al.*, (20), reported that horseradish, which is virtually 100% infected with turnip mosaic virus and also with cauli-flower mosaic virus developed symptoms typical of brittleroot when inoculated with *S. cit-ri*. The results of their study indicated that the viruses had no effect on infection by *S. citri* or on the development of brittle-root disease in horseradish.

Maize Mosaic Virus and Corn Stunt in Maize

Bacilliform virus-like particles as well as

corn stunt spiroplasma were detected in a plant experimentally inoculated with corn stunt spiroplasmas and maize mosaic virus (17). The results suggested that *Dalbulus maidis* could acquire both pathogens and transmit them simultaneously to maize.

Viruses, Rickettsia-like organisms and Mollicutes in White Clover, *Trifolium repens*

Benhamou *et al.,* (8) reported that the prokaryotes caused severe stunting and curling of leaflets of white clover and were limited to sieve elements. The virus particles were localized to parenchyma cells and appeared to be responsible for mosaic symptoms in foliage.

Rayado Fino Virus and Corn Stunt in Maize

Both disease agents were transmitted simultaneously to maize by individual insects of either species of leafhoppers (54). *Rayado fino* virus (RFV) symptoms appeared in 7-12 days and intensified up to 3 weeks. Corn stunt (CS) symptoms appeared after 3-4 weeks and then completely obscured RFV so that the symptoms could not be distinguished from corn stunt alone.

DISCUSSION

The occurrence of viruses and Mollicutes in simultaneous infections of plants may change or intensify symptoms as in OBDV and AY-MLO in flax (6) or as occurs with tristeza virus and greening MLO in Kagzi lime (9). In the later example this association was reported to be synergistic. In flax, plants with dual infections were rarely found in the field and, therefore, the disease had no economic importance. Other cultivated species such as barley, *Hordeum vulgare* are also hosts for both AY and OBDV. When barley was infected with both disease agents in the greenhouse, the symptoms, after 4-6 weeks appeared to be like that of AY alone and were severe. In AY-OBDV infections, both pathogens cause major damage to phloem tissues and, therefore, it would not be unexpected that the effects on the host might be accentuated.

In certain dual infections such as turnip mosaic and brittleroot (*S. citri*) in horseradish (20) and with rayado fino and corn stunt spiroplasma in maize (54), the symptoms caused by the Mollicutes eventually mask the virus symptoms. In maize, particularly, it is apparent that it will be increasingly difficult to make visual diagnoses alone for these kinds of pathogens (10).

The presence of VLP and Mollicutes in plants has prompted differing explanations for their occurrence and significance. In some cases (8,19,30,38,41,52) there is a likely possibility that the VLP were, at the time of the report, unidentified plant viruses that were either part of the disease complex and contributed to symptoms or were simply viruses normally latent in their hosts that happened to be observed in histological studies of the Mollicute component. In certain dual infections in either insect vectors or plants involving VLP and Mollicutes (23,24,25,41) the authors speculated that the VLP were pathogens of the Mollicutes. Such viruses of *Spiroplasma citri* have been identified (2,14,16), thus these explanations are certainly plausible. The occurrence of the VLP in close association with ruptured or degenerating Mollicute bodies were discussed as supportive evidence that these VLP might be pathogens of the Mollicutes.

There is evidence accumulating that there may be interference between AY-MLO and OBDV in its vector, *M. fascifrons* (21,33 unpublished data). These experiments warrant further experimentation and verification, but the data indicate there is interference between these pathogens in their vector. Whether competitive interference for sites or substrates for replication occurs (46) or whether some kind of primitive immune system is triggered by the initial invading pathogen (53) are speculative hypotheses.

REFERENCES

1. Allen, J.C. 1972. Bacilliform particles within asters infected with a western strain of aster yellows. Virology 47: 491-493.

2. Alivizatos, A.S., Townsend R. and Markham P.G. 1982. Effects of infection with a spiroplasma virus on the symptoms produced by *Spiroplasma citri*. Ann. Appl. Biol. 101: 85-91.

3. Banttari, E.E. 1981. Serological assays for the oat blue dwarf virus. Phytopathology 71: 1242-1244.

4. Banttari, E.E. and Zeyen, R.J. 1979. Interaction of mycoplasma-like organisms and viruses in dually infected leafhoppers, planthoppers and plants. Chapter 8. In: Maramorosch K. and Harris, K.F. Leafhopper Vectors and Plant Disease Agents. Academic Press, NY 654p.

5. Banttari, E.E. and Zeyen, R.J. 1976. Multiplication of the oat blue dwarf virus in the aster leafhopper. Phytopathology 66: 896-900.

6. Banttari, E.E. and Zeyen, R.J. 1972. Ultrastructure of flax with a simultaneous virus and mycoplasma-like infection. Virology 49: 305-308.

7. Basu, A.N., Ghosh, A., Mishra, M.D., Niazi, F.R. and Raychaudhuri, S.P. 1974. The joint infection of rice tungro and yellow dwarf. Ann. Phytopath. Soc., Japan 40: 67-69.

8.   Benhamou, N., Ouellette, G.B. and Pause F.
     1983.   Mise en e'evidence d'infections a'
     prokaryotes (rickettsoides et mycoplasmes)
     chez   le   *Trifolium   repens,*   Quebec.
     Phytoprotection 64: 53-59.

9.   Bhagabati, K.N. and Nariani, T.K. 1980. In-
     teraction of greening and tristeza patho-
     gens in Kagzi lime (*Citrus aurantifolia*
     (Christm.)   Swing.) and their effect on
     growth and development of disease symptoms.
     Indian Phytopathol. 33: 292-295.

10. Bradfute,   O.E.,   Tsai, J.H. and Gordon, D.
     T.   1981.   Corn   stunt   spiroplasmas and
     viruses associated with a maize disease
     epidemic in southern Florida. Plant Disease
     65: 837-841.

11. Bradfute,   O.E.,   Robertson, D.C. and Poet-
     hig, R.S. 1979. Detection and characteri-
     zation of Mollicutes in maize and sorghum
     by light and electron microscopy. Proc.
     R.O.C. - U.S. Crop Science Seminar on Myco-
     plasma Diseases of Plants. NCS Symposium
     Series No. 1 R.O.C.

12. Casper,   R.,   Lesemann,   D. and Bartels, R.
     1970.   Mycoplasma-like bodies and viruses
     in *Opuntia tuna* with witches' broom
     disease. Pl. Dis. Reptr. 54: 851-853.

13. Chen, Ming-Hsiúng, Miyakawa, T. and Matsui,
     C. 1972. Simultaneous infections of citrus
     leaves with tristeza virus and mycoplasma-
     like organism. Phytopathology 62: 663-666.

14. Cole,   R.M.,   Tully,   J.G.,   Popkin T.J. and
     Bove, J.M.   1973. Morphology, ultrastruc-
     ture, and bacteriophage infection of the
     helical mycoplasma-like organism (*Spiro-
     plasma citri* gen. nov., sp. nov.) cul
     tured from "stubborn" disease of citrus.
     J. Bacteriology 115: 367-386.

15. deLeeuw, G.N.T. 1975. Virus-like particles associated with mycoplasmas in *Vinca rosea* plants affected by the yellowing disease of *Brassica oleracea*. Acta Bot. Neerl. 24: 252-253.

16. Dickenson, M.J., Townsend, R., and Curson, S.J. 1984. Characterization of a virus infecting the wall-free prokaryote *Spiroplasma citri*. Virology 135: 524-535.

17. Eden-Green, S.J. and Waters, H. 1981. Isolation of corn stunt spiroplasma in Jamaica, and probable dual infection with maize mosaic virus. Ann. Appl. Biol. 99: 129-134.

18. Erasmus, D.S. and VonWechmar, M.B. 1983. Reduction of susceptibility of wheat to stem rust (*Puccinia graminis f. sp. tritici*) by brome mosaic virus. Plant Disease 67: 1196-1198.

19. Fedotina, V.L. 1977. Pupation disease of cereals-a mixed infection. Archiv. Phytopathol. und Pflanzenschultz, Berlin 13 (3): 177-191.

20. Fletcher, J., Schultz, G.A. and Eastman, C.E. 1984. Effect of mosaic viruses on infection of horseradish by *Spiroplasma citri*. Plant Disease 68: 565-567.

21. Fredericksen, R.A 1964. Simultaneous infection and transmission of two viruses in flax by *Macrosteles fascifrons*. Phytopathology 54: 1028-1030.

22. Gamez, R. 1973. Transmission of rayado fino virus of maize (*Zea mays*) by *Dalbulus maidis*. Ann. Appl. Biol. 73: 285-292.

23. Giannotti, J. and Vago, C. 1977. Infec-

tion complexe A, virus, mycoplasmas et Rickettsoides A. L'Echelle D'une meme cellule chez des plantes malades. Travaux dedes A. G. Viennot-Bourgin 393-403.

24. Giannotti, J., deVauchelle, G., Vago C. and Marchoux, G. 1973. Rod-shaped virus-like particles associated with degenerating mycoplasma in plant and insect vectors. Ann. Phytopathol. 5: 461-465.

25. Gourrett, J.P., Maillet, P.L.and Gouranton, U. 1972. Virus-like particles associated with the mycoplasmas of clover phyllody in the plant and in the insect vector. J. Gen. Microbiol. 74: 241-249.

26. Goodell, J.J., Powelson M.L. and Allen, T. C. 1982. Interrelations between potato virus X, *Verticillium dahliae,* and *Collectotrichum atramentarium* in potato. Phytopathology 72: 631-634.

27. Granados, R.R. 1969. Electron Microscopy of plants and insect vectors infected with the corn stunt disease agent. Contrib. Boyce Thompson Inst. 24: 173-188.

28. Hemmati, K. 1979. Localization and pathogenicity of dwarf asters yellows mycoplasma-like agent in the mid-gut of the leafhopper vector *Macrosteles fascifrons.* Iran J. of Plant Pathol. 15: 45-52.

29. Hepperly, P.R., Bowers, G.R. Sinclair, J. B.and Goodman, R.M. 1979. Predisposition to seed infection by *Phomopsis sojae* in soybean plants infected by soybean mosaic virus. Phytopathology 69: 846-848.

30. Hirumi, H., Maramorosch, K.and Hichez, E. 1973. Rhabdovirus and mycoplasma-like organism: Natural dual infection of *Cajanus cajan.* Phytopathology 63: 202.

31. Hirumi, H. and Maramorosch, K. 1972. Natural degeneration of mycoplasma-like bodies in an aster yellows infected host plant. Phytopathology 62: 9-26.

32. Hirumi, H. and Maramorosch, K. 1969. Mycoplasma-like bodies in the salivary glands of insect vectors carrying the aster yellows agent. J. Virology 3: 82-84.

33. Hsu, T.P. and Banttari, E.E. 1979. Dual transmission of the aster yellows mycoplasma-like organism and the oat blue dwarf virus and its effect on longevity and fecundity of the aster leaf hopper. Phytopathology 69: 843-845.

34. Jenns, A.E. and Kuc, J. 1977. Localized infection with tobacco necrosis virus protects cucumber against *Colletotrichum lagenarium*. Physiological Plant Pathology 11: 207-212.

35. Kahn, R.P., Lawson, R.H., Monroe, R.L., Hearon, T.S. 1972. Sweet potato littleleaf (witches' broom) associated with a mycoplasma-like organism. Phytopathology 62: 903-909.

36. Lin, C.P. and Chen, T.A. 1985. Monoclonal antibodies against the aster yellows agent. Science 227: 1233-1235.

37. Maramorosch, K. 1952. Direct evidence for the multiplication of aster yellows virus in its insect vector. Phytopathology 42: 59-64.

38. McCoy, R.E., Tsai, J.H,, Norris, R.C. and Gwin, G.H. 1983. Pigeon pea witches' broom in Florida. Plant Disease 67: 443-445.

39. Nitzany, F.E. 1966. Synergism between *Pythium ultimum* and cucumber mosaic

virus. Phytopathology 56: 1386-1389.

40. Pieczarka, D.J. and Zitter, T.A. 1981. Effect of interaction between two viruses and *Rhizoctonia* on pepper. Plant Disease 65: 404-406.

41. Ploaie, P.G. 1971. Particles resembling viruses associated with mycoplasma-like organisms in plants. Revue Roum. Biol. Ser. Bot. 16: 3-6.

42. Potter, L.R. 1982. Interaction between barley yellow dwarf virus and rust in wheat, barley, and oats, and the effects on grain yield and quality. Ann. Appl. Biol. 100: 321-329.

43. Powell, N.T. 1979. Internal synergisms among organisms inducing disease. Chapt. 6, pp. 113-133. In: Horsfall, J.C. and Cowling, E.B. (Eds.) Plant Disease. An Advanced Treatise Vol. IV Academic Press, NY, 466 p.

44. Pratt, R.G., Ellsbury, M.M., Barnett O.W. and Knight, W.E. 1982. Interactions of bean yellow mosaic virus and an aphid vector with Phytophthora root disease in arrow leaf clover. Phytopathology 72: 1189-1192.

45. Sinha, R.C. 1979. Purification and serology of mycoplasma-like organisms from aster yellows-infected plants. Can. J. Plant Pathol. 1: 65-70.

46. Singer, S.H., Barile, M.F. and Kirschstein, R.I. 1973. Mixed mycoplasma-virus infections in cell cultures. Ann. NYAC 225: 304-310.

47. Stevens, C. and Gudauskas, R.T. 1982. Relation of maize dwarf mosaic virus infection to increased susceptibility of

corn to *Helminthosporium maydis* race O. Phytopathology 72: 1500-1502.

48. Stuckey, R.E., Ghabrial, S.A. and Reicosky, D.A. 1982. Increased incidence of *Phomopsis* sp. in seeds from soybeans infected with bean pod mottle virus. Plant Disease 66: 826-829.

49. Sylvester, E.S. 1985. Multiple acquisition of viruses and vector-dependent prokaryotes: Consequences on transmission. Ann. Rev. Entomol. 30: 71-88.

50. Tanaka, H., Uegaki, R. and Fujimori, T. 1983. Antibacterial activity of sesquiterpenoids from tobacco leaves elicited by *Pseudomonas solanacearum* and tobacco mosaic virus. Ann. Phytopathol. Soc., Japan 49: 501-507.

51. Townsend, R., Markham, P.G. and Plackitt, K.A. 1977. Multiplication and morphology of *Spiroplasma citri* in leafhopper *Euscelis plebejus*. Ann. Appl. Biol. 87: 307-313.

52. Virus Research Group and Division of Mulberry Protection. 1974. Studies on the pathogens of mulberry dwarf disease. I. Virus-like particles and mycoplasma-like bodies associated with mulberry yellow dwarf disease. Scientia Sinica. Vol. 17: 421-427.

53. Whitcomb, R.F., Shapiro, M. and Granados, R.R. 1974. Insect defense mechanisms against microorganisms and parasitoids. In: Rockstein, M. The Physiology of the Insecta Vol. V. Second Edit. Academic Press, NY and London, p. 648.

54. Wolanski, B.S. and Maramorosch, K. 1979. Rayado fino virus and corn stunt spiro-

plasma: Phloem restriction and trans-
mission by *Dalbulus elimatus* and *D. maides*. Fitopatol. Brasileria 4: 47-54.

55. Zummo, N., Bradfute, O.E., Robertson, D.
C.and Freeman, K.C. 1975. Yellow sorghum stunt: A disease symptom of sweet sorghum associated with a mycoplasma-like body in the United States. Pl. Dis. Reptr. 59: 714-716.

12

BIONOMICS OF *DALBULUS MAIDIS*

(DeLONG AND WOLCOTT),

A VECTOR OF MOLLICUTES AND VIRUS

(HOMOPTERA: CICADELLIDAE)

by

James H. Tsai

Fort Lauderdale Research and Education Center

University of Florida

3205 College Avenue

Fort Lauderdale, FL 33314

INTRODUCTION

Nault and DeLong (1980) introduced evidence to support the co-evolution of leafhoppers in the genus *Dalbulus* DeLong (1950) with maize (*Zea mays L.*) and its ancestors. Due to its close relationship with maize, the corn leafhopper, *Dalbulus maidis* (DeLong and Wolcott) has become an economically important pest because it not only causes feeding damage to corn (Bushing and Borton, 1974), but it also can transmit three important corn disease agents, the corn stunt spiroplasma (CSS), maize bushy stunt mycoplasma (MBSM), and maize rayado fino virus (MRFV) (Nault and Knoke, 1981). Two of these diseases, CSS and MRFV, were associated with the outbreak of a disease complex resulting in a loss of the $60 million hybrid seed production and breeding nurseries in southern

Florida in 1979 and 1980 (Bradfute *et al.*, 1981; Niblett *et al.*, 1981).

Nearly four decades have passed since *D. maidis* was first reported as a vector of a corn stunt pathogen (Kunkel, 1946), yet very little is known of the biology and ecology of *D. maidis* (Nault, 1984). Davis (1966) performed a brief study on the biology of *D. maidis* at six temperatures and Pitre *et al.*, (1966) and Pitre (1970) reported *Tripsacum dactyloides (L.)* L. as a new host for *D. maidis* and other plant species susceptible to CSS. This paper reports the biology of *D. maidis* including the development of life stages at four temperatures, adult longevity, fecundity and its hosts.

MATERIALS AND METHODS

Dalbulus maidis eggs were excised from the midribs of corn (*Zea mays L.*) and placed on cut leaf of corn plant (var. Saccharata 'Guardian') and allowed to hatch. Newly hatched nymphs were placed in individual 10 cm petri dishes containing fresh corn leaf pieces measuring 80 x 20 mm. The petri dishes were lined with a layer of moist paper and then covered with a plastic bag to maintain high humidity and placed in an environmental chamber at one of four temperatures, 10 (50 F), 15.6 (60 F), 26.7 (80 F) and 32.2 C (90 F) and a light cycle of 12L:12D. These temperatures were selected to represent the temperatures in south Florida during the corn growing season. Daily observations on molting, and survivalship were made and fresh leaf pieces replaced the old ones every two days until adult emergence. Adults were then removed and placed on young corn plants at 4-5 leaf stage and covered with a cylinder cage 22 cm tall and 5 cm in diameter constructed of butyrate tubing and maintained until their death. The dates of molting, length of stadia, length and width of nymphs, and adult longevity at each of the four temperatures were recorded. The length of nymphs and adults was measured from tip of vertex to tip of abdomen; width was measured across the widest part of the body. Developmental time for each instar at the four temperatures was analyzed using an Analysis of Variance and a Waller/Duncan k-ratio t-test for means.

Upon adult emergence 11 and 15 pairs (one male and one female) of *D. maidis* were maintained at 15.6 and 26.7 C, respectively on individual corn plants at the 5- to 6-leaf stage covered by a cylinder cage. Pairs were removed daily to new plants, and the exposed plants were then dissected for egg counts. Information recorded included the number of eggs laid per female each day and the number of eggs laid per female in a lifetime at both temperatures. Six unmated females were maintained as above at 15.6 and 26.7 C; their adult longevities were then compared to that of the mated females at the same temperatures.

Other plant species and varieties including *Tripsacum dactyloides* (L.) L., *T. dactyloides* (L.) L. var. *meridonale* de Wet and Timothy, *Tripsacum sp.*, *Rottboellia exaltata* L., *Secale cerale* L., and *Avena sativa* L. were tested as alternate hosts. Newly emerged adults were placed singly on the caged plants, and the number of insects tested ranged from 14 to 32. The survivorship was recorded daily.

RESULTS

At all four temperatures, development proceeded with five instars. However, an occasional supernumerary instar VI would occur at all temperatures. The developmental times for instars I-V ranged up to 9-fold over the range of temperatures (Table 1). The ANOVA for the development of instar I was significant with an F.05 [3,33] value of 9.90. The Waller/Duncan K-ratio t-test on the means revealed that development at 10 C was significantly longer than at the other three temperatures. The development of instar II was significant with an F.05 [3,34] value of 86.14. The development of instar III was significant with an F.05 [3,33] value of 23.32. The development of instar IV was significant with an F.05 [3,33] value of 120.03. The development of instar V was significant with an F.05 [3,33] value of 47.98. The Waller/Duncan k-ratio t-test on the means for instars II-V revealed that development at 10. and 15.6 C was significantly longer than at 26.7 and 32.3 C.

The regression analysis also demonstrates the effect of temperature on development with r values and slopes $\pm$ SE of 0.705 (-0.056 $\pm$ 0.097), 0.945 (-0.242 $\pm$ -.100), 0.837 (-0.304 $\pm$ 0.135), 0.961 (-0.360 $\pm$ 0.164), and 0.910 (-0.666 $\pm$ 0.313) for instars I-V, respectively (Fig. 1).

Length and width for instars I-V and adults $\bar{x}$ $\pm$ SD) were as follows: 0.87 $\pm$ 0.05 mm, 0.27 $\pm$ 0.03 mm for instar I; 1.14 $\pm$ 0.11 mm, 0.37 $\pm$ 0.04 mm for instar II; 1.69 $\pm$ 0.21 mm, 0.45 $\pm$ 0.07 mm for instar III; 2.14 $\pm$ 0.24 mm, 0.62 $\pm$ 0.11 mm for instar IV; 2.82 $\pm$ 0.36 mm, 0.81 $\pm$ 0.10 mm for instar V, and 2.79 $\pm$ 0.27 mm, 0.81 $\pm$ 0.07mm for adult male; 3.18 $\pm$ 0.19 mm, 0.88 $\pm$ 0.05 mm for adult female (Table 2).

Adults lived longest at 15.6 C with an average duration ($\bar{x}$ $\pm$ SD) of 106.88 $\pm$ 51.64 days and shortest at 32.2 C with a duration ($\bar{x}$ $\pm$ SD) of 15.67 $\pm$ 9.93 days (Table 1). Oviposition data obtained from 11 and 15 pairs of *D. maidis* at 15.6 and 26.7 C, respectively, showed that the mean number ($\bar{x}$ $\pm$ SD) of eggs per day per female was 3.62 $\pm$ 1.09 at 15.6 and 14.18 $\pm$ 3.55 at 26.7 C. Number ($\bar{x}$ $\pm$ SD) of eggs per female per life at 15.6 and 26.7 C, respectively was 402.33 $\pm$ 140.03 and 611.08 $\pm$ 164.96. Eggs were seldom laid within 24 hours after adult emergence, and failed to hatch at 20 C, but hatched if they were returned to high temperatures. Adult longevity ($\bar{x}$ $\pm$ SD) of mated females was shorter than that of unmated females, respectively, at 15.6 and 26.7 C;

111.00 ± 14.54, 180 ± 26.09 days at 15.6 ; 45.15 ± 15.81, 112.00 ± 16.52 days at 26.7 C.

D. maidis did not survive beyond 20 days on *T. dactyloides, Tripsacum sp., R. exaltata, S. cereale* and *A. sativa* (Table 3). However, *T. dactyloides* var. *meridonale* was suitable for oviposition and continuous reproduction by this species.

DISCUSSION

Introductions of tropical pests and diseases are of great concern to plant protectionists. Southern Florida provides an ideal environment for importations of pest species from the American tropics by either human activities or wind transport. Bradfute *et al.,* (1981) suggested that *D. maidis* and corn stunt pathogens were carried to Florida by hurricane David in 1979. Regardless of the mode of introduction, *D. maidis* and the disease agents including CSS, MBSM. and MRFV were well established in south Florida where they have been found in insect surveys in the last five years (J.H. Tsai and B.W. Falk, unpublished).

In this study, the development of *D. maidis* occurred at each of the four tested temperatures, but, at least 1 supernumary instar VI occurred at all temperatures. This occasional supernumary molt may occur at these temperatures because of an asynchrony in biochemical reactions to the temperature effect on the insect's metabolism (Ratte 1985). Several studies have shown that as temperature increases, insect development time decreases (Butler *et al.,* 1983; Braman *et al.,* 1984; Kilian and Nielson, 1971). In the present study, developmental times decreased as temperatures increased through 26.7 and from 26.7 to 32.2 C the developmental times were essentially the same (Fig. 1). Davis (1966) found that *D. maidis* failed to lay eggs at 12.8 and 18.30 C. However, in this study, development occurred at temperatures of 10 and 15.6 C; the average number of eggs laid at 15.6.C per female per life was 402.33. Davis found that at 21.1 C adult longevity was 26-51 days as compared to an average of 106.9 days in this study. Because his study was based on a very small sample size (3 pairs), it may not reflect longevity accurately. Davis (1966) also found that at 21.1 C the average number of eggs produced per female per life was 151. Our data at 15.6 and 26.7 C, respectively were 402.33 and 611.08 eggs. Again, this large discrepancy may possibly be due to small sample size (3 females) in his test.

In 1966, Pitre *et al.*, reported *Tripsacum dactyloides (L.) L.* a newly discovered host for *D. maidis.* Prior to that only two species of plants, corn and teosinte (*Euchlaena mexicana* Schrad) were reported to be hosts of *D. maidis* (Barnes, 1954). Pitre (1966) successfully reared and maintained a colony of *D. maidis* on *T. dactyloides* through five generations, although insect development was slower and adult longevity was greatly reduced when compared to colonies reared on corn. In the present study, the greatest longevity of this insect on *T. dactyloides* was 20 days which is not in agreement with the earlier findings (Pitre *et al.*, 1966; Pitre, 1970). But D. maidis could be reared continuously on *T. dactyloides* var. *meridonale.* The variation could be due to the different biotypes of *D. maidis* and/or host plant varieties.

Dalbulus maidis is basically a neotropical species and has been reported in maize fields in southern California (Bushing and Burton, 1974) and the southeastern United States (Pitre *et al.*, 1966). It was first found in large numbers in 1978 in southern Florida (Bradfute *et al.*, 1981). Since then, this insect has been collected throughout the year (J.H. Tsai, unpublished). Although this insect has been occasionally found on sticky traps as far north as Ohio (L.R. Nault, personal communication), the potential for *D. maidis* to become established in the major corn growing areas in the U.S. is limited by 1) its host range: the perennial hosts, *Tripsacum dactyloides* and other varieties of *T. dactyloides* are rarely found outside of the southeastern USA (Hitchcock, 1935) which would restrict its ability to survive in areas where corn is not grown continuously, and 2) the inability of this insect to reproduce below 10 C, and apparent absence of winter hardiness or diapause clearly demonstrates that it did not survive the winter in Mississippi (Pitre, 1966), and certainly it would not overwinter in the northern U.S. Likewise, the potential for spread of CSS, MBSM and MRFV from the south to the north is rather limited.

From the temperature study, I have been able to maintain small colonies of *D. maidis* at 15.6 C for long periods without frequent transfers. All other rearing of *D. maidis* and maintenance of CSS, MBSM and MRFV have been done regularly at either 26.7 or 32.2 C depending on the need of the experiment.

ACKNOWLEDGMENT

Appreciation is extended to Grace M. Johns and Paul D. Calvert for their technical assistance and to the American Seed Research Foundation, Washington, D.C. for financial support.

SUMMARY

The corn leafhopper, *Dalbulus maidis* (DeLong and Wolcott), a vector of several economically important pathogens of maize, was reared, from eggs to adults, at four constant temperatures, 10 , 15.6 , 26.7 , or 32.2 C, with a light cycle of 12L: 12D. The average development times for instars I-V ranged from 11.6 to 33.6 days at 10 C, 6.3 to 13.3 days at 15.6 C, 2.5 to 3.8 days at 26.7 C, and 2.4 to 4.4 days at 32.2 C. Both male and female longevities were greatest at 15.6 C and lowest at 32.2 C. Oviposition data obtained from 11 and 15 pairs of *D. maidis* at 15.6 and 26.7 C, respectively, showed that the number ($\bar{x} \pm$ SD) of eggs per female per day was 3.62 $\pm$ 1.09 at 15.6 C and 14.18 $\pm$ 3.55 at 26.7 C, the number ($\bar{x} \pm$ SD) of eggs per female per life was 402.33 $\pm$ 140.03 at 15.6 and 611.08 $\pm$ 164.96 at 26.7 C. Eggs were seldom laid within 24 hours after adult emergence. Adult longevities ($\bar{x} \pm$ SD) between mated females and unmated females at 15.6 and 26.7 C were 111.00 $\pm$ 14.54 days for mated, and 180.00 $\pm$ 26,09 days for unmated at 15.6 C, and 45.15 $\pm$ 15.81 days for mated and 112.00 $\pm$ 16.52 days for unmated at 26.7 C.

Of six other plant species and varieties tested as alternate hosts for *D. maidis*, only *Tripsacum dactyloides* (L.) L. var. *meridonale* de Wet and Timothy was suitable for oviposition and rearing.

The length and width ($\bar{x} \pm$ SD) of eggs and each instar at 26.7 C were measured as follows: 0.74 $\pm$ 0.06mm, 0.15 $\pm$ 0.02mm for 16-hr old eggs; 0.83 $\pm$ 0.05mm, 0.22 $\pm$ 0.03mm for 163-hr old eggs; 0.87 $\pm$ 0.05mm, 0.27 $\pm$ 0.03mm for instar I; 1.14 $\pm$ 0.11mm, 0.37 $\pm$ 0.04mm for instar II; 1.69 $\pm$ 0.21mm, 0.45 $\pm$ 0.07mm for instar III; 2.14 $\pm$ 0.24mm, 0.62 $\pm$ 0.11mm for instar IV; 2.82 $\pm$ 0.36mm, 0.81 $\pm$ 0.10 mm for instar V. The measurements for adults were $\bar{x} \pm$ SD) 2.79 $\pm$ 0.27mm, 0.81 $\pm$ 0.07mm for adult males, and 3.18 $\pm$ 0.19mm, 0.88 $\pm$ 0.05mm for adult females.

REFERENCES

1. Barnes, D. 1954. Biologia, ecologia y distribucion des las chicharritos, *Dalbulus elimatus* (Ball) y *Dalbulus maidis* (DeL. and W.) Officina de Estudios. Especiales de S.A.G. Feletto Tech. No. 11:5-112.

2. Bradfute, O.E., Tsai, J.H. and Gordon,

D.T., 1981. Corn stunt spiroplasma and viruses associated with a maize disease epidemic in southern Florida. Plant Disease 65: 837-841.

3. Braman, S.K., Sloderbeck, P.E. and Yeargan, K.V., 1984. Effects of temperature on the development and survival of *Nabis americoferus* and *N. roseipennis* (Homoptera: Nabidae). Ann. Entomol. Soc. Am. 77:592-596.

4. Bushing, R.W. and Burton, V.E. 1974. Leaf-hopper damage to silage corn in California. J. Econ. Entomol. 67:656-658.

5. Butler, G.D., Jr., Henneberry, T.J. and Clayton, T.E., 1983. *Bemisia tabaci* (Homoptera: Aleyrodidae): Development, oviposition, and longevity in relation to temperature. Ann. Entomol. Soc. Am. 76: 310-313.

6. Davis, R. 1966. Biology of the leafhopper *Dalbulus maidis* at selected temperatures. J. Econ. Entomol. 59:766.

7. DeLong, D.M. 1950. The genera *Baldulus* and *Dalbulus* in North America including Mexico. (Homoptera: Cicadellidae). Bull. Brooklyn Entomol. Soc. 45:105-116.

8. Hitchcock, A.S. 1935. Manual of the grasses in the United States. Government Printing Office, Washington, D.C. Misc. Pub. No. 200. pp. 1040.

9. Kilian, L. and Nielson, M.W. 1971. Differential effects of temperature on the biological activity of four biotypes of the pea aphid. J. Econ. Entomol. 64: 153-155.

10. Kunkel, L.O. 1946. Leafhopper transmission

of corn stunt. Proc. Nat. Acad. Sci. 32: 246-247.

11. Nault, L.R. 1984. A report from chairman of vector subject matter committee. Maize Virus Diseases Newsletter 1:65.

12. Nault, L.R. and DeLong, D.M. 1980. Evidence for co-evolution of leafhoppers in the genus *Dalbulus* (Cicadellidae: Homoptera) with maize and its ancestors. Ann. Entomol. Soc. Am. 73: 349-353.

13. Nault, L.R. and Knoke, J.K., 1981. Maize vectors. pps. 77-84. Gordon, D.T., Knoke, J.K. and Scott, G.E., (eds.). In: Virus and virus-like diseases of maize in the United States. Southern Coop. Series Bull. 247. p. 219.

14. Niblett, C.L., Tsai, J.H. and Falk, B.W., 1981. Virus and mycoplasma diseases of corn in Florida. Proc. 36th Annual Corn and Sorghum Industry Research Conf. 36: 78-88.

15. Pitre, H.N., 1966. Corn virus diseases of recent occurrences in southeastern United States with emphasis on Mississippi. Proc. N. Cen. Br. Ent. Soc. Amer. 21 43:47.

16. Pitre, H.N., 1970. Notes on the life history of *Dalbulus maidis* on gamma grass and plant susceptibility to the corn stunt disease agent. J. Econ. Entomol. 63: 1661-1662.

17. Pitre, H.N., Combs, R.L., Jr., and Douglas, W.A., 1966. Gammagrass, *Tripsacum dactyloides*: a new host of *Dalbulus maidis*, vector of corn stunt virus. Plant Dis. Reptr. 50: 570-571.

18. Ratte, H.T., 1985. Temperature and insect

development, pp. 33-66. Hoffman, K.H. (ed.) In: Environmental physiology and biochemistry of insects. Springer, New York.

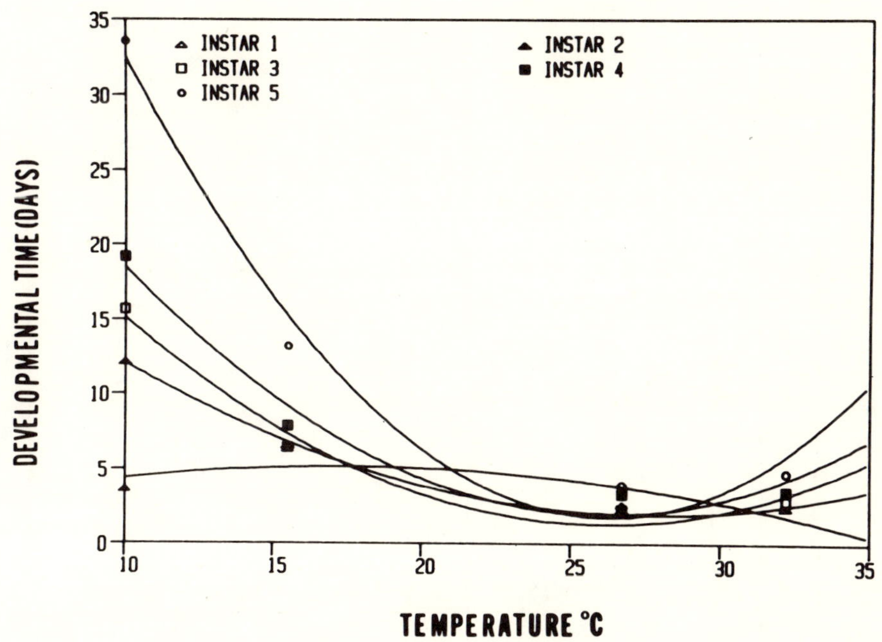

Figure 1. Regression analysis of effect of temperature on instar development.

Table 1. *Dalbulus maidis* instar development (in days $\bar{x} \pm SD$) and adult longevity for four temperatures.

Temp. °C	Instar						Adult Longevity	
	I	II	III	IV	V	VI		
10°								
	11.62	21.64	22.00	26.15	33.55	30.50	66.60	(♂) 30.50 ± 21.92
	±10.06	12.05	7.73	11.00	14.43	0.71	38.21	(♀) 90.67 ± 22.55
n	42	39	33	26	11	2	5	
15.6°								
	6.27	6.50	6.36	7.82	13.33	16.00	106.88	(♂)105.64 ± 49.39
	±2.29	0.96	1.59	1.40	3.60	--	51.64	(♀)135.75 ± 25.04
n	22	22	22	22	21	1	45	
26.7°								
	2.51	2.59	2.53	3.14	3.77	3.00	77.50	(♂) 74.78 ± 36.32
	±1.11	1.19	0.96	0.82	0.62	1.73	29.82	(♀) 87.64 ± 18.14
n	80	80	79	72	52	3	26	
32.2°								
	2.76	2.44	2.76	3.46	4.41	2.00	15.67	(♂) 18.33 ± 7.51
	±0.90	0.76	0.82	1.50	1.23	--	9.93	(♀) 12.67 ± 12.10
n	84	79	72	56	39	1	33	

Table 2. Measurements (mm) of instars and adults of *Dalbulus maidis* reared at 26.7°C.

Stage	Length		Width		# of
	Range	Avg.	Range	Avg.	insects tested
Instar I	0.77 - 0.98	0.87	0.21 - 0.28	0.27	30
Instar II	0.84 - 1.40	1.14	0.28 - 0.42	0.37	33
Instar III	1.26 - 2.10	1.69	0.35 - 0.63	0.49	27
Instar IV	1.75 - 2.73	2.14	0.35 - 0.77	0.62	27
Instar V	1.96 - 3.92	2.82	0.70 - 1.05	0.81	24
Adult male	2.52 - 3.36	2.79	0.70 - 0.91	0.81	20
Adult female	2.73 - 3.50	3.18	0.77 - 0.98	0.88	20

Part III: Diseases of Rice, Potato, Corn, Citrus, Eggplant and Other Plants

13

THE APPARENT YELLOWS DISEASE OF
DODONAEA SPP. IN HAWAII

by

Donald E. Gardner

National Park Service
Cooperative Park Studies Unit

Department of Botany

University of Hawaii at Manoa

Honolulu, Hawaii 96822

INTRODUCTION

The genus *Dodonaea* (Sapindaceae) is widely distributed throughout the islands of the Pacific and other warmer regions of the world. *Dodonaea eriocarpa Sm.* and *D. sandwicensis* Sherff, both commonly known as 'a'ali'i, are woody, highly variable evergreen species with several forms endemic to Hawaii. These plants occur as small to large shrubs or small trees up to 10 m tall. The plants are found throughout the Hawaiian Islands in mid to upper elevation open, dry forests. *Dodonaea spp.* are major floral components of extensive portions of Hawaii Volcanoes and Haleakala National Parks, where they often comprise the dominant woody species in drier habitats. Aside from their prominence in native ecosystems, *Dodonaea spp.* are valued for ornamental purposes in traditional Hawaiian cultural practices, such as use of the colorful fruits in lei making.

A conspicuous and severe disease, with symptoms suggesting a mycoplasma-like organism (MLO)-associated yellows disease, is common in certain stands of both *Dodonaea spp.* The disease appears to be widespread throughout the islands, having been observed on the islands of Hawaii (at several sites), Maui, and Kauai. Since this range includes the most widely separated major islands, site documentations are likely in the future for other islands. Notwithstanding this distribution and prominence, no report of the disease has been found prior to the current observations. Nothing is known, therefore, of the possible origin and history of *Dodonaea* yellows in Hawaii. Until recently, however, relatively little attention has been devoted to native Hawaiian forest disease investigations.

SYMPTOMATOLOGY

The disease is characterized by production of abnormally lengthened pendulous twigs which hang loosely from branches (Fig. 1), or proliferate to form drooping, tightly compacted witches' brooms with reflexed shoot tips (Fig. 2). Internodal elongation may be 10-fold or greater and becomes conspicuous early in disease expression (Fig. 3). Leaves of brooms are typically strikingly yellow, contrasting with dark green healthy foliage, and are often stunted. Young diseased twigs and petioles are red rather than the normal shades of green or gray.

Death and defoliation of brooms results in a hairy or stringy appearance of draped, persistent twigs (Figs. 4 and 5). A single, or several, isolated brooms may occur throughout an otherwise healthy appearing plant (Figs. 3 and 6), with brooms produced on the same branches and arising in immediate proximity to normal tissue.   On the other hand, an entire plant may express symptoms more or less uniformly (Fig. 7), giving the plant an abnormal growth form recognizable at a distance in the field.

Flowering and fruiting may be markedly reduced on symptomatic portions, but production of functional flowers followed by apparently normal fruit development with viable seeds is not uncommon.   Flower and fruit production on asymptomatic portions of broomed plants appears to be fully normal. The disease usually progresses slowly (over a number of years) both within populations of *Dodonaea* plants and in individual infected hosts, but eventually results in the death of symptomatic portions. Death of entire plants has been observed in extreme cases. No conclusive evidence of recovery under natural conditions has been thus far observed; however, diseased plants severly pruned and transplanted from the field to the greenhouse or laboratory often regenerate healthy appearing, rather than symptomatic, shoots.

DIAGNOSIS

In attempts to determine the etiology of the *Dodonaea* yellows disease, standard diagnostic approaches have been followed, although many of these are as yet only preliminary. These approaches have included injection with oxytetracycline hydrochloride solutions, histochemical staining with diagnostic stains, heat treatment, and direct electron microscopic examination of petiole, peduncle, and twig phloem tissue for MLO bodies. Diagnostic attempts, thus far, have not yielded positive results, nor have potential arthropod (particularly leafhopper) vectors been found. Successful graft unions between diseased and healthy tissue have not yet been established.

DISCUSSION

Although no direct evidence of an MLO etiology for the yellows disease of *Dodonaea* is yet available, the distinctive symptoms of leaf yellowing and stunting, and production of witches' brooms suggest that inclusion of *Dodonaea* yellows among the known MLO-associated yellows diseases is justified for the present. No other macro- or microscopic explanation for the symptoms has been found, despite thorough investigation.

Notwithstanding the general similarities, the production of internodally lengthened, pendulous twigs of brooms appears to be a notable variation from the growth form of witches' brooms reported for other yellows diseases, which are characterized by proliferation of stunted, abnormally upright twigs (Seliskar and Wilson, 1981). Ghosh (1981), however, reported that individual infected shoots of sandalwood spike-diseased trees are sometimes pendulous or drooping, a condition associated with the continuous apical growth of the diseased shoot.

It is of possible significance to this discussion that *Dodonaea viscosa* is a suspected alternate host of the MLO associated with sandalwood spike disease in India (Hull *et al.*, 1970; Nayar and Ananthapadmanabha, 1977; Sastri and Narayana, 1931; Venkata, 1935). *Dodonaea viscosa* is closely related to *D. eriocarpa* and has been placed in synonomy with both *D. eriocarpa* and *D. sandwicensis* by some authors. The extent to which yellows symptoms on *Dodonaea spp.* in Hawaii resemble those on *D. viscosa* in India is not entirely clear at present, although production of spike-like growth does not characterize the disease in Hawaii.

REFERENCES

Ghosh, S.K., 1981. Association of mycoplasma and allied pathogens with tree diseases in India. In: Mycoplasma Diseases of Trees and Shrubs. Maramorosch K., and Raychaudhuri (eds.), pp. 231-243. Academic Press, New York.

Hull, R., Plaskitt, A., Nayar, R.M. and Ananthapadmanabha, H.S., 1970. Electron microscopy of alternate hosts of sandal spike pathogen and of tetracycline-treated spike-infected sandal trees. J. Ind. Acad. Wood. Sc. 1 (1): 62-64.

Nayar, R., and Ananthapadmanabha, H.S., 1977. Little leaf disease in collateral hosts of sandal (*Santalum album L.*). Eur. J. Forest Pathol. 7: 152-158.

Sastri, B.N., and Narayana, N., 1931. The spike-disease of *Dodonaea viscosa*. J. Ind. Inst. Sci. 13A: 147-152.

Seliskar, C.E., and Wilson, C.L., 1981. Yellows diseases of trees. In: Mycoplasma Diseases of Trees and Shrubs, Maramorosch, K., and Raychaudhuri, S.P., (eds.) pp. 35-96, Academic Press, New York.

Venkata, R.M.G., 1935. The role of undergrowth in the spread of the spike disease of sandal. Ind. Forest. 61: 169-188.

Figure 1.

Figure 2.

Figure 3.

Fig. 1.   Loosely broomed branch with abnormally lengthened, pendulous twigs.

Fig. 2.   Pendulous, compact broom with reflexed shoot tips.

Fig. 3.   Abnormally lengthened, pendulous twigs of a young broom developing on an otherwise asymptomatic branch.

Figure 4.

Figure 5.

Figure 6.

Figure 7.

Fig. 4.   Dead broom with compact persistent, pendulous twigs.

Fig. 5.   Severely diseased shrub, most brooms of which are dead.

Fig. 6.   Young, isolated broom developing on an otherwise normal shrub.

Fig. 7.   Large, uniformly diseased shrub (left) near a healthy shrub (right).

14

RICE YELLOW DWARF DISEASE

by

V. Muniyappa

Department of Plant Pathology

University of Agricultural Sciences

Hebbal, Bangalore - 560024

and

S.P.Raychaudhuri

Chairman, International Union of Forestry

Research Organization

W.P. Plant Mycoplasma Disease

A-61, Alkananda, Kalkaji, New Delhi - 110 019

INTRODUCTION

Yellow dwarf disease of rice has been reported in most of the rice growing countries of the East and South East Asia. In some areas of Japan and India, the disease has been observed to cause serious damage to the crop (Hayashi, 1961; IRRI, 1969; Ishii *et al.*, 1969; Mishra *et al.*, 1971; Miyahara and Yamaguchi, 1965; Ling,1972; Muniyappa, 1974; Muniyappa and Ramakrishnan, 1976a, 1976b; Murata *et al.*, 1965; Nishio *et al.*, 1964 and Raychaudhuri *et al.*, 1967b). In recent years with continuous cropping, with the release of varieties which are comparatively photo-insensitive, the rice yellow dwarf disease is tending to attain economic importance in India. Details have been discussed earlier by Raychaudhuri, 1977, 1984) and Raychaudhuri and Nariani (1977).

Rice yellow dwarf is a translation of Japanese "Ooi-byo". In Malaysia, the disease was first called "Padi-Jantan" (Lim and Goh,1968). Lim (1970) proposed changing "Padi-Jantan" to yellow dwarf because of the similarities of these two diseases. The disease was first recorded in the South Western region of Japan around 1910. Hashioka (1964) stated that the yellow dwarf disease first appeared in literature in 1919 in the annual report of Kochi Agricultural Experiment Station. In Taiwan, the disease has been known since 1932 (Kurosawa, 1940). The disease has been recorded in different prefectures in Japan, Chiba (Shimazu, 1976), Ehime (Ide and Tanaka, 1968), Fukuoka (Murata *et al.,* 1965), Ibaraki (Komori, 1966; Matsunaga *et al.,* 1973; Takano and Komori, 1961), Kanto and Kinki (Iida and Shinkai, 1950), Kanto-Tosan (Ishii *et al.,* 1969), Kagoshima (Niidome and Itoga, 1962), Kyushu (Suenaga *et al.,* 1965), Matsumoto-daira (Nakamura *et al.,* 1962), Miyazaki (Samejima, 1967), Nagano (Shimoyama and Shibamoto,1966), Okinawa (Shinkai *et al.,* 1963), Shikuko (Anon., 1943), Shizuoka (Mori *et al.,* 1965) and Shiga (Watanabe *et al.,* 1979).

The disease is widely distributed in Sri Lanka (Abeygunawardena *et al.,* 1970), India (Govindu *et al.,* 1968; Muniyappa and Ramakrishnan, 1976b; Pathak *et al.,* 1967; Raychaudhuri and John 1970; Raychaudhuri *et al.,* 1967b), Bangladesh (Galvez-E and Shikata, 1969; Galvez-E and Miah, 1969), the Philippines (Ling, 1969; Rivera and Ou, 1964; Palomar and Rivera, 1967), China southern, China and Zhejiang (Hashioka, 1952; Li *et al.,* 1979), Thailand (Wathanakul and Weerapat, 1969), Indonesia (Satomi *et al.,* 1978) and Java (Satomi *et al.,* 1978). The rice yellow dwarf was noticed in almost all rice growing areas in India, Purnea (Bihar), (Raychaudhuri *et al.*, 1971), Patna (Bihar) (Singh and Chaudhury, 1983), Karnataka Govindu *et al.*, 1968; Muniyappa and Ramakrishnan, 1976b), Orissa, Bihar, West Bengal, Andhra Predesh,Delhi (Raychaudhuri *et al.*, 1976b), Tamil Nadu (Mohan and Natarajan, 1973; Narayanaswamy and Jaganathan, 1973) and Kerala (Mathew and Abraham, 1977).

Amici *et al.*, (1970) observed mycoplasma-like organisms in sieve tubes of phloem of yellows infected rice plants "enrochat" disease in Spain. They suggested that the diseases known as "giallume", yellow dwarf, enanismo amarillo are caused by the same pathogen. If this is confirmed, yellow dwarf is also occurring in Spain. It is likely that it may be present in other countries.

DISEASE INCIDENCE AND LOSSES

Heavy losses in rice yield occurred due to yellow dwarf disease (Komori, 1965; Kureha *et al.*, 1974; Hsieh, 1976). Hashioka (1964) reported that in an epidemic, 70-80 percent of the hills were infected causing 50 percent yield reduction in Japan. He observed that the plants became infected at all stages of growth, the yield decreased most when infection occurred before formation of the panicle primordia. When the rice seedlings were inoculated at the 11 leaf stage or earlier, the plants produced no grains, when inoculated at the 14 leaf stage or later, the yield reduction was not significant (Shinkai, 1962). In the Philippines BPI-76 inoculated 10 or 30 days after sowing failed to yield fertile grains, but there was no significant yield reduction when inoculated at 60 days (Palomar and Rivera, 1967). In Japan, the area severely damaged was reported to be about 20,000 ha, the yield loss amounting to 10,000 tons (Iida, 1969). Abeygunawardena (1969) in Sri Lanka recorded 13.8 to 20.7 percent infection in the ratoon at Peradeniya and 0.08 to 2.2 percent infection at Batalogoda.

In Karnataka (India) the yellow dwarf disease incidence was observed up to 27.34 percent (Var. MR-249) in planted crop and up to 82.75 percent (Selection CR-126-16) in ratoon. There was less incidence in the summer grown crop than in *kharif* grown crop. High incidence was observed in *rabi* season (October-December) (Muniyappa and Ramakrishnan, 1976b).

There was considerable reduction in yield and height when the rice plants were infected up to 50 days after transplanting. Infection 75 days after transplanting had no significant effect on yield and height of the plants (Muniyappa, 1980a). There was no yield when the plants were artificially inoculated at 10 to 45 days after sowing. Plants inoculated at 50 to 65 days after sowing produced very low yields. Normal yield was obtained when the plants were inoculated 85 and 100 days after sowing (Muniyappa and Ramakrishnan, 1976a; Palomar and Rivera, 1967).

The rice yellow dwarf is more serious in the second crop than in the first crop (Chen and Ko, 1975; Chien, 1979; Kobayashi *et al.*, 1975; Yako *et al.*, 1979). An equation is presented for estimating yield reduction based on the plant age at the time when symptoms of rice yellow dwarf infection appear (Chen *et al.*, 1977).

SYMPTOMS

Babaguchi and Hara (1966), Shinkai (1962), Hashioka (1963), Palomar and Rivera (1967), Raychaudhuri *et al.*, 1967a), Lim and Goh (1968), Iida (1969) and Muniyappa (1974) studied the disease in detail and gave a full description of the symptoms. The early symptoms were seen on the newly emerging leaves. The leaves were pale green in the beginning and as the disease progressed, they became severely reduced in size, tillering markedly increased and leaves became soft and slightly drooped (Fig. 1). The young infected plants produced either no panicles or a few small panicles bearing mostly unfilled grains (Fig. 2). Plants infected late in the growth period showed few or no symptoms until harvest and then distinct chlorotic symptoms appeared on ratoon (Fig. 3). Numerous tillers were produced at all stages of infection. Plants infected one to 1-1/2 months before harvest did not show clear symptoms on old leaves and culms. However, such plants produced yellow, stunted tillers from their basal nodes (Muniyappa, 1974; Chen, 1975). Hara and Itoga (1961) noticed abnormal tillering from the upper nodes.

Symptoms in artificially inoculated plants were first observed 19 days after inoculation. Incubation period increased with the age of the plant at the time of inoculation. Plants inoculated 10 to 15 days after sowing produced symptoms within 19 to 20 days and those inoculated later produced symptoms progressively later; those inoculated at 100 days produced symptoms after an incubation of 45 days (Muniyappa and Ramakrishnan, 1976a). Similar observations were reported by Chen and Ko (1975).

No evidence was obtained for the differentiation of strains based on varietal response but possible differences in virulence according to the latent period in the host (longer for one Indonesian isolate) were indicated (Asaga *et al.*, 1977).

VECTORS AND TRANSMISSION

Yellow dwarf disease was transmitted by five species of *Nephotettix* i.e., *N. cincticeps* (Uhler) (Enjoji, 1948); Iida and Shinkai, 1950; Chen, 1970), *N. virescens* (Distant) (=*N. impicticeps* Ishihara) (Shinkai, 1959, 1962; Chen, 1970), *N. nigropictus* (Stal.)(=*N. apicalis*) (Motschulsky) (Ouchi, and Suenaga, 1963; Rivera and Ou, 1964; Shinkai *et al.,* 1963; Chen,1970) *N. parvus* Ishihara and Kawase, (IRRI, 1972) and *N. malayanus* Ishihara and Kawase (IRRI, 1973).

Vectors of yellow dwarf disease were widely distributed in Asia (Harris and Govindu, 1967; Nielson, 1968; Nasu, 1969; Iida and Shinkai, 1969; Chiu and Chien, 1971; Asaga and Furuta, 1974; Muniyappa and Ramakrishnan, 1976b). *N. nigropictus* has been reported as vector of yellow dwarf from India (Raychaudhuri *et al.,* 1976b; Muniyappa and Ramakrishnan,1980), Japan (Ouchi and Suenaga, 1963; Shinkai *et al.,* 1963; Shinkai, 1965), the Philippines (Rivera and Ou, 1964; Palomar and Rivera,1967), Sri Lanka (Abeygunawardena *et al.,* 1970), Taiwan (Chiu, 1964), Indonesia (Satomi *et al.,* 1978), *N virescens* from India (Raychaudhuri *et al.,* 1976b; Muniyappa and Ramakrishnan, 1980), Japan (Shinkai, 1959), Malaysia (Lim and Goh, 1968), Bangladesh, (Galvez-E and Miah, 1969) and *N. cincticeps* in Japan (Iida and Shinkai, 1950) and Taiwan (Chiu, 1964).

Several homopteran insects did not transmit the disease. These include: *Inemadara oryzae, Macrosteles fascifrons, M. quadrimaculatus, Racilia dorsalis and Cicadella viridis* (Cicadellidae); *Laodelphax striatellus, Nilaparvata lugens, Sogatella furcifera* (Delphacidae); and *Nisia atrovenosa* (Meenoplidae) (Shinkai, 1962).

Yellow dwarf infected plants were also infected with tungro through *N. virescens,* producing typical symptoms of the tungro. Individuals simultaneously acquired both pathogens from mixed source and transmitted them effectively (Basu *et al.,* 1974).

Active Transmitters

Efficiency of transmission of vector depends on the percentage of active transmitters. The percentage of active transmitters in *N. nigropictus* was found to be 45 percent (Muniyappa and Ramakrishnan, 1980) to 69 percent (Palomar and Rivera, 1967), *N. virescens*, 55 percent (Muniyappa and Ramakrishnan, 1980; Muniyappa and Veeresh, 1979), 83 percent (Palomar and Rivera, 1967), 94 percent (Shinkai, 1962) and in *N. cincticeps*, 88 percent to 96 percent (Shinkai, 1962). The females were more efficient transmitters compared to males (Muniyappa and Ramakrishnan, 1980).

Acquisition Feeding Period

The minimum acquisition feeding period was 10 min (Lim, 1970; Muniyappa and Ramakrishnan, 1980; Muniyappa and Veeresh, 1979) to 30 min (Shinkai, 1962; Palomar and Rivera, 1967) for *N. virescens* and 10 min for *N. cincticeps* (Shinkai, 1962).

Incubation Period in Vectors

The incubation period in the vectors was generally long. It was 20 to 35 days for *N. nigropictus* (Palomar and Rivera, 1967; Satomi *et al.*, 1978), 23 days (Asaga *et al.*, 1977) to 26 to 40 days for *N. cincticeps* (Shinkai, 1962); and 20 days (Palomar and Rivera, 1967; Raychaudhuri *et al.*, 1967b; Lim, 1970; Abeygunawardena *et al.*, 1970 Satomi *et al.*, 1978) to 55 days (Abeygunawardena *et al.*, 1970) for *N. virescens*.
Shinkai (1962) reported 34 days as incubation period and Muniyappa and Ramakrishnan (1980) reported it as 25 days for *N. virescens*.

Retention Period

The maximum retention period obtained was 38 days for *N. nigropictus* (Palomar and Rivera, 1967), 103 days for *N. cincticeps* (Shinkai, 1962) and 27 days (Lim, 1970) to 104 days (Shinkai, 1962) for *N. virescens*. The pathogen was retained throughout the life of *N. virescens* (Muniyappa and Ramakrishnan,

1980).

Inoculation Feeding Period

The minimum inoculation feeding period was 5 min for *N. cincticeps* (Shinkai, 1962). The minimum inoculation feeding period was found to be 2 min (Palomar and Rivera, 1967; Shinkai, 1962), 5 min (Muniyappa and Ramakrishnan, 1980) and 10 min (Lim, 1970) in *N. virescens*.

Incubation Period in Plants

The incubation period in plants was reported to be 23 days (Palomar and Rivera, 1967), 25 days (Muniyappa and Ramakrishnan, 1980) and 90 days (Shinkai,1962; Abeygunawardena *et al.*, 1970). The incubation period in plants varied with age of the plants at the time of inoculation.

Transovarial Transmission

The yellow dwarf was not transmitted transovarially in *N. cincticeps, N. nigropictus and N. virescens* (Shinkai, 1962; Palomar and Rivera, 1967; Muniyappa and Ramakrishnan 1980), though the infectivity was retained through molts (Shinkai, 1962; Palomar and Rivera, 1967; Muniyappa and Ramakrishnan, 1980).

Effect of Temperature on Transmissive Ability of Vectors

When the temperature during inoculation feeding was 10, 15, and 20 C, the average incubation period was 35, 34 and 28 days respectively. When the temperature during an acquisition feeding period of 4 hours was 5, 10, 15, 20 and 25 C, 0, 7, 27, 64 and 73 percent of the insects became infective, respectively. Viruliferous *N. cincticeps* were incubated at 10, 15, 20 and 25 C, the percentage of infective insects were 60, 60, 100 and 100 respectively. As the temperature decreased, the transmissive ability was also decreased (Ishii *et al.*, 1969).

Optimum temperature for acquisition of rice yellow dwarf from rice by *N. cincticeps* was 25-33 C. Optimum temperature for inoculation was 26-30 C when 40-50 percent of the seedlings developed symptoms. Illumination during inoculation feeding had no significant effect on transmissibility (Chen, 1975).

Cytopathological Changes in the Vectors

Cytopathological changes were observed in *N. cincticeps* when it acquired the causal agent of yellow dwarf (Takahashi and Sekiya, 1962). Ten to 15 days after acquisition feeding, nuclei of the fat body cells enlarged and became irregular in shape. After 20 days, the enlarged nuclei seemed to shrink and 25 days after acquisition feeding shrinkage of the nuclei reached its maximum and the vacuoles in the cytoplasm increased so greatly in number that the fat body cells appeared to be completely reticulated (Maramorosch, 1969; Takahashi and Sekiya, 1962; Takahashi, 1963a).

Disappearance of transmissibility of viruliferous green rice leafhoppers was noticed when they were parasitized by Pipunculidae (Diptera) (Santa, 1965).

Races of *Nephotettix virescens*

Muniyappa and Viraktamath (1981) reported the transmission of rice yellow dwarf with the blue race of *N. virescens*. The percentage of active transmitters were found to be 64, 44, 56, 44 for blue female, blue male, green female and green male races of *N. virescens*, respectively. There was slightly higher transmission of the disease by the females of blue race compared to the females of green leaf hopper.

Seed, Soil and Mechanical Transmission

The disease was not transmitted by seed (Shinkai, 1959; Lim, 1970; Miniyappa, 1974) or soil (Lim and Goh, 1968) or mechanical means (Lim, 1970) or pin pricks through infected leaves (Lim, 1970) or through dodder, *Cuscuta reflexa* L. (Muniyappa, 1974).

HOST RANGE

Yellow dwarf disease was transmitted by *N. virescens* to *Alopecurus aequalis, Glyceria acutiflora, Oryza cubensis* (Shinkai, 1951; 1960; 1962), *O. barathii, O. glaberrima, O. longistaminata, O. minuta, O. nivara, O. perennis, O. punctata, O. rufipogon, O. rufipogon x O. nivara, O. sativa f. spontanea* and *Heteropogon contortus*. All these species belong to the family Poaceae. The incubation period in these plants varied from 45 to 100 days (Muniyappa, 1974; Muniyappa and Ramakrishnan, 1976c, 1981; Muniyappa and Raju, 1981).

The following species of *Oryza* were not infected by the disease, *Oryza alta, O. australiensis, O. brachyantha, O. echingeri, O. grandiglumis, O. latifolia, O. officinalis, O. ridleyi* (Muniyappa, 1974; Muniyappa and Ramakrishnan, 1981; Muniyappa and Raju, 1981).

VARIETAL RESISTANCE

Kurosawa (1940) was the first to observe varietal reaction to yellow dwarf in the field in Taiwan in 1932. He noted that almost all the Japanese varieties examined after harvest were severely infected with yellow dwarf when their tillers grew from the stubble. In field tests, Hashioka (1952) found considerable differences in resistance or susceptibility among 331 varieties tested in Taiwan. Among the varieties tested, 151 varieties showed 0-6 percent infection, 43 varieties 7-10 percent infection, 50 varieties 11-20 percent infection, 32 varieties 21-30 per cent infection and 34 varieties 41-60 percent infection. Chiu (1964) in Taiwan reported that Nan-gai-yu 27 and Hwalien-Yu showed the lowest percentage of infected plants among 46 varieties and selections tested in 1963. Komori and Takano (1964) found that Kaladumai, Loktjan, PeBiHun, Saitama Mochi No. 10 and Tetep were highly resistant in the field in Japan. Of 26 rice varieties inoculated with Japanese isolates, Tadukan and Saitamamochi No. 10 were highly resistant (Asaga *et al.*, 1977). Sakurai (1969), Morinaka and Sakurai (1969, 1970) reported that the following varieties having less than 10 percent infected seedlings were resistant in the field, Bason Takakal, Bellepatna, Blue bonnet, Chiang-nan-tao, Chiem Chank, Karalath (H 33), Keu N 525, Loktjan, Naozane-mochi, Pa-shihtze-sien, Russian No. 25, Russian No. 33, Russian No. 35, Tao-ren-chiao, Tetep, Yang-sien-tao. Of 70 varieties tested, only Taipei 131, Taipei 310, and No-Lin-49 were disease free (Chiu *et al.*, 1968).

In Sri Lanka, H^4 showed the lowest percentage of infected seedlings among 8 varieties inoculated artificially (Abeygunawardena *et al.*, 1970). None of the 34 varieties of rice belonging to *indica, japonica,* and *javanica* groups tested artificially by seedling inoculation was found to be immune to the disease. However, there was lower percentage of infection in the varieties TKM-6, S-317, S-749, S-705, Co.25 and BCP-1 belonging to *indica* group (Muniyappa and Ramakrishnan, 1976c).

Several cultures derived from crosses of T(N)-1X (T-65 X S-705), T-65 x S-705 and Wanar-1X T-65 showed no infection in planted crop and low infection in ratoon under field conditions. The varieties/selections, CH-2, CH-45, CH-126-33-11, IR-127-80-1-10, MR-278, MR-279, IR-442-2-58-2-1-2, S-317, CR-126-53-1 and S-705 did not show any infection, both in *kharif* and *rabi* seasons. Varieties which showed low infection in the field showed high percentage of infection in the seedling test.   Varieties grown in *kharif* and *rabi* seasons showed a higher percentage of infection than in the summer season (Muniyappa and Ramakrishnan, 1976c,. Muniyappa and Raju 1981) tested 80 rice varieties under field conditions.   Fifty-eight varieties showed symptoms in the planted crop and 77 in ratoon. Twenty-two varieties showed no symptoms in the planted crop.   Three varieties MR-363, Gamasolu, Hy-256 showed no symptoms in either the planted crop or in ratoon.

Of the several varieties tested, few showed resistance for yellow dwarf (Chen, 1978a; 1979; Narayanaswamy and Jaganathan, 1973; Chiu and Chen, 1973).

Hashioka (1952) reported that the resistance to disease was a heritable character. Morinaka *et al.*, (1970) and Morinaka (1978) concluded that the resistance to yellow dwarf was controlled by a dominant or incompletely dominant major gene based on the reactions of F 1 to F2 seedlings from crosses between resistant varieties Saitama Mochi No. 10 and two susceptible varieties, Manryo and Sanpuki.

Varieties Tainan 61 and Tainan 5 (both susceptible) were crossed with the resistant Firooz-1 and Kalbara and the F self-pollinated and back crossed to the parents. Seedlings of F 1, and F2, BCPS and BCPR were inoculated with vector insects for 24 hours and 5 to 6 weeks later the degree of resistance was assessed. The resistance in Firooz-1 and Kalbara was controlled by a single dominant gene (Lin *et al.*, 1975).   Five highly resistant lines were selected from progenies of cv. Tainan 61; 4 lines (YD-15, 22, 28 and75) carry the resistant genes from Kalbara and YD-15 (Lin *et al.*, 1980).

RESISTANCE TO VECTORS

Nephotettix cincticeps, vector of rice yellow dwarf showed a non-preference towards the highly resistant rice varieties Firooz-1,

Kalbara and C4-63A. Some vector resistant varieties were susceptible to yellow dwarf disease, while others were moderately susceptible, but under field conditions all were resistant (Chen and Ko, 1976).

Several of 879 rice cultivars and selections were resistant to *N. virescens* and only 0 to 3 percent of first instar nymphs survived on these plants compared with 70 to 90 percent on susceptible plants. Adults lived 4-8 days longer on susceptible plants and laid 15 more eggs (Cheng, and Pathak, 1972) compared to resistant cultivars.

Komori and Takano (1964) did not find any significant differences in the number of vector insects between fields containing resistant and susceptible varieties. The varietal differences in the percentage of diseased plants might be mainly the result of varietal resistance. Inoue (1966) observed higher mortality of nymphs on *indica* varieties than on *japonica* varieties. He further mentioned that the varietal differences in infestation by vector insects in the field may be concerned to some extent, with the apparent varietal resistance.

Out of several varieties tested, MTU-17 was found to be most tolerant towards infestation by *N. virescens* followed by IR-8 and Pankheri-203, while Pikutokumoto showed least tolerance (Mitra and John, 1973).

CAUSAL AGENT

Earlier to 1967, rice yellow dwarf was considered as a virus disease. Takahashi (1964) purified polyhedral virus particles about 55 nm in diameter from diseased leaves by differential centrifugation. The purified virus was infectious when injected into virus free *Nephotettix cincticeps*. Fuyaka and Nasu (1964) observed virus particles in the fat body of the vector 29 days after acquisition feeding.

Nasu *et al.,* (1967) discovered that the rice yellow dwarf is caused by mycoplasma-like organisms.

Electron microscopy

Mycoplasma-like bodies were observed in the phloem sieve tubes of yellow dwarf infected rice plants (Fig. 4) but not from healthy plants. These pleomorphic bodies measured from 80 to 800 nm (Doi *et al.*, 1967; Ishiie *et al.*, 1967; Nasu *et al.*, 1967; Shikata, 1979 *et al.*, 1968; 1979; Galvez-E and Shikata, 1969; Sugiura *et al.*, 1969a; Maramorosch *et al.*, 1970; Singh *et al.*, 1970; Hull, 1972; Maramorosch *et al.*, 1972; Sugiura, 1972; Plavsic-Banjac *et al.*, 1972; 1973; Doi and Okuda, 1973; Chen and Liu, 1974; Raychaudhuri *et al.*, 1977; Chen, 1978a; Ong and Lim, 1978; Satomi *et al.*, 1978; Sugiura, 1978; 1980, 1983; Tiongco *et al.*, 1983).

Mycoplasma-like bodies were observed in the midgut and salivary glands of viruliferous *N. cincticeps* and *N. nigropictus* (Nasu *et al.*, 1967; Sugiura *et al.*, 1969a). The mycoplasma-like bodies were observed in salivary glands 17 days after acquisition feeding (Sugiura *et al.*, 1969a). The bodies were pleomorphic ranging from 80 to 800 nm, devoid of cell walls and bounded by unit membranes. The extracts of salivary glands and midgut of viruliferous insects when injected into disease-free nymphs of *N. virescens*, became infective only when fresh extracts were used (Sugiura *et al.*, 1969b).

Mycoplasma-like bodies, round, oval or elongated and 100-600 nm in size were seen in phloem parenchyma cells of plants with "enrochat" disease in Spain. It is suggested that this disease and the rice diseases known as "giallume", yellow dwarf and "enanismo amarilo" are all caused by the same pathogen (Amici *et al.*, 1970; Amici and Favali, 1972). Further, studies showed that MLO found in the parenchyma and companion cells of the root phloem of rice were rich in ribosomes and incorporated Thymidine 3H (Amici and Favali, 1972).

Chemotherapy

Since animal mycoplasmas are sensitive to some antibiotics, efforts have been made to apply antibiotics to diseased plants to obtain circumstantial evidence of the nature of the causal agent. Sugiura *et al.*, (1968) observed that the development of symptoms was delayed when the seedlings were treated with compounds of tetracycline group at 0 to 5 days after inoculation but treating diseased plants with the compounds had no effect.

When four antibiotics, tetracycline hydrochloride (achromycin), chlortetracycline (aureomycin), dimethyl chlortetracycline (ledermycin) and oxytetracycline (terramycin) at concentrations from 10 to 1000 ppm were applied as foliage spray to diseased plants on alternate day for a period of 20 days, yellow leaves of some diseased plants seemed to be slightly greener and newly grown leaves tended to be symptomless. The recovery was temporary, the symptoms reappeared on the recovered leaves after the termination of spraying (Sakurai and Morinaka, 1970). Spraying the foliage three times immediately before or after inoculation does not suppress the symptom expression (Sakurai and Morinaka, 1970.

When the roots of diseased plants were dipped in antibiotics for two days, remission of symptoms occurred but the symptoms reappeared 48 days after the treatment (Sakurai and Morinaka, 1970). When the diseased plants were immersed in 1000 ppm solution of aureomycin for 30 min and sprayed with the solution at three day intervals for two weeks, plant height increased slightly but tiller number decreased markedly two weeks after treatment (Singh *et al.*, 1970). Diseased plants apparently became normal after dipping the roots in 100 ppm solution of aureomycin before transplanting in pots (Galvez-E and Shikata, 1969).

Muniyappa and Ramakrishnan (1976d) treated yellow dwarf infected plants with oxytetracycline hydrochloride and chlortetracycline at 100, 250, 350 and 500 ppm by wick feeding and 250 to 1500 ppm by spraying. Continuous feeding through cotton wick was the most effective. The remission of symptoms was observed at 250, 350 and 500 ppm. The symptoms reappeared sooner or later. Spraying with these antibiotics for several days also produced remission of symptoms. However, higher concentrations ranging from 750 to 1500 ppm were required to produce remission. Benomyl (Benlate) did not have any therapeutic effect on yellow dwarf diseased rice plants (Muniyappa and Ramakrishnan, 1976d).

Sugiura *et al.*, (1969a) stated that the tetracycline antibiotics effectively suppressed the symptoms only when applied as root dip at a concentration of about 40 ppm for 24 hr shortly before or after inoculation. The yellow dwarf infected plants apparently recovered on dipping the roots in chlortetracycline solution at 100 ppm (Galvez-E and Shikata, 1969).

Transmissive ability of *N. cincticeps* treated with tetracycline antibiotics before or after acquisition feeding, decreased compared to the control insects which had been fed with two percent sucrose solution (Sugiura *et al.*, 1969a); Sakurai and Morinaka, 1970).

Anatomical changes were observed in yellow dwarf infected rice plants after treatment with streptomycin (Shimada *et al.*, 1966).

Heat Therapy

Higher temperature suppressed the symptom development in inoculated seedlings and the transmissive ability of viruliferous *N. cincticeps* (Takasaki *et al.*, 1970).

Culturing of the Pathogen

Attempts to isolate and cultivate the pathogen on artificial media were unsuccessful (Sugiura *et al.*, 1969; Nasu and Sugiura,1974). Raju and Nyland (1978) isolated and cultured a *Spiroplasma* from rice plants with yellow dwarf symptoms in the University of California, Davis, USA. They were not sure about the identity of the disease. Su *et al.*, (1978) isolated *Spiroplasma* from the green leaf bug (*Trigonotylus rubornis*) and rice plant. They did not indicate its relationship with the yellow dwarf disease of rice.

Serological Detection of the Pathogen

Seki and Onitsuka (1965) reported the detection of rice yellow dwarf virus by hemagglutination test. In the light of the discovery of MLOs in yellow dwarf infected plants, this has to be re-examined.

Evidences indicate that the rice yellow dwarf disease is caused by mycoplasma-like organisms. In order to identify the exact causal organism, the pathogen has to be isolated and Koch postulates proven (Doi *et al.*, 1967; Ishiie *et al.*, 1967; Nasu *et al.*, 1967; Shikata *et al.*, 1968; 1969; Sugiura *et al.*, 1969a; Whitcomb and Davis, 1970; Sugiura, 1972; 1977; Hull, 1972; Ghosh and Raychaudhuri, 1974; Chiu, 1978; Okuda, 1978; Quebral, 1978; Saleh *et al.*, 1978).

PHYSIOLOGY OF YELLOW DWARF INFECTED RICE PLANTS

Infection of rice plants with yellow dwarf enhanced respiration which reached a maximum with the appearance of symptoms (Tschen, 1976; Tschen *et al.*, 1977; Tschen and Chang, 1982). When symptoms of rice yellow dwarf appeared on the first leaf, the activities of glucose-6-phosphate dehydrogenase (G-6-PDH),

6-phosphogluconate-dehydrogenase (6-PGDH), phosphoglucomutase (PGM) and phosphoglucoisomerase (PGI) were lower than in healthy leaves. Reducing sugars, glucose and fructose accumulated in infected leaves while photosynthesis was reduced. The lowered activity of G-6-PDH and 6-PGDH suggested that the pentose phosphate pathway had not increased in infected plants but the glycolysis pathway still appeared to be active (Tschen, 1976; Tschen *et al.*, 1979).

EPIDEMIOLOGY

Active Transmitters in the Field

Collections of leafhoppers (*N. nigropictus* and *N. virescens*) from the field showed a small percentage (4%) of leafhoppers occurring in the field carried the disease and were capable of infecting rice seedlings under artificial conditions in the glass house. These active transmitters would carry the disease from one season to the other wherever there is continuous cropping (Muniyappa and Ramakrishnan, 1976b). Ishii *et al.*, (1969) collected leafhoppers from the field at different times and tested for active transmission. The active transmitters were 1 to 21 percent at first adults, 0 percent of nymphs in the first generation, 0 percent of second adults and 9 to 32 percent of third adults in 1962-64. The percentage of the disease carrying individuals in the over-wintered population was in some cases as high as 70 percent (Iida, 1966). In epidemic fields, the percentage of infective leafhoppers was found to be 80 percent in autumn in Japan (Shinkai, 1961, 1962) to 90 percent in Faizabad, India (Mishra *et al.*, 1971.

Spread and Seasonal Occurrence of Disease in Relation to Vector Population in the Field

The experiments conducted at the University of Agricultural Sciences, Bangalore, India, indicated that there was a steady increase in the incidence throughout the growth of first ratoon reaching a peak in the second ratoon. It was obvious that there was active transmission within the crop (Muniyappa and Ramakrishnan,1976b).

In Japan in the field, over-wintered leafhoppers which are the primary vectors of the yellow dwarf disease were able to survive until the end of May or the beginning of June (Shinkai, 1960; Mori *et al.*, 1965 and Babaguchi and Hara, 1966). The secondary vectors of the disease appeared continuously after the beginning of August (Mori *et al.*, 1965; Babaguchi and Hara, 1966). In early cultivation, the disease appeared first in the middle of July, reached maximum occurrence from the end of July to the beginning of August and then decreased from the middle of August to September (until harvesting time). Rice plants were hardly infected in nursery (early cultivation) from the beginning of April to the beginning of May, but in the paddy fields they were severely infected from the beginning of May to the middle of June. Most of the diseased hills showing symptoms before harvest were infected before the middle of June (primary infection was introduced by the first adult and the secondary infection was caused by the third adult (Takano *et al.*, 1963; Yasuo *et al.*, 1963; Mori *et al.*, 1965; Babaguchi and Hara, 1966; Ishii *et al.*, 1969). Ishii *et al.*, (1969) could not recover the disease from the weeds indicating that the disease might overwinter not in diseased plants but mainly in the leafhoppers. Disease incidence depended much on overwintered leafhoppers which acquired the disease in the previous autumn from diseased shoots growing from cut stubbles (Sato and Sugino, 1965; Iida, 1966). Higher disease incidence was noticed when artificial light was provided during the night in the field (Kurosawa, 1940). The difference in the occurrence of yellow dwarf disease in manual and mechanised transplanting paddy was noticed (Fujisawa and Shibamoto, 1975). Vectors, *Nephotettix spp.* collected from various localities in Taiwan transmitted the yellow dwarf (Asaga, 1978).

In India, the leafhopper population (*N. virescens* and *N.nigropictus*) in the field was high during July to September in Delhi (Mishra *et al.*, 1971), April to September in Bihar (John, 1970), May to June and October to November in Karnataka (Muniyappa and Ramakrishnan, 1976b; Prabhuswamy, 1972). The incidence of disease was low in the months of January to April and early August transplanted crops and high incidence was noticed from September to December transplanted crops. The incidence of disease appeared to be correlated with the vector population (Muniyappa and Ramakrishnan, 1976b). The average maximum and minimum temperatures were low, when the incidence of the disease was high. The humidity also was high during the months of September to December. A combination of low temperature and high humidity has been shown to favor the leafhopper population (Muniyappa and Ramakrishnan, 1976b).

Many reports regarding ecological and epidemiological aspects of yellow dwarf disease of rice have appeared in literature; most of them were in the Japanese language (Shinkai, 1960; 1962; Hayashi, 1961; Niidome *et al.*, 1961; Komori *et al.*, 1962; Sameshima and Nagai, 1962; Itoga and Babaguchi, 1963; Takai, 1963; Takahashi, 1963b; Takano *et al.*, 1963; Yasuo *et al.*, 1963; Hara *et al.*, 1964; Iwahashi and Goto, 1964; Nagai *et al.*, 1964; Nishio *et al.*, 1964; Shigenaga and Koyama, 1964; Babaguchi *et al.*, 1965; Goto *et al.*, 1965; Hasuko *et al.*, 1965; Miyahara and Yamaguchi,1965; Mori *et al.*, 1965; Murata *et al.*, 1965; Satomi and Mizuta, 1965; Suenaga *et al.*, 1965; Miyahara, 1966, 1969; Yoshimeki, 1966; Iwaki *et al.*, 1967; 1969; Samejima, 1967; Takahashi and Kihune, 1967; Yamanaka, 1969; Kimizaki and Komori,1971; Komori *et al.*, 1972; Kobayashi *et al.*,

1975; Chen, 1977, 1978b; Hirao and Inoue,1978).

Forecasting of the occurrence of the disease has been tried in Japan (Sato and Sugino, 1965; Suenaga *et al.,* 1965). Mishra *et al.,* (1971) suggested the use of serodiagnostic tests to detect, rapidly, the infective *Nephotettix spp.*

CONTROL

Cultural

The yellow dwarf disease was reduced by field sanitation and management practice (Abeygunawardena *et al.,* 1970).

Chemical

The occurrence of the yellow dwarf disease in the paddy field was decreased to some extent by applying NAC (1-Naphthyl N-methyl-carbomate) or Vamidothion (n-methyl-O, O-dimethyl-triolophosphoryl-5-thia-3-methyl-2-valeramide) two times during the rice nursery period (Mori *et al.,* 1965). Attempts to control the insect vectors by aerial application of insecticides over large areas have been tried in Japan. Large scale cooperative control of the disease was done by dusting or spraying various insecticides from helicopters. Malathion, carbaryl dust and other chemicals were used usually soon after transplanting to kill the over-wintering insects (Enjoji, 1948; Miyazawa *et al.,* 1961; Hayashi *et al.,* 1962; Nakamura *et al.,* 1962; Ichikawa,1963; Shinkai *et al.,* 1963; Hayashi and Komatsu, 1964; Iwamato and Kimizaki, 1964; Miyazawa and Hayakawa, 1964; Nasu, 1964; Toyoda *et al.,* 1964; Goto *et al.,* 1965; Yoshimura and Sakai, 1965; Hara *et al.,* 1966; Iida, 1966; Komori, 1966; Shimoyama and Shibamoto, 1966; Aguiero *et al.,* 1966; Hosino *et al.,* 1967; Iida, 1969; Iwaki *et al.,* 1967; Samejima, 1967; Takita *et al.,* 1967; Ide and Tanaka, 1968;

Oda *et al.*, 1968a; 1968b; Lim, 1969; Miyahara, 1969; Nagai *et al.*, 1969; Yamanaka, 1969; Komori, 1971; Komori *et al.*, 1972; Lim, 1972; Ishii, 1973; Yamazaki and Hariya, 1978). Insecticides were applied into boxes for mat type of seedlings at the second leaf stage for yellow dwarf control (Takai and Ino, 1975).

Sevin (1-naphthyl-N-methyl-carbamate) when applied to soil at 3 kg a.i. per ha controlled *Nephotettix spp.* effectively (John, 1966). Seed treatment with carbofuran, lannate, thimet or propoxure before sowing the seeds protected the rice seedlings from the leafhoppers (Mitra *et al.*, 1970; Raychaudhuri *et al.*, 1970, 1971, 1981). The seed treatment with carbofuran or lannate repelled the leafhoppers in seed bed stage for 3-4 weeks and subsequently, one application of carbofuran granules in rice fields after transplantation was effective in repelling the vectors from feeding on the plants for 3-4 weeks (Raychaudhuri *et al.*, 1981).

The mortality of leafhoppers were found to be more in dimecron (0.03%), carbofuran (0.04%), dimethoate (0.0125%) and methyl demeton (0.025%) treated plots reducing the incidence of yellow dwarf disease from 5.8 to 0.94 percent (Muniyappa, 1980b).

Applying propaphos to the soil in the nursery box reduced the incidence of yellow dwarf in the field (Asaka *et al.*, 1979).

SUMMARY

In this chapter an attempt has been made to review the various aspects of the rice yellow dwarf disease. Yellow dwarf disease of rice has been reported in most of the rice growing countries of the East and South East Asia. Plants became infected at all stages of growth, the yield decreased most when infection occurred before formation of the panicle primordia. In Karnataka, the yellow dwarf disease incidence was observed up to 27 percent in planted crop (var. MR-249) and up to 82 percent (selection CR-126-6) in ratoon. Yellow dwarf disease was transmitted by leafhoppers, *Nephotettix cincticeps, N. nigropictus, N. virescens, N. malayanus,* and *N. parvus.* Vectors of yellow dwarf disease were widely distributed in Asia. The minimum acquisition feeding period was 10 to 30 min. for *N. virescens* and 10 min. for *N. cincticeps.*. The incubation period in the vectors varied from 20-40 days. The pathogen was retained throughout the life of *N. virescens.* The minimum inoculation feeding period was found to be 2 - 10 min. Yellow dwarf disease was transmitted by leafhoppers to *Alopecurus aequalis, Glyceria acutiflora, Oryza cubensis, O. barathii, O. glaberrima, O. longistaminata, O. minuta, O. nivara, O. perennis, O. punctata, O. rufipogon, O. rufipogon x O. nivara* and *O. sativa f. spontanea* and *Heteropogon contortus.* Mycoplasma-like bodies were observed in the phloem sieve tubes of yellow dwarf infected rice plants, but not of healthy plants. Mycoplasma-like bodies were also observed in the midgut and salivary glands of viruliferous *N. cincticeps* and *N. nigropictus.* Remission of symptoms occurred when the yellow dwarf infected plants were treated with tetracyclines; oxytetracycline hydrochloride, chlortetracycline and dimethyl chlortetracycline. Attempts to isolate and cultivate the pathogen on artificial media were unsuccessful. In India, the leafhopper population (*N. nigropictus* and *N. virescens*) in the field was high during July to September in Delhi, April to September in Bihar, May to June and October to December in Karnataka. The disease incidence was high from September to December transplanted crops and low incidence was noticed in the months of January, February, March, April and early August transplanted crops. There was a positive correlation between the leafhopper population and disease incidence. A combination of low temperature and high humidity has been shown to favor the leafhopper population. The occurrence of the yellow dwarf disease in the field was decreased to some extent by the application of insecticides. The resistance sources have been identified for the yellow dwarf disease.

REFERENCES

Abeygunawardena, D.V.W., 1969. The present status of virus diseases of rice in Ceylon. In: Proceedings of a symposium on The virus diseases of the rice plant. p. 53-57, 25-28 April, 1967, Los Banos, Philippines, Johns Hopkins U., Press Baltimore, Maryland, pp. 354.

Abeygunawardena, D.V.W., Bandaranayaka, C.M. and Karandawela, C.B., 1970. Virus diseases of rice and their control. Trop. Agriculturist, 126: 1-13.

Aguiero, V.M., Ixconde, D.R. and Quintana, I. J., 1966. Insecticidal control of *Nephotettix spp.* the vector of tungro and yellow dwarf disease of rice in the Philippines. Philippine Agric., 49: 871-878.

Amici, A., Favali, M., 1972. Thymidine-H labelling of mycoplasma-like bodies in roots of rice plants affected by a yellows-type disease. Protoplasma, 75: 37-43.

Amici, A., Ballesteros, R., Batalla, J.A. and Belli, G., 1970. Mycoplasma-like bodies in Spanish Rice Plants with "Enrochat" disease. Riv. Pathol. Veg. (Ser.4), 6 (4): 247-253.

Anonymous, 1943. Studies on rice yellow dwarf disease: Results of experiments made in 1941 and 1942 (In Japanese). Kochi Agricultural Experiment Station, pp. 39. (Mimeographed).

Asaga, K., 1978. Comparison between rice yellow dwarf pathogens and vectors from different localities. Tech. Bull. ASPAC Food Fert. Technol. Cent., 42: 16-25. Also in Plant Diseases due to mycoplasma-like organisms, p. 116-125. Taiwan, Food and Fertilizer Technology Center for the Asian and Pacific Region, 1978. (FFTC Book Ser. No. 13).

Asaga, K. and Furuta, T., 1974. Affinity between pathogen of rice yellow dwarf and green rice leafhopper from several districts in Japan (in Japanese. Proc. Kanto-Tosan Plant Protect. Soc., 21: 19-20.

Asaga, K. and Furuta, T., and Yamada, M., 1977. Studies on establishment of standard experimental system in rice yellow dwarf (in Japanese). J. Central Agric. Exp. St., 26: 105-132.

Asaka, S., Fujisawa, T., Nagata, K. and Harada, T., 1979. Control of yellow dwarf by the application of propaphos to the soil in the nursery box (in Japanese). Proc. Kanto-Tosan Plant Prop. Soc., 26: 27.

Babaguchi, K. and Hara, K., 1966. Infection period of the yellow dwarf virus of early growing rice plants. (in Japanese). Proc. Assoc. Plant Protect. Kyushu, 12: 20-25.

Babaguchi, K., Hara, K., and Waki, K., 1965. Infection period of the yellow dwarf disease in early seasonally cultured rice (in Japanese). Kyushu Agr. Res., 27: 141-143.

Basu, A.N., Ghosh, A., Mishra, M.D., Niazi, F. R. and Raychaudhuri, S.P., 1974. The joint infection of rice tungro and yellow dwarf. Ann. Phytopath. Soc., Japan, 40: 67-69.

Chen, C.C., 1970 Comparative transmission of

rice yellow dwarf by three *Nephotettix* leafhoppers in Taiwan. (in Chinese). Plant Protect. Bull. (Taiwan). 12: 160-165.

Chen, C.C., 1975. Studies on the varietal resistance of rice plant to yellow dwarf. I. Factors affecting the varietal resistance test method (in Chinese). Plant Protect. Bull. Taiwan, 17: 263-271.

Chen, C.C., 1977. Some epidemiological studies on rice yellow dwarf disease especially emphasized on the transmission cycle (in Chinese). Natl. Sci. Counc. Mon. (Taiwan), 5: 98-105.

Chen, C.C., 1978a. Studies on varietal resistance of rice to yellow dwarf in Taiwan. In: Plant diseases due to mycoplasma-like organisms. p. 126-134, Taiwan, Food and Fertilizer Technology Center for the Asian and Pacific Region 1978. (FFTC Book Ser. No. 13).

Chen, C.C., 1978b. Epidemiological studies on rice yellow dwarf (in Chinese). In: Diseases and insect pests of rice: ecology and epidemiology. p. 139-166. Taipei, Joint Commission on Rural Reconstruction.

Chen, C.C., 1979. Varietal resistance to yellow dwarf in rice. In: Proc. R.O.C.-United States Co-operative Science Seminar on Mycoplasma Diseases of plants. p. 153-159. Taipei, Taiwan, National Science Council.

Chen, C.C. and Ko, W.H., 1975. Effect of yellow dwarf disease on the agronomic characters of rice plant (in Chinese). Plant Protect. Bull. Taiwan, 17: 250-262.

Chen, C.C. and Ko, W.H., 1976. Studies on the varietal resistance of rice plant to yellow dwarf. II-III. II. Screening test for varietal resistance to yellow dwarf in rice plant III. Types of yellow dwarf resistance (in Chinese). Plant Protect. Bull., Taiwan, 18: 207-217, (4): 346-353.

Chen, C.C., Ko, W.H., Wang, E.S. and Hu, D. Q., 1977. Effect of yellow dwarf on rice plants as determined by the time of symptom appearance (in Chinese). Plant Protect. Bull., Taiwan, 19: 251-256.

Chen, M.J., 1978. Electron microscope observations of several plant-infecting mycoplasma-like organisms. Tech. Bull. ASPAC Food Fert. Technol. Cent., 42: 5-15. Also in Plant diseases due to mycoplasma-like organisms, p. 13-23. Taiwan, Food and Fertilizer Technology Center for Asian and Pacific Region, 1978 (FFTC Book Ser. No. 13).

Chen, M.J., 1979. Electron microscopic observations of several plant infecting mycoplasma-like organisms in Taiwan. In: Proc. R.O.C.-United States Cooperative Science Seminar on Mycoplasma Diseases of plants. p. 25-36. Taipei. National Science Council.

Chen, M.J. and Liu, H.Y., 1974. Electron microscopic study of rice yellow dwarf (in Chinese). Plant Protect. Bull., Taiwan, 16: 42-55.

Cheng, C.H. and Pathak, M.D., 1972. Resistance to *Nephotetix virescens* in rice varieties. J. Econ. Entom., 65: 1148-1153.

Chien, C.C., 1979. Occurrence and influence of rice diseases on yield in the first and second rice crops in Taiwan. Taipei, Taiwan, National Science Council, p. 177-189.

Chiu, R.J., 1964. Virus diseases of rice in

Taiwan--A general review. FAO-IRC working party on Rice production and protection, 10th Meeting, Manila, Philippines (Mimeo).

Chiu, R.J., 1978. Plant diseases in Taiwan due to mycoplasma-like organisms. In: plant diseases due to mycoplasma-like organisms. p. 5-12. Taiwan Food & Fert. Tech. Center for the Asian and Pacific Region (FFTC Book Ser. No. 13).

Chiu, R.J. and Chien, C.C., 1971. Yellow dwarf of rice (in Chinese). In: Proc. Symp. Rice Diseases, p. 135-154. Taipei, 1969.

Chiu, R.J. and Chen, C.C., 1973. Varietal tests for resistance to yellow dwarf in rice. International Rice Research Conference, Los Banos, Laguna, IRRI, 1973, pp. 5.

Chiu, S.M., Lin, M.H. and Huang, C.S., 1968. A screening test for rice varieties resistant to yellow dwarf disease (in Chinese). J. Taiwan Agr. Res., 17: 19-23.

Doi, Y. and Okuda, S., 1973. Electron microscopic study of mycoplasma-like organisms (MLO) in yellows-diseases host plants (in Japanese). Shokubutsu Boeki Kenkyu, 8: 203-221.

Doi, Y., Terenaka, M., Yora, K. and Asuyama, H., 1967. Mycoplasma or PLT group-like microorganisms found in the pholem elements of plants infected with mulberry dwarf, potato witches' broom, aster yellows, or paulownia witches' broom (in Japanese). Phytopathol. Soc., Japan Ann., 33: 259-266.

Enjoji, S., 1948. Experimental results in the control of rice yellow dwarf virus. Chiba Agr. Breed. Sta. Rep. (Mimeographed).

Fujisawa, T. and Shibamoto, T., 1975. Difference in the occurrence of yellow dwarf in manual and mechanical transplanting paddies (in Japanese). Proc. Kanto-Tosan Plant Prot. Soc. 22: 19.

Fukaya, M. and Nasu, S., 1964. Distribution of plant viruses in insect vector (1). Multiplication of rice dwarf and yellow dwarf viruses in fat body of green rice leafhopper (in Japanese). Phytopathol. Soc.,Japan, Ann. 29: 72.

Galvez-E, G.E., and Miah, M.S.A. 1969. Virus and mycoplasma-like diseases of rice in East Pakistan. Int. Rice Comm. Newslett., 18: 18-26.

Galvez-E, G.E. and Shikata, E., 1969. Microorganismos similares a los micoplasmas o al grupo PLT (Psittacosis-lymphogranuloma-trachoma), probables agentes causales del enanismo amarillo del Arroz. Agricultura Trop., 25: 109-115.

Ghosh, S.K. and Raychaudhuri, S.P., 1974. Investigations on mycoplasmal diseases of plants in India. In: Raychaudhuri, S.P. and Verma, J.P. Eds. Current trends in Plant Pathology, p. 112-121. Dept. Botany, University of Lucknow.

Goto, S., Iwahashi, I. and Nagai, K., 1965. Studies on the ecology and control of rice yellow dwarf. V. On the annual fluctuation of the disease appearance in early seasonally cultured rice (in Japanese). Kyushu Agr. Res., 27: 144-145.

Govindu, H.C., Harris, H.M., and Yaraguntaiah, R.C., 1968. Possible occurrence of tungro and yellow dwarf viruses on rice in Mysore. Mysore J. Agri. Sci., 2: 125-127.

Hara, K. and Itoga., 1961. Studies on the yellow dwarf of earlier planted paddy. I. On the abnormal tillering from upper nodes (in Japanese). Proc. Assoc. Plant Protect. Kyushu, 7: 32-34.

Hara, K., Babaguchi, K. and Horikiri, M., 1966. Studies on the control of virus disease of rice by aerial dusting in autumn (in Japanese). Kyushu Agri. Res., 28: 117-119.

Hara, K., Babaguchi, K. Horikiri, M., Fukamachi, S. and Yanagita, Y., 1964. Problems in controlling virus diseases of rice in regions where two crops of rice are annually harvested (in Japanese). Proc. Assoc. Plant Protect. Kyushu, 10: 101-103.

Harris, H.M. and Govindu, H.C., 1967. Insect vectors in relation to virus diseases of rice. Mysore J. Agric. Sci., 1: 303.

Hashioka, Y., 1952. Varietal resistance of rice to brown spot and yellow dwarf. Studies on pathological breeding of rice, VI. Jap. J. Breed., 2: 14-16.

Hashioka, Y., 1963. Rice blast varietal resistance in relation to the local environments with nodes on stem nematode and other diseases of rice in Thailand. Final Report to FAO, pp. 66.

Hashioka, Y., 1964. Virus diseases of rice in the world. IRISO, 13: 295-309.

Hasuko, E., Kaida, H. and Goto, S., 1965. Studies on the ecology and control of rice yellow dwarf. VI. Period of the virus acquirement and the transmission of the green rice leafhopper, *Nephotettix cincticepts* Uhler in autumn (in Japanese). Kyushu Agr. Res., 27: 146-147/

Hayashi, K., 1961. Expansion of area affected

with rice yellow dwarf and its cause (in Japanese). Proc. Kanto Tosan Plant Protect. Soc., 8: 14.

Hayashi, S. and Komatsu, S., 1964. Injury of yellow dwarf on rice plant in the infected area (in Japanese). Proc. Kanto-Tosan Plant Protect. Soc. 11: 24.

Hayashi, K., Miyazawa, T., Nakamura, T. and Yanagi, T., 1962. Large scale cooperative control of rice yellow dwarf disease using helicopter, in Matsumotodaira, Nagano Prefecture: Several problems encountered (in Japanese). Proc. Kanto-Tosan Plant Protect. Soc., 9: 17.

Hirao, J. and Inoue, H., 1978. Bionomics of the green rice leafhopper, *Nephotettix cincticeps* in relation to the incidence of rice yellow dwarf disease in Japan. In: Plant diseases due to mycoplasma-like organisms. p. 143-157.

Hoshino, M., Takita, Y. and Kamei, K., 1967. Studies on chemical control of virus diseases by aerial spraying. 2. Effects of aerial spraying to smaller brown plant hopper and green rice leafhopper in early spring season (at low temperature period). Proc. Kanto-Tosan Plant Protect. Soc., 14: 18.

Hsieh, S.P.Y., 1976. Assessment of yield losses due to rice transitory yellowing and yellow dwarf (in Chinese). Plant Protect. Bull., Taiwan, 18: 91-105.

Hull, R., 1972. Mycoplasma and plant diseases. PANS 18: 154-164.

Ichikawa, H., 1963. Autumn control test of leafhopper as mediator of rice yellow dwarf disease (in Japanese). Plant Protect. (Japan), 17: 313-316.

Ide, A. and Tanaka, T., 1968. Occurrence and control of yellow dwarf disease in Ehime Prefecture (in Japanese). Agr. Hort., 43: 1682-1686.

Iida, T.T., 1966. Ecology and control of rice virus diseases in Japan. Jarq, 1: 1-4.

Iida, T.T., 1969. Dwarf, yellow dwarf, stripe and black streaked dwarf diseases of rice. In: Proceedings of a symposium on the virus diseases of the rice plant. p. 3-11. 25-28 April, 1967, Los Banos, Philippines, Johns Hopkins U. Press, Baltimore, Maryland, pp.354.

Iida, T.T. and Shinkai, A., 1950. Transmission of rice yellow dwarf by green rice leafhopper (in Japanese). Ann. Phytopathol. Soc., Japan 14: 113-114.

Inoue, H., 1966. Larvae development of the green rice leafhopper, *Nephotettix cincticeps* Uhler and the White back plant hopper, *Sogatella furcifera* Horvath on japonica and indica rice varieties. Odokon-Chugoku, 8: 17-19.

IRRI (Int. Rice Res. Inst.), 1969. The virus diseases of the rice plant. Proceedings of a symposium, 25-28 April, 1967. Los Banos, Phillipines. The Johns Hopkins U. Press, Baltimore, Maryland, pp.354.

IRRI, 1972. International Rice Research Institute Annual Report for 1971. p. 112-113. Los Banos, Philippines, pp. 238.

IRRI, 1973. International Rice Research Institute Annual Report for 1972. p. 132-133. Los Banos, Philippines, pp. 246.

Ishii, M., 1973. Control of rice virus diseases. Japan Pesticide Information. 17:

11-16.

Ishii, M., Yasuo, S. and Ono. K., 1969. Epidemiological studies on rice yellow dwarf diseases in Kanto-Tosan District, Japan (in Japanese). J. Cent. Agr. Exp. Sta., 13: 1-21.

Ishiie, T., Doi, Y., Yora, K. and Asuyama, H., 1967. Suppresive effects of antibiotics of tetracycline group on symptom development on mulberry dwarf disease. Phytopathol. Soc. Japan Ann., 33: 267-275.

Itoga, S. and Babaguchi, K., 1963. On the occurrence of yellow dwarf disease on early rice crop (in Japanese). Proc. Assoc. Plant Protect. Kyushu, 9: 58-60.

Iwahashi, T. and Goto, S., 1964. Ecological and therapeutical studies on the yellow dwarf disease of rice plant (in Japanese). Proc. Assoc. Plant Protect. Kyushu, 10: 103-105.

Iwaki, H., Katayama, E., Tobita, T. and Saito, S., 1967. Characteristics of rice yellow dwarf and its control by aerial spraying (in Japanese). Proc. Kanto-Tosan Plant Protect. Soc., 14_ 30-31.

Iwaki, H., Tobita, T. and Takita, Y., 1969. Damage analysis of rice yellow dwarf (in Japanese). Proc. Kanto-Tosan Plant Protect. Soc., 16: 27.

Iwamoto, S. and Kimizaki, K., 1964. Control of rice yellow dwarf in Ibaraki Prefecture (in Japanese). Proc. Kanto-Tosan Plant Protect. Soc., 11: 25.

John, V.T., 1966. Insecticidal control of *Nephotettix spp.* the vectors of tungro and yellow dwarf diseases of rice in the Philippines. Indian Phytopathology, 19: 150-154.

John, V.T., 1970. Yellowing disease of paddy. Indian Farming, 20 (3): 27-30.

Kimizaki, K. and Komori, N., 1971. On the transition of the occurrence of yellow dwarf and smaller planthopper of rice plant after readjustment of paddy field (in Japanese). Proc. Kanto-Tosan Plant Protect. Soc., 18: 20.

Kobayashi, S., Kureha, Y. and Shibamoto, T., 1975. On the importance of post-transplanting infection with rice yellow dwarf (in Japanese). Proc. Kanto-Tosan Plant Protect. Soc., 22: 20.

Komori, N., 1965. The damage caused by rice yellow dwarf virus (in Japanese). Proc. Kanto-Tosan Plant Protect. Soc., 12: 16-17.

Komori, N., 1966. Occurrence of yellow dwarf disease and its control in Ibaraki Prefecture (in Japanese). Plant Protect. (Japan), 20: 285-288.

Komori, N., 1971. Problem of occurrence of rice yellow disease and its control in Ibaraki Prefecture (in Japanese). Plant Protect. (Japan), 20: 285-288.

Komori, N. and Takano, S., 1964. Varietal resistance of rice plants to rice yellow dwarf in field (in Japanese). Proc. Kanto-Tosan Plant Protect. Soc., 11: 12.

Komori, N., Iwamoto, S. and Takano, S., 1962. Infection period of rice yellow dwarf disease (in Japanese). Proc. Kanto-Tosan Plant Protect. Soc., 9: 15.

Komori, N., Takano, T., Iwamoto, S. and Kimizaki, K., 1972. The epidemiology and the control of yellow dwarf of rice plant in Ibaraki Prefecture (in Japanese). Bull.

Ibaraki Agr. Exp. Sta., 12: 85-134.

Kureha, Y., Kobayashi, S. and Shibamoto, T., 1974. Relation between the degree of yellow dwarf of rice plant and yield (in Japanese). Proc. Kanto-Tosan Plant Protect. Soc., 21: 18.

Kurosawa, E., 1940. On rice yellow dwarf disease of rice occurring in Taiwan (in Japanese). J. Plant Protect., 27: 161-166.

Li, D.B., Wang, G.C. and Cheng, F.J., 1979. Epidemiological study on rice virus diseases and their control in Zeejiang province. Acta Phytopathologica Sinica, 9: 73-82.

Lim, G.S., 1969. The bionomics and control of *Nephotettix impicticeps* Ishihara and transmission studies on its associated viruses in West Malaysia. Malaysian Min. Agr. Coop. Bull. No. 121, pp. 62.

Lim, G.S., 1970. Transmission studies on yellow dwarf disease of rice in West Malaysia. Malaysian Agric. J. 47: 517-523.

Lim, G.S. 1972. Chemical control of rice insects and diseases in Malaysia. 27-36, Japan Pesticide Information No. 10, 6-102

Lim, G.S., and Goh, K.G., 1968. Leafhopper transmission of a virus disease of rice locally known as Padi Jantan in Kiran, Malaysia. Malaysian Agric. J., 46: 435-450.

Lin, M.H., Chiu, C.L. and Chien, C.C., 1975. Inheritance of resistance to yellow dwarf disease in rice (in Chinese). J. Agric. Res., China, 24: 1-7.

Lin, M.H., Chang, T.M., Chien, C.C. and Chang, Y.C., 1980. Breeding for the resistance of yellow dwarf disease of rice. J. Agric. Res. China, 29: 259-264.

Ling, K.C., 1969. Virus diseases or rice in the Philippines and the effect on yield. Sci. Rev., 10: 23-30.

Ling, K.C., 1972. Rice virus diseases. International Rice Research Institute, Los Banos, Philippines, pp. 134.

Maramorosch, K., 1969. Effects of rice pathogenic viruses on their insect vectors. In: Proceedings of a symposium on the virus diseases of the rice plant. p. 179-203. 25-28 April 1967, Los Banos, Philippines, Johns Hopkins Press U., Baltimore, Maryland, pp. 354.

Maramorosch, K., Granados, R.R. and Hirumi, H., 1970. Mycoplasma diseases of plants and insects. Adv. Virus Res., 16: 135-193.

Maramorosch, K., Plavsic-Banjac, B., John, V.T. and Raychaudhuri, S.P., 1972. Etiology of rice yellow dwarf in India. Phytopathology, 62: 776.

Mathew, J. and Abraham, J., 1977. Occurrence of rice yellow dwarf in Kerala. Agri. Res. J. Kerala, 15: 172-173.

Matsunaga, M., Kimizaki, Y. and Komori, N., 1973. Occurrence of yellow dwarf and smaller planthopper of rice plant in northern part of Ibaraki Prefecture (in Japanese). Proc. Kanto-Tosan Plant Protect. Soc., 20: 18-19.

Mishra, M.D., Raychaudhuri, S.P., Everett, T.R. and Basu, A.N., 1971. Possibilities of forecasting outbreaks of tungro and yellow dwarf of rice in India and their control. Proc. Indian Natn. Sci. Acad., 37 (B): 352-

356.

Mitra, D.K. Raychaudhuri, S.P., Everett, T. R., Ghosh, A. and Niazi, F.R., 1970. Control of rice green leafhopper with insecticidal seed treatment and pre-transplant seedling soak. J. Econ. Ent., 63:1958-1961.

Miyahara, K., 1966. Studies on the preference of the green leafhopper to the yellow dwarf infected rice plant (in Japanese). Kyushu Agr. Res., 28: 119-120.

Miyahara, K., 1969. On the studies on ecology of green rice leafhopper (*Nephotettix apicalis cincticeps* Uhler) referring to the control of yellow dwarf of rice plant (in Japanese). Bull. Saga Agr. Exp. Sta., 9: 127-135.

Miyahara, K. and Yamaguchi, N., 1965. Analytical studies on the causal factors of outbreak of the yellow dwarf in 1964 (in Japanese). Kyushu Agr. Res., 27: 148-150.

Miyazawa, T. and Hayakawa, H., 1964. Relation between the effect and the number of times of dusting by helicopter for the control of yellow dwarf (in Japanese). Proc. Kanto-Tosan Plant Protect. Soc., 11: 27.

Miyazawa, T., Muraga, Y., Miyairi, Y. and Ichikawa, H., 1961. Large scale cooperative control of rice yellow dwarf using helicopter (in Japanese). Proc. Kanto-Tosan Plant Protect Soc., 8: 17.

Mohan, J.C. and Natarajan, T., 1973. Occurrence of yellow dwarf virus disease on rice at Coimbatore. Madras Agric. J., 60: 578.

Mori, K., Makino, T. and Osawa, T., 1965. Ecological investigations and control measures of yellow dwarf of rice plant in Shizuoka

Prefecture (in Japanese). B. Shizuoka Pref. Agr. Expt. Sta., 10: 33-42.

Morinaka, T., 1978. Inheritance of resistance to rice yellow dwarf disease. In: Plant diseases due to mycoplasma-like organisms. p. 135-142.

Morinaka, T. and Sakurai, Y, 1969. Varietal resistance to yellow dwarf of rice plant in field (in Japanese). Chugoku Agr. Res., 40: 16-17.

Morinaka, T. and Sakurai, Y., 1969. Varietal resistance to yellow dwarf of rice plant and the method of testing resistance. Bull. Chugoku Agric. Exp. Stn. Serv. E., 6: 57-79.

Morinaka, T. and Sakurai, Y., 1970. Inheritance of resistance to yellow dwarf disease in rice. Jap. J. Breed., 20: 22-28.

Muniyappa, V., 1974. Status of rice yellow dwarf disease in rice production in Karnataka State. Ph.D. Thesis, Univ. Agril. Sci. Bangalore, India, pp. 125.

Muniyappa, V. 1980a. Effect of yellow dwarf disease on the yield of rice under field conditions. ILRISO, 29; 223-226.

Muniyappa, V. 1980b. Control of yellow dwarf disease of rice using insecticides. ILRISO, 29: 309-312.

Muniyappa, V. and Raju, B.C., 1981. Response of cultivars and wild species of rice to yellow dwarf disease. Plant disease. 65: 679-680.

Muniyappa, V. and Ramakrishnan, K., 1976a. Effect of yellow dwarf disease on growth and yield of rice crop. Curr. Res., 5: 30-31.

Muniyappa, V. and Ramakrishnan, K., 1976b. Epidemiology of yellow dwarf disease on rice. Mysore J. Agric. Sci., 10: 259-279.

Muniyappa, V. and Ramakrishnan, K., 1976c. Reacion of cultures and varieties of rice and species of *Oryza* to yellow dwarf disease or rice. Mysore J. Agric. Sci., 10 270-279.

Muniyappa, V. and Ramakrishnan, K., 1976d. Chemotherapy of yellow dwarf disease of rice. Mysore J. Agric. Sci., 10 431-439.

Muniyappa, V. and Ramakrishnan, K., 1980. Transmission studies on yellow dwarf disease of rice. Mysore J. Agric. Sci., 14: 55-59.

Muniyappa, V. and Ramakrishnan, K., 1981. Rice yellow dwarf and host range. p. 22-26. In: Mycoplasma and allied pathogens of plants, animals and human beings (Eds. H.C. Govindu, K. Maramorosch, S.P. Raychaudhuri and V. Muniyappa). UAS, Bangalore, pp. 117.

Muniyappa, V. and Veeresh, G.K., 1979. Leafhoppers as vectors of plant mycoplasma diseases. Proc. Indian Natn. Sci. Acad., B46: 827-831.

Muniyappa, V. and Viraktamath, C.A., 1981. Transmission of rice yellow dwarf with the blue race of *Nephotettix virescens*. Intl. Rice Res. Newsl., 6: 15.

Murata, T., Takashi, T., Nado, M., Uchida, N. and Tateishi, I., 1965. On the severe epidemics of rice yellow dwarf disease in Fukuoka Prefecture in 1964 (in Japanese). Proc. Assoc. Plant Protect. Kyushu, 11: 6-7.

Nagai, K., Iwahashi, T. and Goto, S., 1964. Ecological and control studies on the yellow dwarf disease of rice plants. I-

III. Kyushu Agri. Res. 26: 153-158.

Nagai, K., Kayashima, S. and Goto, S., 1969. On the occurrence and control of the rice yellow dwarf. IX. Controlling effect by paddy water application of insecticides. Proc. Assoc. Plant Protect. Kyushu, 15: 107-109.

Nakamura, T., Hayashi, K., Miyazawa, T. and Yanagi, T., 1962. Large scale cooperative control of rice yellow dwarf disease using helicopter in Matsumotodaira, Nagano Prefecture: Outline of the result (in Japanese). Proc. Kanto-Tosan Plant Protect. Soc., 9: 16.

Narayanaswamy, P. and Jaganathan, T., 1973. Assessment of varietal reaction of rice yellow dwarf disease. Madras Agric. J. 60: 1809-1810.

Nasu, S., 1964. Taxonomy, distribution, host range, life cycle and control of rice leafhoppers. Symposium on the major insect pests of rice. International Rice Research Institute, Philippines.

Nasu, S., 1969. Vectors of rice viruses in Asia. In: The virus diseases of the rice plant. p. 93-109. International Rice Research Institute, Philippines.

Nasu, S. and Sugiura, M., 1974. Research for culture media of the rice yellow dwarf mycoplasma-like organism (in Japanese). Jap. J. Appl. Entomol. Zool., 18: 145-146.

Nasu, S., Sugiura, M., Wakimoto, S. and Iida, T.T., 1967. Pathogen of rice yellow dwarf disease (in Japanese). Phytopathol. Soc. Japan, Ann., 33: 343-344.

Nielson, M.W., 1968. The leafhopper vectors of

phytopathogenic viruses (Homoptera:Cicadellidae): taxonomy, biology and virus transmission. U.S. Dept. Agr. Tech. Bull., 1382, pp. 386.

Nishio, Z., Takita, Y., Iwaki, H., Toyoda, F., Aoki, M. and Katayama, E., 1964. The outbreak of rice yellow dwarf in Tochigi Prefecture (in Japanese). Proc. Kanto-Tosan Plant Protect. Soc., 11: 23.

Niidome, I. and Itoga, S., 1962. Occurrence of rice yellow dwarf disease and its damage in Kagoshima Prefecture (in Japanese). Plant Protect. (Japan), 16: 159-162.

Niidome, I., Yanagida, Y. and Itoga, S., 1961. On the damage of rice yellow dwarf in earlier planted paddy rice. Proc. Assoc. Plant Protect. Kyushu, 7: 29-32.

Oda, K., Takita, Y. and Takahashi, S., 1968a. Studies on chemical control of rice virus diseases by aerial application. I. Effects on aerial dusting in early spring as the control measure for rice yellow dwarf disease (in Japanese). Bull. Tochigi. Agr. Exp. Sta. 12: 117-120.

Oda, K., Iwaki, H., Takita, Y., Saito, S. and Takahashi, S., 1968b. Effect on green rice leafhopper of low volume spraying of malathion by helicopter for the control of rice yellow dwarf disease (in Japanese). Proc. Kanto-Tosan Plant Protect. Soc., 15: 35.

Okuda, S., 1978. Plant diseases due to mycoplasma-like organisms in Japan. In: Plant diseases due to mycoplasma-like organisms, p. 24-28. Taiwan, Food & Fert. Tech. Center for the Asian and Pacific Region 1978 (FFC Book Ser. No. 13).

Ong, C.A. and Lim, G.S., 1978. Plant diseases

due to mycoplasma-like organisms in Malaysia. In: Plant diseases due to mycoplasma-like organisms. p. 52-60. Taiwan, Food and Fertilizer Tech. Center for the Asian and Pacific Region. (FFTC Book Ser. No. 13).

Ouchi, Y. and Suenaga, H., 1963. On the transmissibility by the leafhopper, *Nephotettix apicalis* of rice yellow dwarf virus (in Japanese). Proc. Assoc. Plant Protect. Kyushu, 9: 60-61.

Palomar, M.K. and Rivera, C.T., 1967. Yellow dwarf of rice in the Philippines. Philippine Phytopathol., 3: 27-34.

Pathak, M.D., Ling, K.C., Lowe, J.A. and Yoshimura, S., 1967. A survey of insects and diseases of rice in India. Report by International Rice Research Institute, Philippines, p. 1-10.

Plavsic-Banjac, B., and Maramorosch, K., 1972. Electron microscopic diagnosis of Indian rice yellow dwarf, sandal wood spike and African maize streak. Acta Biol. Yugosl. Ser. B. Microbiol. 9 (2): 201-211.

Plavsic-Banjac, B., Maramorosch, K., John, V.T. and Raychaudhuri, S.P., 1973. Rice yellow dwarf in India: Mycoplasma-like etiology. FAO Plant Protect. Bull., 21: 1-4.

Prabhuswamy, H.P., 1972. Rice leafhoppers and planthoppers in Bangalore area with emphasis on the biology and control of *Zygina maculifrons* (Homoptera:Cicadellidae). M. S. (Agri.) Thesis, Univ. Agric. Sci., Bangalore, pp. 73.

Quebral, F.C., 1978. Plant diseases in the Philippines presumed to be due to mycoplasma-like organisms. In: Plant diseases due to mycoplasma-like organisms p. 46-51. Taiwan, Food and Fert. Tech. Cent. Asian and Pacific Region (FFTC Book Ser. No. 13).

Raju, B.C. and Nyland, G., 1978. Isolation and culturing of a spiroplasma from rice plants with yellow dwarf symptoms. Phytopathol. News., 12: 267 (Abstr.).

Raychaudhuri, S.P. 1977. A manual of virus diseases of Tropical plants. MacMillan Co. of India, Ltd, pp. 299.

Raychaudhuri, S.P. 1984. Virus diseases of cereal crops, maize and millets in India, Res. Trop. Plant Patholog. Vol I, pp. 519-531.

Raychaudhuri, S.P. and John, V.T., 1970. Progress in rice insect and disease investigation. (c) Virus diseases. FAO-IRC working party on rice production and protection, 13th Meeting, Teheran, Iran (Mimeo).

Raychaudhuri, S.P., Mishra, M.D. and Ghosh, A., 1967a. Preliminary note on transmission of virus disease resembling tungro of rice in India and other virus-like symptoms. Plant Dis. Reptr. 51: 300-301.

Raychaudhuri, S.P., Mishra, M.D. and Ghosh, A., 1967b. Preliminary note on the occurrence and transmission of rice yellow dwarf virus in India. Plant Dis. Rptr., 51: 1040-1041.

Raychaudhuri, S.P., Mitra, D.K. and Everett, T.R., 1970. Control of rice green leafhopper vectors of rice tungro virus and yellow dwarf in India. Proc. 7th International Cong. Plant Prot., Paris, pp. 171.

Raychaudhuri, S.P., Mishra, M.D. and Basu, A.N., 1971. Investigations on tungro and yellow dwarf diseases of rice. *Oryza* 8 (2, Suppl.), : 361-363.

Raychaudhuri, S.P., Mishra, M.D. and Basu, A. N., 1981. Chemical control of vectors of rice yellow dwarf. p. 27-30. In: Mycoplasma and Allied Pathogens of plants, Animals and Human Beings. (Eds. H.C. Govindu, K. Maramorosch, S.P. Raychaudhuri and V. Muniyappa), Univ. Agril. Sciences Bangalore. pp. 117.

Raychaudhuri, S.P., and Nariani, T.K. 1977, Virus and mycoplasma diseases of plants in India, Oxford, 1BH, Publ. pp. 102.

Raychaudhuri, S.P., Varma, A. and Verma, J.P., 1977. Mycoplasmal diseases of plants in India. In: Physiology of microorganisms. p. 437-446. Today and Tomorrow's Printers and Publishers, New Delhi.

Rivera, C.T. and Ou, S.H., 1964. Studies on the virus diseases of rice in the Philippines. The International Rice Res. Inst., Los Banos, Laguna, Philippines, (Mimeographed).

Sakurai, Y. and Morinaka, T., 1970. Effects of tetracyclines on symptom development and leafhopper transmission of rice yellow dwarf disease. Bull. Chugoku Agric. Exp. Stn. Ser. E., 5: 1-14.

Saleh, N., Hibino, H., Roechan, M. and Tantera, D.M., 1978. Plant diseases associated with mycoplasma-like organisms, p. 61-67. Taiwan, Food and Fertilizer Technology Center for Asian and Pacific Region (FFTC Book, Ser., No. 13).

Samejima, T., 1967. Occurrence and control of yellow dwarf disease of rice in Miyazaki Prefecture (in Japanese). Plant Protect. (Japan), 21: 47-50.

Sameshima, T. and Nagai, K., 1962. Relation

between the life history of green rice leafhopper, (*Nephotettix cincticeps* Uhler) and rice yellow dwarf propagation (in Japanese). Jap. J. Appl. Exp. Zool., 6: 267-273.

Santa, H., 1965. Disappearance of virus--transmissibility of viruliferous green rice leafhopper (yellow dwarf virus) parasitized by Pipunculidae (in Japanese). Proc. Kanto-Tosan Plant Protect. Soc., 12: 66.

Sato, M. and Sugino, T., 1965. Studies on the forecasting method of rice yellow dwarf disease (in Japanese). Kanto-Tosan Plant Protect. Soc., 12: 18.

Satomi, H. and Mizuta, T., 1965, Comparison of the seasonal abundance of rice insects and virus diseases in direct sown paddy field and in the transplanted one. Proc. Assoc. Plant Protect. Kyushu, 11: 90-92.

Satomi, H., Roechan, M., Iwaki, M. and Saito, Y., 1978. Occurrence of rice yellow dwarf in Indonesia. Contr. Cent. Res. Inst. Agric. Bogor, No. 40, pp. 6.

Seki, M. and Onitsuka, S., 1965. Detection of rice yellow dwarf virus by hemagglutination test (in Japanese). Kyushu Agr. Res., 27: 125-126.

Shigenaga, T. and Koyama, S., 1964. On the transmission of rice yellow dwarf disease by over-wintered generation of green rice leafhopper (*Nephotettix cincticeps* Uhler) (in Japanese). Proc. Assoc. Plant Protect. Kyushu, 10: 8-10.

Shikata, E., 1979. Rice viruses and MLOs and leafhopper vectors. In: Leafhopper vectors and plant disease agents, Maramorosch, K., and Harris, K.F., (eds.) p. 515-527.

Academic Press, New York.

Shikata, E., Maramorosch, K., Ling, K.C. and Matsumoto, T., 1968. Mycoplasma-like structures in diseased plants with American aster yellows, corn stunt, Philippine rice yellow dwarf and Taiwan sugarcane whiteleaf. Ann. Phytopath. Soc. Japan, 34: 208-209.

Shikata, E., Maramorosch, K., Ling, K.C. 1969. Presumptive mycoplasma etiology of yellows diseases. Pl. Prot. Bull. F.A.O., 17: 121-128.

Shimada, H., Kawai, T. and Nishitani, S., 1966. Anatomical observation of recovery process of infected rice plant with yellow dwarf by treatment of streptomycin (in Japanese). B. Shiga Pref. Agr. Expt. Sta., 9: 23-26.

Shimazu, H., 1976. Occurrence and control of rice yellow dwarf disease in northern part of Chiba Prefecture. Proc. Kanto-Tosan Plant Prot. Soc., 23: 22-23.

Shimoyama, M. and Shibamoto, T., 1966. Occurrence of yellow dwarf disease of rice and its control in Nagano Prefecture (in Japanese). Plant Protect., Japan, 20: 289-292.

Shinkai, A., 1951. Host range and problem of virus transmission through seeds of rice yellow dwarf (in Japanese). Ann. Phytopathol. Soc., Japan 15: 176.

Shinkai, A., 1959. Transmission of rice yellow dwarf by "Taiwan tsumaguro-yokobai" (in Japanese). Ann. Phytopathol. Soc., Japan, 24: 36.

Shinkai, A., 1960. Over-wintering of rice yellow dwarf virus (in Japanese). Ann. phytopathol. Soc., Japan 24: 36.

Shinkai, A., 1961. Percentage of viruliferous leafhoppers collected from rice field affected with yellow dwarf (supplement) (in Japanese). Proc. Kanto-Tosan Plant Protect. Soc., 8: 12.

Shinkai, A., 1962. Studies on insect transmission of rice viruses in Japan (in Japanese). Nat. Inst. Agr. Sci. Bull. Ser. C., 14: 1-112.

Shinkai, A., 1965. Transmission of four rice viruses by leafhoppers (in Japanese). Ann. Phytopathol. Soc., Japan, 31: 380-383.

Shinkai, A., Miyanaga, T. and Tobechi, K., 1963. Infection and control of rice yellow dwarf disease in Okinawa (in Japanese). Okinawa Agr., 2: 40-42.

Singh, K.G., Saito, Y. and Nasu, S., 1970. Mycoplasma-like structures in rice plant infected with "padi jantan" disease. Malaysian Agric. J., 47: 333-337.

Singh, S.K. and Chaudhury, R.C., 1983. Occurrence of rice yellow dwarf disease in Patna (Bihar) India. Int. Rice Res. Newsl., 8 (6): 11.

Su, H.J., Lei, J.D. and Chen, T.A., 1978. Spiroplasmas isolated from green leaf bug (*Trigonotylus rubornis*) and rice plant. Proc. 3rd Intl. Cong. Plant Pathol., Munich, Germany, 16-23 August, 1978. Dtsch. Phytomed. Ges. p. 61 (Abstr.).

Suenaga, H., Yoshimeki, M. and Fujiyoshi, M., 1965. Studies on the formulae for forecasting the occurrence of important insect pests in Kyushu Districts II. Relation between climatic factors and the numerical population of rice planthoppers and leaf-hoppers in the light trap catch. Proc. Assoc. Plant Protect. Kyushu, 11: 82-84.

Sugiura, M., 1972. Mycoplasma and virus (in Japanese). Noyaku Kenkyu, 19: 6-11.

Sugiura, M., 1977. Present status and problems of studies on mycoplasma-like microorganisms of plant (in Japanese). Plant Prot. Japan, 31: 47-52.

Sugiura, M., 1978. Plant disease due to mycoplasma-like organisms, recent study results and their trend (in Japanese). Agric. Hortic., 53: 613-620.

Sugiura, M., 1980. Mycoplasma and mycoplasma diseases of plants. In: Rice Protection in Japan, Part 1, p. 27-35. Kobe, International Co-operation Agency.

Sugiura, M., 1983. Rickettsia-like bacteria, spiroplasmas and mycoplasma-like organisms as a plant pathogen (in Japanese). Plant Prot., (Japan), 37 (1): 11-18.

Sugiura, M., Kaida, H. and Osawa, T., 1969a. Rice yellow dwarf and antibiotics of tetracycline group (in Japanese). Shokubutsu Boeki, 23: 293-297.

Sugiura, M., Kaida, H., Nasu, S., Wakimoto, S. and Iida, T.T., 1969b. Location of rice yellow dwarf pathogen in the body of viruliferous vector (in Japanese). Phytopathol. Soc., Japan, Ann., 35: 130.

Sugiura, M., Nasu, S., Wakimoto, S. and Iida, T.T., 1968. Studies on rice yellow dwarf. Ann. Phytopath. Soc. Japan, 34: 205-206.

Takahashi, Y., 1963a. Globules in the adipose cells of the green rice leafhopper (*Nepho-*

tettix cincticeps) infected with the virus of yellow dwarf disease of the rice plant (in Japanese) Jap. J. Appl. Ent., Zool., 7: 350-351.

Takahashi, Y., 1963b. Detecting method of the viruliferous green rice leafhopper, *Nephotettix cincticeps* Uhler, a vector of the yellow dwarf disease of the rice plant (in Japanese). Jap. J. Appl. Ent. Zool., 7: 200-206.

Takahashi, Y., 1964. Purification of rice yellow dwarf virus (in Japanese). Ann. Phytopathol. Soc. Japan, 29: 73.

Takahashi, Y. and Kihune, 1967. On the second infection of rice yellow dwarf virus in paddy field. Proc. Kanto-Tosan Plant Protect. Soc., 14: 28-29.

Takahashi, Y. and Seikiya, I., 1962. Adipose tissue of the green rice leafhopper, *Nephotettix cincticeps* Uhler, infected with virus of the yellow dwarf disease of the rice plant (in Japanese). Jap. J. Appl. Ent. Zool., 6: 90-94.

Takai, A., 1963. On the distribution pattern of the yellow dwarf disease of the rice plant (in Japanese). Jap. J. Ecol., 13: 151-156.

Takai, A. and Ino, M., 1975. Control of yellow dwarf by the application of insecticides into boxes for mat type seedlings at second leaf stage (in Japanese). Proc. Kanto-Tosan Plant Prot. Soc., 22: 21.

Takano, S. and Komori, N., 1961. Occurrence of rice yellow dwarf in Ibaraki Prefecture (in Japanese). Proc. Kanto-Tosan Plant Protect. Soc., 8: 13.

Takano, S., Takano, T., Komori, N., Iwamoto, S., Yasuo, S., Ishii, M. and Kawamori, N., 1963. Studies on the occurrence of rice yellow dwarf disease. I-II (in Japanese). I. Influence of transplanting period. II. Some experiments on primary and secondary infection. Proc. Kanto-Tosan Plant Protect. Soc., 10: 1-2.

Takasaki, T., Sugiura, M. and Iida, T.T., 1970. Effect of high temperature treatment on inoculated plants of yellow dwarf and viruliferous insects (in Japanese). Phytopathol. Soc., Japan Ann., 36: 190.

Takita, Y., Oda, K., Iwaki, H., Toyoda, F. and Saito, K., 1967. Studies on chemical control of virus disease by aerial spraying. 3. Effects of aerial spraying for control of rice virus diseases in spring season (at low temperature period)(in Japanese). Proc. Kanto-Tosan Plant Protect. Soc., 14: 19-20.

Tiongco, E.R., Cabunagan, R.C., Lapiz, D.B., Flores, Z.M. and Ling, K.C., 1983. Common diseases of rice. 3. Virus and MLO associated diseases of rice. In: Rice production Manual, Philippines, Rev. Ed. p. 353-368. University of the Philippines at Los Banos, College of Agriculture.

Toyoda, F., Ichikawa, T. and Kimura, M., 1964. Effect of malathion dusting by helicopter for the control of rice yellow dwarf (in Japanese). Proc. Kanto-Tosan Plant Protect. Soc., 11: 26.

Tschen, J., 1976. Changes in respiration and carbohydrate metabolism of rice plants infected by mycoplasma. In: Biochemistry and cytology of plant parasite interaction, p. 236-238.

Watanabe, K., Hasegawa, Y. and Tanaka, T.,1979.

Real conditions of the first occurrence of yellow dwarf in Shiga Prefecture (in Japanese). Bull. Shiga Protect. Agric. Expt. Stn., 21: 101-104.

Wathanakul, L. and Weerapat, P., 1969. Virus diseases of rice in Thailand. In: Proceedings of a symposium on the virus diseases of the rice plant, p. 79-85. 25-28 April, 1967. Los Banos, Philippines, The Johns Hopkins U. Press, Baltimore, Maryland, pp. 354.

Whitcomb, R.F. and Davis, R.E., 1970. Mycoplasma and phytarboviruses as plant pathogens persistently transmitted by insects. Ann. Rev. Entomol., 15: 405-464.

Yako, N., Onodera, S. and Saito, M.,1979. Study on yellow dwarf control. IV. Annual changes in disease occurrence; especially occurrence in regenerated rice plant in Hawadori, Fukushima Prefecture (in Japanese). Ann. Rep. Soc. Plant Prot. North Japan, 30: 78.

Yamanaka, I., 1969. Ecology and controlling method of yellow dwarf disease (in Japanese). Agr. Hort., 44: 87-90.

Yamazaki, F. and Hariya, N., 1978. Control of yellow dwarf of rice plants by the application of chemicals in the seedling case (in Japanese). Proc. Kanto-Tosan Plant Prot. Soc., 25: 27.

Yasuo, S., Ishii, M., Komori, N. and Iwamoto, S., 1963. Studies on the occurrence of rice yellow dwarf disease. II. Some experiments on primary and secondary infection. Proc. Kanto-Tosan Plant Protect. Soc., 10: 2.

Yoshimeki, M., 1966. Ecological and physiological studies on the dynamics of the migratory population of rice planthoppers and leafhoppers.   Bull. Kyushu Agr. Exp. Sta., 12: 1-78.

Yoshimura, S. and Sakai, H., 1965. Studies on the control of virus diseases of rice plants in direct sowing culture. I. A test on the control effect of insecticidal soil treatment against vectors, *Nephotettix* and *Delphacodes* (in Japanese).   Kyushu Agr. Res., 27: 155-157.

15

MYCOPLASMA-ASSOCIATED POTATO DISEASES AND THEIR CONTROL IN INDIA

by

S.M. Paul Khurana, R.A. Singh and

D.M. Kaley

Scientist S-3, (Plant Pathology), Scientist S-2 (Plant Pathology, Central Potato Research Institute, Simla-171001, H.P., Scientist S-2 (Plant Physiology), Central Potato Research Station, Rajgurunagar--410 505, Pune, Maharashtra, India

INTRODUCTION

Six potato yellows diseases vis, marginal flavescence (MF), potato phyllody (PP), potato stolbur (stolbur), purple-top-roll (PTR), purple-top wilt (PTW) caused by aster yellows (AY) and witches' broom (WB), have been found to be associated with mycoplasma-like organisms (MLOs) (2,24,18,10). Of these MF, PTR, PP and WB have been reported from India and have been found to be responsible for development of hairy sprouts (HS) in tubers (17,20). Though WB has universal distribution (37) PTW and stolbur have been recorded only from Australia, parts of Europe, Israel and the United States of America (14,23,36,38). Recently, PTR has also been reported from France and Israel (1, 38).

Because of the national importance of the

potato yellows diseases, particularly the PTR, work was intensified on disease losses and control in the '70s (20,21). The present review endeavors to collate and critically compare the available information on the Indian potato yellows diseases for a better understanding of their nature, vectors, interrelationship, and possible control.

Although the incidence of WB may not exceed 30% in the Deccan plateau, up to 90% and 35% of PTR and MF have been recorded in exceptional years in certain varieties/localities of Himachal Pradesh hills (11,20). PP was discovered from the WB branded material obtained from Rajgurunagar, Pune, Maharashtra. No data are available about its actual incidence in nature.

Inadequate data are available on the losses due to these diseases. But in light of the information available on the comparative yields in experimental infections vs. 'healthy' controls under field conditions, these diseases have been found to cause very high losses (Table 1). Severe and early infections of PTR and MF have been observed to cause losses of 50-75% and 50-90% respectively. However, mild or delayed infections may cause only 25-35% losses. Though yield reductions due to severe WB may be as high as the disease incidence itself, mild or late infections caused losses of 17-29% in 1985. PP depressed the yields up to 64% in the variety Kufri Kuber (20,22).

SYMPTOMATOLOGY

Purple Top Roll (PTR)

Its symptoms comprise rolling of the basal part of the leaflets of young top leaves, usually with light to intense pink or purple pigmentation. It may be preceded by reducing the size of the leaflets and by marginal chlorosis. Infected plants become stiff, have shortened internodes and develop many axillary shoots often having aerial tubers. Such plants are invariably stunted, have general chlorosis and lack in vigor. There is very little use of the nutrients from the PTR infected mother tubers. These tubers remain unshriveled and firm until harvest. The infected plants produce small tubers aggregated around the main stem or short stolons (17). On storage, hairy sprouts (HS) develop in a large number of PTR infected tubers (18,20).

There are at least three distinct isolates of PTR according to variation in symptom severity, intensity of color development and yield depressions (6).

Marginal Flavescence (MF)

Initial symptoms of MF comprise mild chlorosis on the margins of leaflets of upper leaves which remain very small. The chlorosis intensifies progressively, leaf blades become slightly thick, rough puckered, and the growth of infected plants continues to be arrested i.e. they remain extremely stunted because of short internodes with small leaves having narrow leaflets partly overlapping each other. In some varieties, viz. Kufri Jyoti, the chlorosis may be accompanied by purplish pigmentation in cool weather. MF infected mother tubers also remain unshriveled over 2-3 months. Plants with early and severe attack of the disease either produce none or a few small tubers in clusters i.e. on very short thin stolons (16). A mild strain of MF was described recently (7), which does not induce thick, rough and puckered leaflets but causes acute yield depressions and almost all the tubers produced develop hairy sprouts (33).

Witches' Broom

Plants infected by typical or several strains of WB develop many axillary and basal shoots simulating a bushy 'broom' bearing small, round, simple chlorotic leaves and often producing none or only very minute tubers. Infected tubers may be pigmented and develop many sprouts soon after harvest (37,9,11). Five different strains of the pathogen that occur in India vary in their symptoms and effect on the crop (9).

Potato Phyllody (PP)

Enlarged phylloid flowers, extreme hairyness, flattening of stems with prominent wings and development of chlorotic small yet compound leaves having small leaflets but enlarged folioles are the characteristic symptoms of PP. Phylloid flowers often proliferate into vegetative shoots. After a prolonged period of infection, the potato plants develop several naked androecia as small yellow 'rosettes' on stem,and also axillary aerial tubers (20,34).

COMPARATIVE CHARACTERISTICS

The characteristic, distinctive and common symptoms of potato yellows diseases, listed in order of their importance in India, are summarized below and compared in Tables 2-4 (Fig. 1).

MF, PTR, PP and WB

MF and PTR occur mainly in the high hills, mid-hills and northern plains while PP and WB are restricted to the Deccan plateau (16,20). Incidence of PTR in nature is the highest (up to 90-95%) followed by MF (up to 1-70%); it is very low for WB and PP (<7%).
Tuber transmission of MF and PTR is comparatively low (0.1-66% and 8.3-69%) but up to 100% for PP and WB (20). MF and PTR induce HS up to 16-65% while PP and WB produce HS and finely divided multi-sprouts in 79-100% tubers (20). MF affected potato plants show premature senescense and normally bear no flowers; PTR does not affect flowering but WB increases the vegetative age and growth of plants and arrests the terminal reproductive phase while PP in-

vokes phyllody (20). Such plants mature only slightly later than the healthy plants. Absolute infections of MF and PTR can cause higher yield depressions (30-92%) than PP and WB (64, 17-69%) (21,11,6,33). MF and PTR pathogens are more sensitive to the chemotherapeutants Benlate and Bavistin than are PP and WB (20, 22).

MF and PTR

The MF symptoms are more severe than PTR in potato. In many other hosts, PTR causes severe symptoms (20). *Trifolium repens* is a collateral host for PTR but not MF. Similarly, periwinkle and calendula plants, with chronic infections, develop diagnostic symptoms for PTR but not MF, while tobacco var. NC-95 is susceptible to PTR but immune to MF (32,12). MF symptoms are expressed best during the period of low humidity (R.H. 40-60%) and higher temperature (20-28 C) while PTR requires high humidity (R.H. 50-90%) and lower temperatures (15-20 C) for symptom expression (20).

As observed in ultrathin sections, the MLOs of MF and PTR are typical of other MLOs in size and morphology (19).

Losses in yield due to MF (60-92%) are higher than due to PTR (30-82%) (16,11,20). The tuber transmission of MF alone is usually higher than for the PTR. Tuber transmission of MF from PTR infected plants is neglibible compared to that of PTR from MF (MF + PTR) - infected plants. MF induces up to 65% HS, compared to 16-20% due to PTR (20,33). Potato produce manifests a negative interaction between PTR MF but not MF PTR and PTR + MF (21, 20). *Alebroides dravidanus*, the vector of PTR is not a vector of the MF pathogen (26,21). MF induces extensive necrosis of phloem and regeneration in the form of a pouch in infected potato plants (13).

Eight tobacco varieties, viz Ambalema, Burley-21, Delcrest, Kanak Prabha and McNair-12, susceptible to MF, are immune to PTR (32). The

MF pathogen is more susceptible to cycloheximide and antiamoebin than PTR (31,20). MF and
PTR induce distinguishable phloem florescence
in infected plants (25).

PP and WB

Current year infection losses in yield due
to WB are lower (17-29%) in contrast to PP (up
to 64%)(20,33). The WB symptoms are expressed
sooner (25-35 days) upon infection in host
plants than PP (60-90 days) (11). *A. dravidanus* and *O. albicinctus*, the vectors of
WB, do not transmit the PP agent (20,21). The
pathogens of PP and WB do not cross protect in
potato or tomato (20). Besides the PP pathogen
is more thermolabile than the WB (11). Induction of large phylloid flowers by PP is in contrast to WB which normally arrests the reproductive phase (20,11).

COMMON CHARACTERS

MF and PTR

PTR and MF can cause simultaneous or combined infections in potato plants (20,21,22).
Both have common leafhopper vectors, *Orosius
albicinctus* and *Seriana equata* (20). In
nature, the development of the PTR disease, in
a potato crop, very often follows the MF. Diagnostic floral anomalies in periwinkle and calendula have been observed due to these pathogens (12). Tuber transmissions of PTR from MF
(MF+PTR) infected plants is fairly high (20,
11).

PP and WB

Both PP and WB are restricted to the Deccan
plateau (20,11). Up to 100% tuber transmission
for PP and WB has been reported (21,11). A mild
strain of WB can also induce phylloid flowers
in tomato resembling symptoms induced by PP

(10). *Trifolium repens* is the common host for PP and WB showing similar symptoms, viz. smalling, chlorosis and extreme narrowing of the leaves and leaflets (20,35). Both PP and WB pathogens are equally susceptible to chemotherapeutants, including tetracyclines.

TRANSMISSION

TUBER TRANSMISSION

The tuber transmission of these potato pathogens seems to be of great significance in nature. A maximum of 70,66,64, 100% tuber transmission was recorded for PTR, MF, PP and WB under field conditions (Table 5). In fields most plants infected with MF developed PTR symptoms suggesting that they can cause complex infections. However, plants initially infected with PTR normally did not express MF symptoms indicating that PTR has a negative interaction with MF i.e. PTR infection results in retardation of MF symptoms and also its tuber transmission. The tuber transmission of PTR from produce of PTR infected plants is high but MF incidence in such cases was negligible (20). On the other hand, the transmission of MF (as well as PTR) from the MF (and/or MF+PTR) infected plants produce is relatively high (11, 22) indicating negative interaction between MF and PTR (Table 5).

Incidence of PTR and MF in Crop from HS Tubers

In the plants obtained from HS tubers collected from farmers' stores, either without distinction of variety or from six known varieties, the highest incidence of PTR and MF was recorded in Kufri Shakti (PTR 22.5%) and Kufri Chandramukhi (MF 24%). The other varieties had relatively lower incidence of PTR and/or MF except Great Scot which had 16% MF but 0.0% PTR. The mixed lot of HS tubers gave rise to plants having an incidence of 23% PTR and 12% MF

(Table 6). A number of plants from HS tubers, which did not have PTR/MF, had either mosaic or leafroll symptoms, yet a fairly high degree of plants were apparently healthy indicating that not all HS tubers produce PTR/MF in succeeding generation.

Almost all the tubers from WB infected Kufri Kuber plants were very small and developed either HS or finely divided sprouts and plants with WB symptoms. But in var. Kufri Lauvkar tuber transmission of WB was low and about 50% of the tubers either failed to develop HS or failed to germinate and grow into plants.

INSECT TRANSMISSION

Apart from tuber and dodder (*Cuscuta hyalina* Roth.) transmission (20,11), the yellows pathogens are readily transmitted in nature through their leafhopper vectors (Table 7). PTR is efficiently transmitted through *Alebroides dravidanus* (26) and inefficiently by *Orosius albicinctus* and *Seriana equata* while *S. equata* is a good vector for MF. Similarly, WB is transmitted by both *O. albicinctus* and *A. dravidanus*. A vector for PP is yet to be established.

Single inoculative leafhopper per plant resulted in low (10-30%) degrees of transmission (Table 7) but a group of five leafhoppers per plant provided 50-100% transmission. *A. dravidanus*, the potent vector of PTR, proved to be merely a laboratory vector of WB since it has not been recorded from the Deccan plateau where WB is prevalent. However, *O. albicinctus* frequently occurs there (20,22,29).

Vector-pathogen relationship has been studied with *O.albicinctus*-PTR and it has been determined to belong to the persistently (or circulative) transmitted yellows pathogens (29, 30). On the other hand, the work on PTR-*A. dravidanus* indicates that it can be readily acquired and transmitted in a manner similar to that reported for semi-persistent type of pathogens. However, such a difference may have resulted from the fact that the former type of vector does not colonize on potatoes and has to be forced to feed while potato is the natural host for the latter. Further, the viruliferous leafhoppers remain inoculative for almost their whole lifespan. Percent transmission increases with an increase in acquisition, incubation or inoculation periods. This indicates the possible multiplication of the pathogen within the vector(s). However, not all individuals are vectors but usually nymphs and females are efficient vectors. Female vector insects survive longer than the males and are consistent transmitters. The incubation of the pathogen is 6-14 days in the vector. Optimum inoculation feeding for 9-10 days by 4-5 individuals/plant gives maximum transmission of PTR on white clover plants (30).

Pronounced lethality of PTR pathogen on its vector, *A. dravidanus* has also been observed. The pathogen, when acquired at the nymphal stage, reduced the longevity, delayed the urge for mating of infective females, enhanced the hatching time of eggs, mortality of nymphs, and even induced sterilization in many females (28).

Etiology

In 1963, identification of the yellows diseases of potatoes in India was initiated with studies on the causes of hairy sprouts in tubers (15). Hairy sprout of potatoes is prevalent under ordinary storage conditions in the hills, in several parts of Indo-Gangetic plains and also in the Deccan plateau. Hairy sprouts occur as a result of both biotic and abiotic factors such as fungi, viruses, mycoplasmas and physiological or environmental conditions. Association of the potato leafroll virus, mycoplasmal diseases, viz. WB, PTR, MF and PP have also been found important in inciting hairy sprouts (18,20).

Antiserum Production

By injecting partially purified and milli-pore filter-sterilized WB and PTR antigens into rabbits, in the initial trial, 4-weekly, intra-muscular, 1 ml. injections with Freund's complete adjuvant were given to rabbits which were bled after 8 weeks. A weak but specific anti-serum was produced against the WB pathogen. An antiserum for detection of PTR was not produced in this manner (20).

Morphology of MLOs

MLOs have been observed in different con-centrations in sieve elements of many parts of PTR, WB and PP infected tomato and tobacco plants (Fig. 2). All forms were bounded by a unit membrane and contained ribosome-like gra-nules. They varied in size (100-800 nm), shape and internal structure. MLOs of smallest size (100-150 nm) were often found in the case of WB and PP in great concentration i.e. packing the sieve elements. Usually there were small, spher-ical MLOs but few were elongated, beaded, or budding (lobed/protruding) (19,11). In MF, rather long bodies (700-1800 nm) with very dark thick membrane were observed (Fig. 2). Most of these bodies were uniformally granulated but some had fibrallar material in the center. Com-monly, these bodies were elipsoid to oval in outline, measured 800-1000 nm, and some were either vacuolate or contained electron-dense centers. At times, elongated or tailed MLOs were also found, particularly in the vector (20). On the basis of above mentioned mor-phology, Prof. L.M. Black (1975, pers. comm.) doubted whether they were MLOs despite the susceptibility of the MF agent to tetracy-clines, but not to penicillin. Recently, we have been able to observe association of rigid-rod and tadpole-like viruses within and outside the pleomorphic bodies in the case of MF in potatoes (I.D. Garg, S.M. Paul Khurana and R.A. Singh, 1985, unpublished). These have been suspected to be similar to viruses of spiroplasmas (R.E. Davis and Alan Liss, 1985, pers. comm.).

Physiological Studies

Total carbohydrates are high in PTR, MF and WB infected plant leaves while total chlorophyll content was high in the healthy controls. In potatoes, infected with PTR and MF, the total carbohydrate and starch contents are high in tubers and low in the sprouts (4, 3). Similarly in WB infected plants, starch content is high in the whole tubers but low in the sprouts. Reducing sugars and sucrose, however, are considerably increased in the sprouts of potatoes infected with PTR, MF and WB pathogens (20,3).

Protein-N increased in the sprouts and tubers infected with these pathogens. PTR and MF incidence was high in the plants raised from HS tubers but their yield was decreased greatly. Application of Parthenin (200 ppm) to the tubers increased the yield of plants from HS tubers. Incidence of PTR and MF in the plants of HS tubers was reduced by treatment of IAA and IPA (dip for 24 hrs plus foliar spray, 50 ppm each). Proline treatment of normal healthy tubers (1 ppm tuber dip for 2 hrs plus 10 ppm foliar spray) did not induce development of HS in their produce (20). IAA dip treatment to WB tubers and/or foliar spray (50 ppm each) reduced the severity of WB symptoms and also increased the yields of tubers (20). However, treatment of HS tubers from apparently healthy and PTR plants with IAA and proline reduced the incidence of HS in the produce (3).

STRAINS AND HOST RANGE

The pathogens of yellows diseases have a tendency for developing strains as indicated by pecularity of symptoms, variation in virulence, insect transmissibility, etc. By the use of a set of differential hosts, at least five strains of the WB pathogen have been established (9,11,20). Three distinct isolates of PTR have been collected which differ not only in symptoms (especially color intensity) but also in their virulence as reflected by reduction in yield (6). Similarly, a 'mild' strain/isolate of MF was also encountered requiring longer time for producing disease symptoms and causing less reduction in yield of infected potatoes as compared to the typical MF (7).

Host range of these yellows pathogens is very broad. Over 40 species in seven families, viz Apocyanaceae, Compositae, Caryophyllaceae, Leguminosae, Solanaceae and Umbelliferae, have been found susceptible. Among these PP agent has the widest host range, followed by WB, MF and PTR.

A good degree of interaction has been reported to occur between strains of other yellows agents, but only limited work has been done with the causal agents of these four diseases. More than one strain of WB may infect potato, tree-tomato and tomato plants but tomatoes produce distinct symptoms for each strain (9). Similarly, MF and PTR, which were earlier thought to be related as primary and secondary symptom phases of the same pathogen, have also been found to infect potato, tobacco and datura plants simultaneously. However, the tuber transmission of MF from MF and PTR infected plants is negligible indicating thereby a negative interaction (20,22,11), esp. between MF following the PTR infection within the plant.

CONTROL

The symptoms and effects of the yellows diseases are significantly suppressed both in host plants and the vectors by tetracycline group of antibiotics and also systemic fungicides (Tables 8,9). This masking of symptoms in plants is also accompanied by a reduction in the quantity of MLOs, as has been observed by electron microscopy or by grafting for bioassay. Similarly, high (around 35 C) or low (less than 10 C) temperatures result in masking of symptoms. Unfortunately, such remission of symptoms is temporary and application of antibiotics does not result in eradication or inactivation of the MLOs. Nevertheless, the chemicals have been used for suppressing the yellows diseases. Extensive field experiments with antibiotics, systemic fungicides and their combinations against PTR and MF at Simla revealed that these diseases can be economically "controlled" by application of systemic fungicides and even tetracycline. Maximum symptom remission (96%) of PTR has been achieved by Lenticlin/Oxytetracycline followed by Benlate, Cycloheximide, BP, etc. (80-70%), while MF was suppressed efficiently by Cycloheximide, Tetracycline or Antiamoebin + Benlate or Benlate + BP (80-60%). Similarly, maximum increase in yield of PTR affected plants was achieved by sprays of Bavistin, Benlate, Tetracycline and Cycloheximide (4.1-3.1 times). Untreated MF causes heavy losses in yield. The chemical treatments increase yields to a much greater extent, viz. Tetracycline + Benlate (9.5 times) followed by Cycloheximide (7.0 times) and Antiamoebin + Benlate (6.8 times) or Benlate (up to 4.3 times). However, a large degree of variation in the efficacy of the chemicals/antibiotics for MF/PTR control in different varieties during different years is suggestive of the existence of a host-pathogen environment interaction (22).

The net increase in yield minus cost of chemical(s) and labor indicates that treatment of PTR (under 75-90% infection) with Bavistin, Benlate and Tetracycline resulted in a net profit of about Rs. 5,200/-Rs. 4,300/- and /Rs. 1,400/-per ha, respectively. The antibiotic therapy (due to high cost) for PTR cannot compete with Bavistin or Benlate. However, a higher degree of incidence and detection of these diseases in early stages are important to achieve maximum benefit with these chemicals.

Similar field trials have not been tried for control of WB but under laboratory conditions WB has not been affected by antibiotics/systemic fungicides (22).

EFFECT OF CHEMOTHERAPY ON THE INCIDENCE OF HS, PTR AND MF IN THE PRODUCE

In general, the tuber produce of chemically treated plants upon storage was found to have a lower incidence of HS indicating that the chemotherapy of diseased plants also affected development of HS in the produce (Table 10). However, a lower degee of PTR and/or MF expression was observed in the succeeding crops(s). It was observed that the maximum incidence of MF was 5.0% and 33.3% in PTR-infected Kufri Jyoti and Kufri Lauvkar upon Tetra + Anti and Cycloheximide treatments, respectively, while it was 26.3% and 58.8% in their respective untreated (diseased) controls. Likewise, the incidence of MF in MF-infected Kufri Chandramukhi was highest (70.0%) with treatment of BP and Benlate followed by Tetra + BP (63.0%) as against 66.4% in diseased control (Table 10).

Chemical Control of WB through Tuber Treatment

By dipping the WB infected Kufri Kuber tubers in the solutions of BP, Benlate, Tetracycline (500 ppm) and Urea (10,000 ppm) and six combinations for 90 min., maximum germination was recorded in the case of Benlate followed by Tetracycline and BP. Such germinated plants had relatively larger compound leaves. Although a great reduction in the symptom intensity of WB was noted with the above treatements, the yields obtained from treated plants were not at par with the healthy controls. This may be on account of the low-vigor of the plants, after masking of symptoms yet carrying the WB pathogen. Unlike PTR and MF, the three systemic

fungicides, viz Bavistinr, Plantvax and Vita-
vax, were poor in causing symptom remission and
increase in yields of WB infected plants.

THERMOTHERAPY

Hot-water Treatment for WB

WB-infected tubers of var. Kufri Kuber were
treated for 10, 20 and 30 minutes at 45 C and
for 10, 15 and 20 minutes at 50 C. Germination
of the tubers was not affected at 45 C while
treatment at 50 C for 15 min. and 20 min. totally arrested the germination. The plants from
untreated tubers and also from tubers treated
at 45 C developed WB symptoms. Tubers exposed
at 50 C for 10 min. produced plants that were
apparently healthy up to about 6-7 weeks. Later
on, a few axillary/basal thin sessile leaves developed in such plants, but their yields were
very high as compared to the diseased control.
In another set of experiments, where 10, 12 and
14 minutes duration at 50 C were tried, the
symptom remissions were at par for a treatment
of 10 and 12 minutes (35). Elimination of PP
pathogen was achieved from the infected tubers
by hot water treatment at 50 C for 10-15 min.
(8).

SOURCES OF RESISTANCE

A number of varieties/hybrids were thought
to be tolerant or resistant in the initial screening for resistance to these diseases. However, rigorous and long term (2-3 years in
succession) screening, under a high-inoculum-
potential, revealed that they have merely "escaped" the infection. Accordingly, out of 20
potato varieties/clones/accessions of wild
Solanums and hybrids, which were earlier
indicated to be resistant, only 3 i.e. HW 1-16,
CP 1447 (var. Mexican Conchita) and CP 1448
(var. Mexican Elinita) proved resistant to PTR
and WB (20).

DETECTION

The commonest way of detection of these diseases, in the course of seed production, had been visual inspection. However, fluorescent and light microscopy techniques have also been standardized for indexing and laboratory confirmation of these infections for quarantine purposes.

Fluorescence Microscopy

Hand-cut transverse sections of the stems of potato plants, individually infected with PTR, MF and WB, were fixed by boiling in water for 4-5 minutes followed by staining with 1% aqueous aniline-blue. Under fluorescent light, the diseased stem sections revealed the presence of a green fluorescence throughout the phloem zone, restricted to outer and inner phloem for PTR and WB and MF, respectively. In the case of healthy stem sections, no fluorescence was observed in the phloem zone. The fluorescence was detected in plants about 15-30 days after graft inoculation even when the plants did not show initial symptoms (25).

Light microscopy

In PTR and MF infected potato plants regeneration of the xylem perenchyma, adjacent to necrotic phloem was observed. This resulted in development of a pouch-like structure consisting of thin-walled, small parenchyma (13). In the case of some of PTR infected plants, such pockets may be developed by both internal and external phloems. In the case of WB, however, the cortex lacked collenchyma and was composed entirely of the thin-walled parenchyma. Xylem parenchyma was more lignified and pith cells were larger than in the healthy plants (13).

DISCUSSION

Comparison of symptoms, host range, transmission through tubers and vectors and their efficacy in transmitting these pathogens, indicate that PP is a distinct disease; PTR and MF appear similar in some respects, but they are different individually as well as distinct from WB, PTW and stolbur. The only common characters varying in degree of their expression, are chlorosis of infected plants and development of hairy sprouts in the tubers produced. The PTW and stolbur agents are regarded as distinct pathogens (38) yet they have some characters in common, viz. both can incite wilt in infected plants and tuber necrosis and flaccidity. Lack of tuber-necrosis and wilt syndrome due to PTR, MF, WB and PP agents make them different from that of PTW (AY) and stolbur (20, 21). Besides, PTW, AY and stolbur are characterized by a rather low rate of tuber transmission compared to WB and PP and even MF and PTR which have distinct insect vectors. Great differences in various symptoms and host range as well as susceptibility of PTR and MF to Benlate and Bavistin, in contrast to AY, also suggest that they are incited by distinct pathogens that might have co-evolved in different parts of the world at the same or different times depending on the availability of host plants, vectors and alternate plant and insect hosts in their environment.

It is also worthwhile to mention that the clover phyllody pathogen, recorded in India (5, 27) and also in North America and Europe does not infect potatoes and hence is distinct from the potato yellows pathogens described here. Furthermore, the southern MF, known only in Andigena potatoes from Peru, is caused by the aphid-borne potato leafroll virus while the MF is a leafhopper-borne mycoplasma-associated disease producing identical symptoms on Tuberosum and Andigena potatoes (20).

REFERENCES

1. Cousin, M.T. 1975.   (Purple top roll of potato) L enroutement violace de la pomme de terre.   Annals Phytopathologie 7: 167-173.

2. Doi, Y., Teranaka, M., Yora, K. and Asuyama, H. 1967.   Mycoplasma or PLT group-like microorganisms found in the phloem elements of plants infected with mulberry dwarf, potato witchs' broom, aster yellows or paulownia witches' broom. Ann. Phyto-path. Soc. Japan. 33: 259-266.

3. Kaley, D.M., and Nagaich, B.B. 1975. Bio-chemical changes in potato tubers showing hairy sprouts and in purple top roll af-fected plants.   Intern. Symp. on Plant Pathology, Budapest, June 24-27.

4. Kaley, D.M., Nagaich, B.B. and Sangar, R.B. S. 1977.   Carbohydrate content of purple top roll and marginal flavescence affected potatoes.   Proc. All India Symposium on Physiology of Host Pathogen Interaction, Madras, Nov. 25-26, pp. 323-326.

5. Khurana, S.M. Paul, Singh, S. and Nagaich, B.B. 1975. Clover phyllody. Ann. Sci. Rept. CPRI, Simla, pp. 150.

6. Khurana, S.M. Paul, Singh, M.N. 1978. Losses in yields: Purple top roll. Ann. Sci. Rept., CPRI, Simla, pp. 79.

7. Khurana, S.M. Paul, and Garg, V.K. 1978. Mild marginal flavescence.   Ann. Sci. Rept., CPRI, Simla, pp. 79.

8. Khurana, S.M. Paul, Singh, V. and Nagaich, B.B. 1979a.   Hot water treatment of tubers for elimination of potato phyllody patho-gen. Indian Phytopath. 32: 646-648.

9. Khurana, S.M. Paul, Singh, V. and Nagaich,
    B.B. 1979b. Identification of five strains
    of potato witches' broom pathogen in India.
    J. Indian Potato Assoc. 6: 119-129.

10. Khurana, S.M. Paul, Arai, K., Nagaich, B.
    B., Singh, V. and Singh, S. 1981a. Myco-
    plasma-like organisms associated with pota-
    to phyllody disease. Indian Phytopath. 34:
    532-533.

11. Khurana, S.M. Paul, Nagaich, B.B., Singh,
    S. and Singh, V. 1981b. Inter-relationship
    between the potato yellows pathogens. pp.
    111-112.   In: S.C. Verma (Ed.) Contemp.
    Trends in Plant Sciences, p. 506, Kalyani
    Publishers, New Delhi.

12. Khurana, S.M. Paul, Singh, V. and Nagaich,
    B.B. 1981c. Floral anomalies in periwinkle
    and calendula induced by potato yellows
    pathogens. Indian J. Hort. 38: 282-283.

13. Khurana, S.M. Paul, Shekhawat, G.S. and
    Nagaich, B.B. 1983.   Light microscopy of
    potato plants for detection of three myco-
    plasmal diseases.   J. Indian Potato Assoc.
    10: 60-63.

14. Klinkowski, M. 1957. Beitrage zur Kenntnis
    der Stolburkrankheiten der Kartoffel. Proc.
    3rd Conf. Potato Virus diseases (Lisse
    Wageningen, 24-26 June, 1957) pp. 264-277.

15. Nagaich, B.B., Lal, S.B. and Upreti, G.C.
    1963.   Hairy sprout diseases of potatoes.
    Indian Potato J. 5: 71-75.

16. Nagaich, B.B. and Giri, B.K. 1971. Margin-
    al Flavescence -- leafhopper transmitted
    disease of potato. Indian Phytopath. 24:
    824-826.

17. Nagaich, B.B. and Giri, B.K. 1973. Purple top roll disease of potato. Amer. Potato J. 50: 79-85.

18. Nagaich, B.B., Kaley, D.M., Giri, B.K. and Bhardwaj, V.P. 1973. Investigations on hairy sprout disease of potato. Final Tech. Report, CPRI, Simla, p. 35.

19. Nagaich, B.B., Puri, B.K., Sinha, R.C., Dhingra, M.K. and Bhardwaj, V.P. 1974. Mycoplasma-like organisms in plants affected with purple top roll, marginal flavescence and witches' broom diseases of potato. Phytopath. Z. 81: 273-279.

20. Nagaich, B.B., Khurana, S.M. Paul, Kaley, D.M., Singh, S., Bhardwaj, V.P., Sunaina, V., Dhingra, M.K., Singh, V., Singh, M.N. and Sangar, R.B.S. 1978. Investigations on purple top roll and witches' broom diseases of potato. Final Tech. Report, CPRI, Simla, pp. 48.

21. Nagaich, B.B. 1979. The yellows diseases of potato. J. Indian Potato Assoc. 6: 1-11.

22. Nagaich, B.B., Khurana, S.M. Paul, Singh, S., Singh, V., Singh, M.N. Sunaina, V., and Parihar, S.B.S. 1982. Studies on the transmission and control of potato yellows and their effect on crop production. In: Potatoes in Developing Countries. B.B. Nagaich *et al.*, (eds.) IPA & CPRI, Simla pp. 294-303.

23. Norris, D.O. 1954. Disease of potato caused by tomato big bud virus. Australian J. Agric. Res. 5: 1-8.

24. Ploaie, P. and Maramorosch, K. 1969. Electron microscopic demonstrations of particles resembling mycoplasmas or Psittacosis lymphogranuloma-Trachoma Group on plants

infected with European yellows type
diseases. Phytopathology 59: 536-544.

25. Shekhawat, G.S., Khurana, S.M. Paul and
Nagaich, B.B. 1981. Flourescence microscopy of potato and tomato infected with
three mycoplasmal diseases. Z. Pflanzenkr.
Pflanzenschutz 87: 148-151.

26. Singh, S. and Nagaich, B.B. 1977. *Alebroides dravidanus*: A new vector for
purple top roll and witches' broom of the
potato. J. Indian Potato Assoc. 4: 71-73.

27. Singh, S., Nagaich, B.B., Arai, K. and Khurana, S.M. Paul. 1983. Clover phyllody in
Simla hills. Indian Phytopath. 36: 771-775.

28. Singh, S., Bhargava, K.S. and Nagaich, B.B.
1983a. Lethality of the potato purple top
roll pathogen to its vector *Alebroides
nigroscutellatus*. Indian Phytopath. 36:
646-650.

29. Singh, S., Bhargava, K.S. and Nagaich, B.B.
1983b. *Orosius albicinctus* and the potato purple top roll: vector pathogen relationship. J. Indian Potato Assoc. 10:
34-41.

30. Singh, S., Bhargava, K.S. and Nagaich, B.B.
1984. Factors affecting transmission of potato purple top roll pathogen by *Orosius
albicinctus*. J. Indian Potato Assoc. 11:
85-92.

31. Singh, V., Singh, M.N. and Nagaich, B.B.
1976. Chemotherapeutic control of purple
top roll and marginal flavescence diseases
of potatoes. J. Indian Potato Assoc. 3: 60-
64.

32. Singh, V., Nagaich, B.B. and Khurana, S.M.
Paul. 1977. *Nicotiana* species and

cultivars: additional hosts of the potato purple top roll and marginal flavescence pathogens. J. Indian Potato Assoc. 4: 46-49.

33. Singh, V. and Nagaich, B.B. 1979. Potato phyllody yield loss. Ann. Sci. Rept. CPRI, Simla, pp. 87.

34. Singh, V., Khurana, S.M. Paul, Nagaich, B. B. and Singh, R.A. 1981. Studies on the symptomatology and host range of potato phyllody disease. Indian Phytopath. 34: 410-413.

35. Singh, R.A., Sangar, R.B.S. and Prasad, S. M. 1982. Hot water treatment of witches' broom infected tubers. In: Potato in Developing Countries. B.B. Nagaich *et al.*, (eds.). IPA and CPRI, Simla, pp. 390-393.

36. Younkin, S.G. 1943. Purple top wilt of potatoes caused by the Aster yellows Virus (Abstract). Phytopathology 33: 16.

37. Wright, N.S. 1952. Studies on the witches' broom virus disease of potatoes in British Columbia. Can. J. Bot. 30: 735-742.

38. Zimmerman, L. Gries, S., Klein, M. and Sneh, B. 1975. Incidence of yellows diseases in potato fields grown for certified seed in Israel. Proc. 6th EAPR Trien. Conf., Wageningen (The Netherlands), Sept. 15-19, pp. 196-197.

TABLE 1. Yield depressions due to 100% infection of potato purple
top roll (PTR), marginal flavescence (MF) and witches' broom (WB)
pathogens

Potato variety	% loss due to		
	PTR	MF	WB
Great Scott	–	–	22.6 (–)*
Kufri Chandramukhi	50–70	60–65	– (30.0)
Kufri Jyoti	55–75	80–92	– (30.0)
Kufri Kuber	–	–	29.0 (69.0)
Kufri Lauvkar	60–70	–	– (63.0)
Kufri Naveen	–	50–60	25.0 (–)
Kufri Red	–	–	17.0 (40.0)
Up-to-Date	60–65	80–85	– (64.0)

– Data not available, *Loss due to one-year old infection.

TABLE 2. Comparison of losses*, hairy sprouts and
conditions for potato yellows

Disease	% yield loss	% (Max.) Hairy sprouts	Favorable weather, crop age at symptom expression
PTR	30–82	20	Humid cool, 90–100 days
MF	60–92	65	Dry cool, 30–40 days
WB	17–29 (49–100)**	100	Dry warm, 25–45 days
PP	64	79	Dry warm, 60–90–130 days
PTW	6% at 14.5% incidence	17	Humid cool, 100–140 days
Stolbur	100	10	Dry warm, 35–50 days

* On absolute infection basis, ** Due to chronic infection.

TABLE 3. Comparison of transmission of potato yellows
through tuber, dodder and leafhopper

Disease	Tuber*(%)	Transmission through		
		Dodder (sp.)	Vector sp.	Vector-pathogen** relationship
PTR	8-69(-35)	*Cuscuta hyalina*	*Alebroides dravidanus* *Orosius albicinctus* *Seriana equata*	Acq. 7 days Tr. 9 days (10-50%)
MF	0.1-56	*C. hyalina*	*O. albicinctus* *S. equata*	Acq. 27 days Tr. 18 days (5%)
WB	100	*C. hyalina*	*A. dravidanus* *O. albicinctus*	Acq. 5 days Tr. 25 days (40-50%)
PP	100	*C. hyalina*		
PTW	3-8	*C. campestris*	*Macrosteles fascifrons* *Elymana virescence*	Acq. 7 days Tr. 21 days (47-78%)
Stolbur	14	*C. campestris* *C. trifolli* *C. epilinum*	*E. virescence* *Hyalesthes obsoletus* *Aphrodes bicinctus* *Euscelis plebejus*	Acq. 2 days Tr. 2-7 days (below 10%)

* Average percentage of 3-5 years trial.

** For the most efficient vector.

TABLE 4. Differential hosts, phloem flourescence and
chemotherapeutic response of potato yellows

Disease	Differential Hosts	Phloem fluorescence	Chemotherapeutic response
PTR	*Trifolium repens* Potato var. Conchita, Mexican Hybrid HW 1-16 Tobacco var. NC-95 Dahlia	Outer & inner	Tetracyclines, Benlate, Bavistin
MF	Tobacco vars. Ambalema, Burley-21, Delcrest, Kanak Prabha, McNair-12, Samsun, Xanthi, Xanthi nc.	Outer only	Tetracyclines, Antiamoebin & Cycloheximide
WB	Potato var. Elinita Mexican	Inner only	Tetracyclines
PP	*Physalis floridana*, *Petunia hybrida*, *N. stolonifolia*	–	Tetracyclines
PTW (AY)	Aster, celery	–	Tetracyclines but not Benlate
Stolbur	*Capsicum annuum* *Anagallis arvensis*, *Stellaria media*	–	Tetracyclines

TABLE 5. Tuber transmission of the potato yellows pathogens

Potato variety		% Tuber transimssion of		
		PTR	MF	WB
Kufri Chandramukhi	PTRa	69.6 (30.3*)	0.19 (0.79)	–
	MF	8.3 (0.0)	66.4 (0.0)	–
Kufri Jyoti	PTR	27.6 (21.6)	1.47 (0.25)	–
	MF	13.1 (5.0)	19.2 (0.0)	–
Kufri Naveen	MF	25.3 (8.3)	33.3 (0.0)	–
Kufri Kuber	WB	–	–	100 (7.0)
Kufri Lauvkar	WB	–	–	50 (10)
Up-to-Date	PTR	18.0 (12.0)	0.0	–

– = Data not available, a= Original infection, *= % Infection in healthy control.

TABLE 6. Percent incidence of PTR and MF in crop from HS tubers

Potato varieties	% plants		
	PTR	MF	Healthy
Kufri Chandramukhi	5	24	61*
Kufri Jyoti	8	7	85
Kufri Kuber	2	1.5	84*
Kufri Kundan	6	5	86*
Kufri Shakti	22.5	11	66*
Great Scot	0	16	84
Others	23	12	65

* Rest were showing mosaic/leaf roll

TABLE 7. Efficiency of vectors in the transmission of potato yellows pathogens

| Vector | % Transmission to and from potato by a single jassid | | |
	PTR	MF	WB
Orosius albicinctus	6–10(20.0)*	3–6(10.0)	10.0(15.0)
Alebroides dravidanus	35–70(100.0)	0–0	15–35(50.0)
Seriana equata	5–10(14.0)	6–15(35.0)	0–0

* = % Transmission through group of five leafhoppers/plant.

TABLE 8. Percent PTR symptom remission/increase in
potato yield after chemotherapy*

| Treatment** | Varieties | | |
	Kufri Jyoti	Kufri Lauvkar	Kufri Chandramukhi
Antiamoebin (Anti)	65/93	64/47	–
BP 101	80/117	68/92	80/55
Benlate (Ben)	60/137(370)	59/90	83/212
Cycloheximide	80/57	54/95	65/38
Bavistin (Bavi)	65/172(390)	–	70/236
Tetracycline (Tetra)	85/313	70/01	81/56
Lenticlin	96/63	–	96/315
Plantvax	65/61	–	94/72
Vitavax	80/51	–	64/177
Anti + BP	56/82	71/130	–
Anti + Ben	68/111	64/80	–
Tetra + Bavi	82/355	–	82/281
Tetra + Ben	72/93	55/123	70/95
Ben + Bavi	92/250	–	86/(410)a

* All chemicals were used @ 500 ppm except Cycloheximide (100 ppm); average of three or five years data except a.

The data within parentheses indicate the maximum increase obtained in one year.

** Only the effective promising ones are listed.

Numerator/Denominator = % Symptom remission/increase in yield.

– = Data not available.

TABLE 9. Percent MF remission/increase in
          yield after chemotherapy*

Treatment**	Varieties			
	Kufri Jyoti	Kufri Chandramukhi	Kufri Naveen	Up-to-Date
Antiamoebin	48/316	50/126	-	60/80
BP	28/245	75/124	43/106	50/100
Benlate	41/180	50/162	12/28	40/119
Cycloheximide	80/706	75/200	-	-
Tetracycline	75/225	-	66/74	40/9
Anti + BP	35/245	60/44	-	60/44
Anti + Ben	80/684	66/50	-	10/87
Anti + Tetra	60/304	33/134	-	30/191
BP + Ben	30/283	80/108	43/39	-
Tetra + BP	28/187	-	-	30/34
Tetra + Ben	35/551(956)	50/120	43/85	30/81

*/** = Same as for TABLE 5.

Numerator = % symptom remission.

Denominator = % increase in yield over control

The datum within parenthesis indicates the maximum increase obtained in one year.

TABLE 10. Incidence of PTR and MF in the produce of antibiotic/chemical sprayed potato plants

Treatment**	Kufri Jyoti (PTR produce)		Kufri Lauvkar (PTR produce)		Kufri Jyoti (MF produce)		Kufri Chandramukhi (MF produce)	
	PTR %	MF %	PTR %	MF %	PTR %	MF %	PTR %	128 %
Tetra	11.4	0	29.2	7.6	0	13.0	12.4	12.4
Anti	7.2	3.1	14.0	15.4	0	7.6	28.7	0
Ben	3.3	0	13.2	6.6	0	0	12.4	18.6
BP	14.8	0	9.2	13.2	0	4.3	18.0	63.0
Tetra + BP	23.0	0	19.1	13.2	0	0	30.0	20.0
Tetra + Anti	17.0	5.0	77.0	30.4	7.1	0	21.3	28.4
Tetra + Ben	9.3	0	10.8	19.8	12.5	12.5	8.3	33.2
Anti + Ben	14.3	0	25.1	30.0	0	0	37.5	12.5
Anti + BP	11.5	2.8	26.6	30.7	5.3	0	-	-
Cycloheximide	5.0	3.3	9.4	33.3	0	0	-	-
Urea	23.0	0	13.2	26.4	-	-	20.7	17.7
Urea + BP	11.7	0	8.3	14.5	-	-	23.0	40.0
BP + Ben	14.2	3.5	28.5	9.8	-	-	30.0	70.0
Diseased control	26.3	3.3	58.8	34.5	11.1	33.3	8.3	66.4
Healthy control	0	0	0	0	0	0	0	0

EXPLANATION OF PLATES

1. Comparative diagrammatic representation of salient features of the symptom syndrome of yellows diseases of the potato (after Khurana *et al.*, 1981). PTR: Many upright stems (a) rolling/pigmentation of top leaves (b,d) axillary poliferation (b) and normal flowers/fruits (b). Note the unshriveled mother until harvest (b) and aerial tuberisation (c).

MF: Chlorotic margins, rough lamina and overlapping leaflets of the top stunted leaves, (a). Unshriveled mother tuber with few small tubers (b) and tubers developing hairy sprouts (c).

PTW: Upright habit of stem devoid of axillary growth/aerial tubers (a) leaflets rolling basally with pigmentation and normal flowers/fruits (b). Wilting phase of plant with pigmentation/necrosis of stem/tubers (c).

STOLBUR: Upright shoots with extreme axillary proliferation (a) 'spooning' of leaflets (b) and wilting plant/shriveled rubbery mother tuber (c). Flowering/aerial tuberisation absent (c).

WB: Typical (a) and mild (b) witches' broom strain infected plants with simple and trifoliate leaves; aerial tubers (b) and hairy sprouts on large/small tuber (c).

PP: Cupped calyx (a) polysepalous phylloid flower with rosetted androecea (b) sessile rosette of androecea (devoid of calyx/corolla) on stem (c); proliferation of a shoot from phylloid flower (d). Leaf showing smalling of leaflets, extreme hairyness but prominent folioles (e) and a tuber lacking apical-dominance with many hairy sprouts (f).

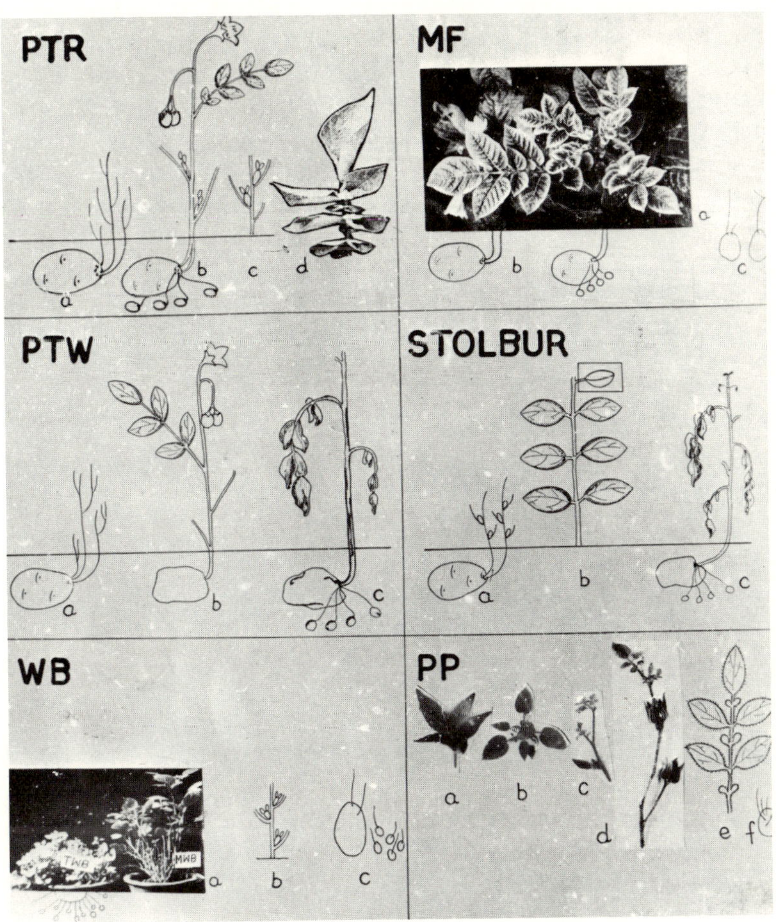

Plate 1.

2. Variations in morphology of mycoplasma-like organisms associated with the potato yellows diseases (i) Purple top roll, (ii) Marginal flavescence, (iii) Witches' broom, and (iv) Potato phyllody.

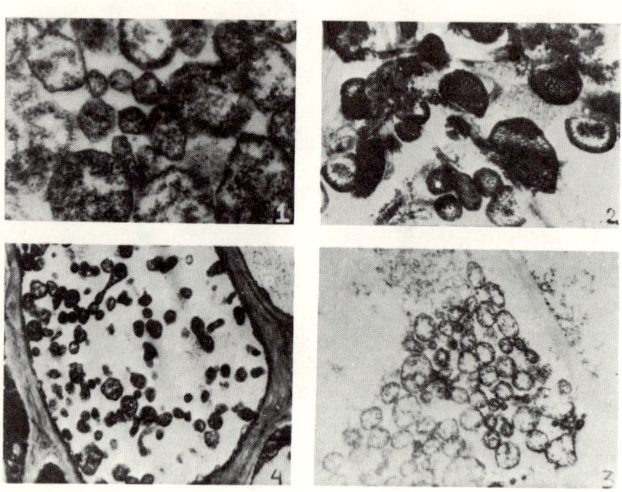

Plate 2.

16

MYCOPLASMA DISEASES OF CORN

IN FLORIDA

by

James H. Tsai

Ft. Lauderdale Research
and Education Center

University of Florida, IFAS

3205 College Avenue

Ft. Lauderdale, FL 33314

INTRODUCTION

Corn stunt is one of the most economically important diseases of corn in the United States, Mexico and Central and South America (Bradfute *et al.*, 1981; Nault and Bradfute, 1979). Prior to the first report of MLO association with mulberry dwarf and other yellows diseases (Doi *et al.*, 1967), corn stunt was thought to be caused by several strains of a virus based on symptomatology and vector transmission (Gordon *et al.*, 1981; Maramorosch *et al.*, 1965; Nault and Bradfute, 1979). Corn stunt was first named by Kunkel (1946) and later described by Maramorosch (1955) as Rio

---

Florida Agricultural Experiment Station
Journal Series No 7080.

Grande corn stunt. Upon re-examination of the fixed materials, researchers at the Boyce Thompson Institute for Plant Research discovered MLO in corn plants and leaf hopper vectors by electron microscopy (Maramorosch *et al.*, 1968). The helical morphology of the causal agent of Rio Grande corn stunt was subsequently established (Chen and Lio, 1975; Davis and Worley, 1973; Williamson and Whitcomb, 1975) and was named as corn stunt spiroplasma (CSS). In 1955, Maramorosch described a strain of corn stunt as Mesa Central corn stunt, which is now believed to be caused by the maize bushy stunt mycoplasma (MBSM)(Bradfute *et al.*, 1977; Nault, 1980).

During 1978-1980, a corn stunt-like disease became epidemic in south Florida: at least five viruses including maize dwarf mosaic virus strain A and B, maize stripe virus, maize mosaic virus, maize rayado fino virus, an unidentified straight tubular virus-like particle, and two prokaryotes (CSS and MBSM) were found in the disease complex (Bradfute *et al.*, 1981; Davis *et al.*, 1984, J.H. Tsai, unpublished). However, the CSS played a dominant role in the epidemic during that period as the infection of CSS in the diseased field samples in 1979 and 1980 reached 68.4 and 98.5%, respectively (Bradfute *et al.*, 1981).

MATERIALS AND METHODS

From Fall 1978 through 1981, leaf samples of field grown corn (*Zea mays L.*) with corn stunt-like symptoms were collected in Dade County FL. They were checked by dark-field microscopy for presence of spiroplasmas in expressed sap before the initiation of transmission experiments. In 1978, two CSS isolates were obtained from the plants which had extensive leaf reddening and maize rayado fino-like symptoms: they were designated as "Red Leaf" and "R-type" isolates, respectively. From 1979 through 1981, only one isolate was obtained for each year and designated as 1979 CSS, 1980 CSS, and 1981 CSS. One isolate of MBSM originated in 1980 from a field grown corn plant with bushy stunt symptoms in Dade County, Florida. This isolate was designated as FL-MBSM. Another isolate of MBSM was obtained from L.R. Nault (Ohio Agric. Res. and Dev. Ctr., Wooster, OH) and designated as TX-MBSM.

All CSS isolates were routinely maintained in a series of healthy sweet corn plants (*Zea mays* var. *saccharata* Guardian) using laboratory reared *Dalbulus maidis*. Test insects were reared on Guardian sweet corn with 12 hrs. of light per day at 25 ± 1C. Guardian sweet corn was used for source and test plants of CSS throughout the experiments whereas "Aristogold Bantam Evergreen" sweet corn was used for MBSM source and test plants. During this study, 156 adults of *D. maidis* from the stock colonies were placed on test plants to check for the presence of transmissible pathogens; none was found to carry CSS or MBSM.

To determine the median incubation period (IP_{50}) and retention of CSS and MBSM in *D. maidis*, a group of 80 leafhoppers was given a three-day acquisition access period (AAP) on the infected plants. After AAP, the leafhoppers were transferred singly to a series of test plants in the two-leaf stage. Transfers were made daily until the death of the insects. Test insects were kept at 25 ± 1 C throughout this study. Test plants were labelled and kept for six weeks at 28 ± 2 C in a greenhouse for symptom development. The procedure for determining IP_{50} values were the same as described previously (Tsai and Zitter, 1982.)

RESULTS

Symptomatology

Symptoms on test plants for all five isolates of CSS were indistinguishable. The initial symptoms of plants inoculated with CSS showed characteristic small chlorotic stripes that developed at the leaf bases of new leaves after about 25-30 days (Fig. 1). The chlorotic stripes became fused and extended further toward the leaf tips in the later leaves which were characterized by green spots and stripes on a chlorotic background (Fig. 2). The infected plants had much shorter internodes and a proliferation of secondary shoots in leaf axils.

Plants inoculated with MBSM initially developed a marginal yellow and orange color of the older leaves and marginal chlorosis along the whorl leaves. The symptoms on subsequently developed leaves were characterized by marginal chlorosis, tearing, shortening and twisting of young leaf tips (Fig. 3). Numerous tillers at the base of the plant and at leaf axils became evident (Fig. 4). Infected "Aristogold Bantam Evergreen" sweet corn developed extensive leaf reddening, and more basal and auxillary shoots as compared to Guardian sweet corn. Symptoms observed for the Texas isolate of MBSM were essentially identical to those observed for the Florida isolate. The height of plants inoculated with MBSM at maturity rarely exceeded 1/2 of that inoculated with CSS.

Transmission characteristics

Five isolates of CSS were transmitted by *D. maidis* at rates ranging from 44.2 to 56%. The IP_{50} ranged from 17.5 to 21.2 days. The average retention period ranged from 27.5 to 32.2. days for all five isolates (Table 1). The Florida isolate of MBSM was transmitted by *D. maidis* at a lower rate than the Texas isolate (43.9% vs 50%). The former had a slightly longer IP_{50} and retention period than the latter (Table 2).

DISCUSSION

Corn stunt spiroplasma and MBSM produce entirely different diseases in corn. Symptoms of

CSS are very similar to those described previously for Rio Grande CSS (Maramorosch, 1955; Nault, 1980). Initially, all five CSS isolates used in this study exhibited a variety of symptoms in the field. However, after a few passages in Guardian sweet corn in the laboratory, no discernible differences could be detected among them. The absence of extensive leaf reddening in the laboratory of 1978 red leaf isolate could well be due to reaction of genotypes of the host plants in the field. The absence of rayado fino-like symptoms for "R-type" isolate in the laboratory could have been due to double infection of the original diseased plants in the field. Mixed infections of CSS and maize rayado fino virus in *D. maidis* had previously been reported (Wolanski and Maramorosch, 1979; Nault, *et al.*, 1979; Gamez, 1973; Bradfute *et al.*, 1981). Symptoms of MBSM noted in the field and laboratory are indistinguishable to those previously reported for the Mesa Central corn stunt (Maramorosch, 1955; Nault, 1980). "Aristogold Bantam Evergreen" sweet corn again proved to be a valuable plant for diagnosing MBSM (Nault, 1980).

There were no appreciable differences in transmission characteristics of the five CSS isolates by *D. maidis* (Table 2). The IP_{50} values for all five isolates were similar to other CSS isolates (Nault, 1980; Nault and Bradfute, 1979). Based on enzyme-linked immunosorbent assay and deformation tests, these five isolates are closely related to E-275 isolate of CSS (Davis *et al.*, 1984). It is therefore reasonable to assume that these five CSS isolates from Florida belong to one genotype. The transmission characteristics of Florida isolate MBSM by *D. maidis* were different from those of Texas isolate (Table 2). Although the symptomatology and morphology between these two isolates are indistinguishable, their relatedness awaits the determination of serological tests.

SUMMARY

The transmission characteristics of five isolates of corn stunt spiroplasma (CSS) (1978 Red Leaf, 1978 R-type, 1979 CSS, 1980 CSS and 1981 CSS) and two isolates of maize bushy stunt mycoplasma (MBSM) (FL-MBSM and TX-MBSM) were studied. The transmission efficiencies for 1978 Red Leaf, 1978 R-type, 1979 CSS, 1980 CSS and 1981 CSS by *Dalbulus maidis* after a three-day acquisition access period (AAP) were 44.2, 46.0, 50.0, 56.0 and 53.3% respectively. The respective median incubation periods (IP_{50}) in the *D. maidis* were 19.0, 18.0, 21.2, 20.3 and 17.5 days, and the mean retention periods by *D. maidis* were 28.5, 30.4, 32.2, 31.5 and 27.5 days, respectively. The transmission efficiencies for FL-MBSM and TX-MBSM by *D. maidis* after a three-day AAP were 43.9 and 50.0%. The IP_{50} for FL-MBSM and TX-MBSM were 28.0 and 25.5 days. The mean retention periods for *D. maidis* were 34.5 and 32.6 days. Based on the transmission characteristics, these five Florida isolates of CSS were indistinguishable and they were similar to other known CSS isolates.

REFERENCES

Bradfute, O.E., Nault, L.R., Robertson, D.C., and Toler, R.W. 1977. Maize bushy stunt -- a disease associated with non-helical mycoplasma-like organism. (Abstr.) Proc. Am. Phytopathol. Soc. 4: 171.

Bradfute, O.E., Tsai, J.H., and Gordon, D.T. 1981. Corn stunt spiroplasma and viruses associated with a maize disease eipdemic in southern Florida. Plant Disease 65: 837-841.

Chen, T.A. and Liao, C.H. 1975. Corn stunt spiroplasma: isolation, cultivation and proof of pathogenicity. Science 188: 1015-1017.

Davis, R.E., and Worley, J.F. 1973. Spiroplasma: motile, helical microorganism associated with corn stunt disease. Phytopathology 63: 403-408.

Davis, M.J., Tsai, J.H., and McCoy, R.E. 1984. Isolation of the corn stunt spiroplasma from maize in Florida. Plant Disease 68: 600-604.

Doi, Y., Teranaka, M., Yora, K., and Asuyama, H. 1967. Mycoplasma or PLT group-like microorganisms found in the phloem elements of plants infected with mulberry dwarf, potato witches' broom, aster yellows, or paulownia witches' broom. [In Japanese, with English abstract.] Ann. Phytopathol. Soc. Japan 33: 259-266.

Gamez, R., 1973. Transmission of rayado fino virus of maize (*Zea mays*) by *Dalbulus maidis*. Ann. Appl. Biol. 73: 285-292.

Gordon, D.T., Bradfute, O.E., Gingery, R.E., Knoke, J.K., Louis, R., Nault, L.R., and Scott, G.E. 1981. Introduction: History, geographical distribution, pathogen characteristics, and economic importance. Pages 1-13 In: Virus and virus-like diseases of maize in the United States. D.T. Gordon, J.K. Knoke, and G.E. Scott, eds. South. Coop. Ser. Bull. pp. 247, 248.

Kunkel, L.O. 1946. Leafhopper transmission of corn stunt. Proc. Nat. Acad. Sci. 32: 246-247.

Maramorosch, K. 1955. The occurrence of two distinct types of corn stunt in Mexico. Plant Dis. Rep. 39: 896-898.

Table 1. Summary of incubation period (IP) and retention of five CSS isolates in *D. maidis* after a 3-day acquisition acess period.

Isolate	Infectivity[a] %		IP (days)			Retention (days)		
			Min.	Max.	IP_{50}	Min.	Max.	Mean
1978 Red leaf	19/43	44.2	14	28	19.0	10	34	28.5
1987 R-type	23/50	46.0	15	25	18.0	9	41	30.4
1979 CSS	24/48	50.0	13	27	21.2	7	45	32.2
1980 CSS	28/50	56.0	14	29	20.3	11	42	31.5
1981 CSS	23/45	53.3	13	26	17.5	6	37	27.5

[a] Numerator = number of infections; denominator = number of trials

Table 2. Summary of incubation period (IP) and retention of two MBSM isolates in *D. maidis* after a 3-day acquisition access period.

Isolate	Infectivity[a] %		IP (days)			Retention (days)		
			Min.	Max.	IP_{50}	Min.	Max.	Mean
FL-MBSM	29/66	43.9	21	35	28.0	15	49	34.5
TX-MBSM	30/60	50/0	23	33	25.5	17	45	32.6

[a] Numerator = number of infections; denominator = number of trials

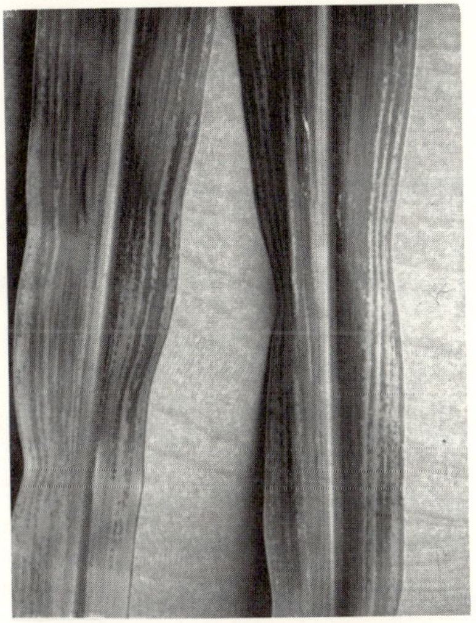

Figure 1. 'Guardian' sweet corn showed characteristic chlorotic stripes at the leaf base 3 weeks after inoculation with CSS.

Figure 2. Advanced stage of corn stunt symptoms on 'Guardian' sweet corn.

Figure 3. Symptoms of marginal chlorosis, tearing, and twisting of young leaf tips on 'Guardian' sweet corn inoculated with MBSM.

Figure 4. Multiple tillers at the base of 'Guardian' sweet corn plant inoculated with MBSM.

17

STUBBORN DISEASES OF CITRUS

CAUSED BY

SPIROPLASMA CITRI

by

D.J. Gumpf

Department of Plant Pathology

University of California

Riverside, CA 92521

INTRODUCTION

The mycoplasma disease of citrus caused by *Spiroplasma citri* called "stubborn" was first observed about 1915 in Washington navel orange trees in orchards near Redlands, California (Fawcett *et al.*, 1944). From 1915-1917 several non-productive trees were observed and some were top-worked using carefully selected "healthy" buds. In spite of the carefully chosen budwood, the growth of these scion buds was slow and exhibited growth characteristics shown by the original tree. This type of growth was described at the time as "stubborn." The first report of stubborn outside California and Arizona was from Palestine where it caused heavy damage when it appeared in epidemic during the summer of 1982 (Reichert and Perlberger, 1931). This disease, "little leaf," was thought to have been he result of drought. Subsequent work showed many similarities between little leaf and s ubborn and at present both names apply to th same disorder.

GEOGRAPHICAL DISTRIBUTION
AND ECONOMIC IMPORTANCE

The occurrence of stubborn disease of citrus in California was once spotty and scattered except in a few orchards where a very high percentage of uniformly affected trees would indicate the use of diseased budwood. In recent years, the disease has become a major problem in all the important warm inland valley orange and grapefruit producing areas of California and Arizona. The total damage caused by stubborn disease in California has never been fully assessed. Calavan (1969) did, however, estimate that in 1969 about one million trees were affected.

Numerous studies have established that stubborn disease is a serious problem widely distributed in many Mediterranean and Mideast countries including Algeria, Corsica, Cyprus, Egypt, Iran, Iraq, Jordan, Lebanon, Morocco, Sardinia, Sicily, Syria, and Tunisia (Anonymous, 1970; Bruno, 1964; Chapot, 1959; Nour-Eldin, F.,1967; Scaramuzzi and Catara, 1968; Vogel and Bové, 1974). In Peru and northwestern Mexico, trees with stubborn-like symptoms have been seen but positive confirmation by culturing the organism has not been attempted. In all countries affected, the percentage of diseased trees varies greatly among individual orchards with many diseased trees difficult to detect. Visual surveys indicated from less than 1% to over 50% of trees in affected orchards appear stubborn diseased in California and Morocco (Calavan and Carpenter, 1965).

SYMPTOMS

Symptoms of stubborn disease vary with host variety, nutritional health, age, season, wind, humidity, temperature, and isolate of *S. citri* involved. *Spiroplasma citri* is a warm temperature organism which, in culture, grows well at 28° -30° C (Saglio *et al.*, 1971; Fudl-Allah *et al.*, 1972). Citrus also grows well and reacts strongly to infection by *S. citri* at these temperatures. Our observations in California indicate that symptoms of stubborn disease are most severe in the hot southern desert, slightly less severe in the San Joaquin Valley and least severe in the cool coastal plain.

Leaf and shoot symptoms include small, abnormally upright leaves, bunchy or rosetted foliage and numerous short leafy branches, growth from multiple axillary buds, foliar chloroses resembling deficiency patterns, stunted growth, premature leaf drop and twig dieback. Symptoms associated with fruiting include heavy normal and unseasonal flowering, excessive fruit drop, acorn-shaped or lopsided fruits with a curved columella, mature fruits paler than normal and sometimes retaining green color on the stylar end, excessive seed abortion, insipid sour or bitter flavor, small fruits and low yield (Gumpf and Calavan, 1981).

CASUAL AGENT

The first indication that stubborn disease might be caused by a mycoplasma rather than a virus came in 1968 when Igwegbe and Calavan (1970) discovered that tetracycline suppressed the disease symptoms in sweet orange seedlings. Subsequent electron microscopy of ultrathin sections of tissue from stubborn affected and healthy citrus by Igwegbe and Calavan (1970) and Laflèche and Bové (1970) revealed mycoplasma-like structures only in the sieve tubes of stubborn infected tissue. In 1970, groups in France (Saglio *et al.*, 1971) and in California (Fudl-Allah *et al.*, 1972), cultured in liquid and on agar media a pleomorphic prokaryote from young leafy shoots of stubborn infected sweet orange. Saglio *et al.*, (1973) fully characterized the organism and proposed for it the name *Spiroplasma citri*. It is characteristically a helically filamentous, motile, cell wall-less prokaryote which is a recognized representative of the family Spiroplasmataceae, placing it in the order Mycoplasmatales and class Mollicutes.

DIAGNOSIS

The symptoms of stubborn disease in affected field trees are variable and often so obscure that positive identification may require confirmation by experimental means. Graft transmission to highly sensitive citrus varieties such as Marsh grapefruit, Sexton tangelo, and Madam Vinous sweet orange provides a fairly reliable method of diagnosis (Calavan and Christiansen, 1966). Electron microscope investigations reveal the presence of typical mycoplasma-like bodies in sieve tubes of stubborn diseased but not healthy citrus (Igwebge and Calavan, 1970). Stubborn diseased seedlings will also show recovery after holding them in hydroponic solutions containing tetracycline (Igwebge and Calavan, 1970).

The stubborn organism, *S. citri*, has been characterized in detail and its biological and physical properties are well studied, therefore, cultivation and subsequent serological identification provides the most rapid and sensitive diagnostic test.

EPIDEMIOLOGY

Natural Spread

Natural spread of the stubborn agent was suspected for years by research workers and growers. Trees surrounded by apparently healthy trees as well as experimental seedling trees inexplicably acquired stubborn disease. In addition, diseased plants were found in nurseries under circumstances that strongly suggested natural spread. As evidence accumulated that the stubborn agent was being spread naturally in the field, an effort was made to discover a possible vector. During the 1973 epidemic of stubborn disease Cartia and Calavan (1974), and Lee *et al.*, (1973) cultured *S. citri* from beat leafhoppers *Circulifer tenellus* (Baker) collected in a stubborn disease field plot at the University of California Moreno Farm, east of Riverside. Subsequently, *S. citri* was also isolated from two other leafhoppers, *Scaphytopius nitridus* (DeLong) and *S. acutus delongi* (Young), and several other species of insects (Kaloostian *et al.*, 1979b; Oldfield and Kaloostian, 1979). This work demonstrated that at least the three named species of leafhoppers could acquire *S. citri* naturally. Later, experimental leafhopper transmission studies also determined that in California these three leafhopper species could transmit *S. citri* (Calavan *et al.*, 1979; Kaloostian, *et al.*, 1979b; Oldfield and Kaloostian, 1979).

Citrus Hosts

All important commercial varieties of citrus appear to be susceptible to stubborn when graft inoculated but in California natural infection occurs mostly in sweet orange, grapefruit, and tangelo trees (Carpenter, 1959; Calavan and Christiansen, 1961). Navel orange and grapefruit trees are most severely damaged but stubborn Valencia orange trees too may have crop losses exceeding 50 percent.

Non-Citrus Hosts

The existence of non-citrus hosts of the

stubborn pathogen was strongly indicated in 1970 when sweet orange seedlings grown far from other citrus developed stubborn disease (Calavan *et al.,* 1974). Oldfield and Kaloostian (1979) were also able to culture *S. citri* from wild *C. tenellus* collected from non-citrus hosts in areas far from citrus. In 1975, *C. tenellus* from south central Washington yielded *S. citri* in pure culture. *Scaphytopius nitridus* fed on these cultures, transmitted the pathogen to sweet orange plants, producing symptoms indistinguishable from stubborn disease originating from other sources (Rana *et al.,* 1975). Groups of *C. tenellus* collected in 1976 in the Mojave Desert about 40 miles from commercial citrus transmitted *S. citri* directly to sweet orange plants.

Environmental Effects

Temperature effects are the only environmental conditions that have been examined. There is very little natural spread of stubborn disease on the cool coastal plain of southern California (Calavan *et al.,* 1979). This fact is probably due to two factors: (1) the poor growth made by *S. citri* below 28°C and (2) the low populations of stubborn disease vectors normally present in this cool area. High populations of *C. tenellus* are most commonly found in warm or hot arid or semi-arid regions.

GRAFT TRANSMISSION AND PERPETUATION IN SCION

Even though early citrus propagators frequently transmitted stubborn disease inadvertently into healthy scions by grafting them onto diseased navel orange trees in unsuccessful attempts to improve fruit quality and production of affected trees (Fawcett *et al.,* 1944), results of more recent graft transmission studies with stubborn disease have been highly variable. Transmissions ranged from 1%

or 2% to nearly 100%, depending on inoculum, variety, season, temperature and other factors. The low percentages of transmission are most likely due to the erratic distribution of *S. citri* in citrus phloem. Distribution of the pathogen in sweet orange and grapefruit trees at various seasons was studied by Calavan *et al.*, (1968), who obtained transmission only from graft pieces containing phloem and the most consistent year-round transmission from leaf patches excised from affected young leaves. *Spiroplasma citri* is sometimes difficult or impossible to culture from many parts of infected field-grown citrus trees, including the fruit, especially during the cool season.

SUMMARY

Natural spread by leafhoppers especially *Circulifer tenellus*, the beet leafhopper, has been observed and monitored for years. As a result, the following observations have been repeatedly verified: 1) spread of stubborn is particularly noticeable and prevalent in young trees; 2) the pattern of early disease spread in a field indicates the disease moves into the citrus from surrounding fields; 3) tree to tree spread may only be important in a very young orchard and only when there are large numbers of infected trees, on the other hand, the presence of an infected tree in a mature grove presents little or no hazard to neighboring trees; and 4) *Spiroplasma citri* can be transmitted by grafting using infected budwood or the organism can move from infected rootstocks into healthy scion grafts.

CONTROL OR DISEASE MANAGEMENT

Satisfactory control of citrus stubborn disease in susceptible varieties has been achieved only in orchards established with disease free trees in areas where populations of the beet leafhopper appeared to be consistently low for several years after planting. Any abnormally low incidence of stubborn disease in highly susceptible varieties is therefore related to a low degree of vector activity or to inadequate inoculum of *S. citri* during the first six years of tree growth. In most areas where control is not possible, a program of disease management may be effective in maintaining productive orchards. What follows is a discussion of several management practices that should be considered as important in maintaining an orchard threatened by stubborn.

Antibiotics

Although *S. citri* is highly sensitive to tetracyclines *in vitro* (Saglio *et al.*, 1973; Bowyer and Calavan, 1974), tetracycline injected into the xylem of large trees apparently is not translocated to the sieve tubes in amounts sufficient to inhibit the pathogen enough for affected trees to improve. Prolonged root immersion, while effective, for small plants (Igwegbe and Calavan, 1970), has no practical value in the field.

Clean Propagative Material

The use of propagative material free of stubborn disease (clean stock) was recommended by Fawcett *et al.*, (1944) and often has been advised as a preventive measure (Chapot and Delucchi, 1964; Calavan and Carpenter,1965; Calavan *et al.*, 1978, 1979). In some areas where considerable natural spread of *S. citri* occurs, clean stock has little value for prevention of stubborn disease during epidemic years (Calavan *et al.*, 1979) because young sweet orange, grapefruit, and tangelo trees are especially susceptible to infection with *S. citri* from leafhopper vectors. Thus, clean propagative material cannot, by itself, assure prevention of stubborn disease in endemic areas but must be supplemented by other cultural practices.

Quarantine

The presence of stubborn disease in commercial citrus growing areas would make any quarantine measure against infected plants and budwood ineffective, particularly since the pathogen is present in other plant hosts and, with adequate leafhopper vector populations, can be transmitted from citrus or other hosts to young citrus trees. Areas, however, where stubborn disease is not present in citrus should exclude stubborn infected propagative material.

Stubborn free nursery trees can be produced in greenhouses from clean propagative materials even when epidemic conditions prevail in outdoor nurseries. This reduces tree exposure to possible infection during the highly susceptible period by about two years, although the trees remain susceptible several years after being planted in the field (Calavan *et al.*, 1976, 1979).

Control of Non-Citrus Hosts

Non-citrus hosts contribute to seasonal build up of *S. citri*, which may be transmitted to citrus by leafhopper vectors (Calavan *et al.*, 1979). Although there is no experimental evidence that controlling weed hosts of *S. citri* decreases the incidence of stubborn disease in young citrus trees, clean cultivation in and near citrus plantings has been suggested (Wallace, 1978) and may be useful in some situations. However, in or near citrus orchards the destruction of non-citrus host plants infested with vectors may cause many of the vectors to feed on citrus which under normal circumstances they might avoid.

Trap Plants

We have observed field plots of citrus

seedlings and periwinkle plants designed to determine the effect of bordering trap plants on the incidence of stubborn disease. Using sugarbeet which is leafhopper attractive and a non-host for *S. citri* as the trap plant, the incidence of stubborn disease can be reduced. This reduction may be as much as 53% in periwinkle when compared to control plots without a surrounding border of trap plants. Although the use of trap plants does not provide complete protection from stubborn, the incidence was reduced sufficiently to merit consideration of its use in a nursery situation.

Vector Control

In field trials also using citrus seedlings and periwinkle plants, the effect of systemic insecticide application for direct control of leafhopper vectors was evaluated. Two systemic insecticides, Temik and Di-Syston, evaluated in our tests were of little value in reducing the incidence of stubborn even though leafhopper feeding tests clearly indicated that the application rate used (10 lb/acre) was sufficient to cause 100% mortality within a 24-hour feeding period 40 days after treatment. Observations made from these studies and confirmed by laboratory experiments indicate that transmission of the stubborn organism by infective leafhoppers can occur in substantially less than 24 hours (Liu *et al.*, 1983). Therefore, it is unlikely that we will be able to kill the insect on the plant before it transmits *S. citri*.

Roguing

Careful observations made over a number of years on commercial plantings indicate the importance of removing stubborn affected trees from young orchards (i.e., six years and under) as soon as a positive diagnosis is made. These trees are not removed because they present a disease hazard to healthy neighboring trees, but because of their early infection, they will never be productive. Trees which become infected when mature in a mature grove need not necessarily be pulled out until economics dictates this action. These trees present minimal hazard to the rest of the grove because there is little tree to tree spread. Usually, only a part of the mature tree will become infected and exhibit symptoms of stubborn, leaving the remainder of the tree normally productive.

Summary and Recommendations

The recommendations that follow, I consider to be useful in a practice of citrus orchard management in lowering the incidence of stubborn disease or minimizing losses from it:

1) Use budwood from disease-free trees for all propagations including top-working.

2) Topworking should be done only on trees which are totally free of stubborn.

3) Survey young orchards annually up to about six years; remove all infected plants and replant with disease-free plants.

4) All replants, whether in young or mature orchards should be removed if found to be stubborn affected and again replanted with disease-free plants.

5) If possible, produce nursery trees in a greenhouse or in areas where there is a low incidence of stubborn.

6) Where nursery trees are produced in high incidence or risk areas, substantial borders of vector attrac-

tive trap plants immune to *S. citri* can be used. These trap plants can also be treated with insecticide to kill the attracted leafhoppers.

7)   Avoid the use of cover crops that are highly susceptible to *S. citri* in orchards less than six years old in areas subject to high populations of stubborn disease vectors.

REFERENCES

Anonymous. 1970. Citrus production problems in the Near East and North Africa. U.N. Devel. Program, FAO Rep. TA 2870, pp. 95.

Bower, J.W. and Calavan, E.C. 1974. Antibiotic sensitivity *in vitro* of the mycoplasma-like organism associated with citrus stubborn disease. Phytopathology 64: 346-349.

Bruno, A. 1964. Citrus virus symptoms (psorosis, xyloporosis, exocortis, and stubborn disease) in Sardinia. Nature (London) 202, 932.

Calavan, E.C. 1969. Investigations of stubborn disease in California: indexing, effects on growth and production, and evidence for virus strains. Proc. 1st, Int. Citrus Sym., 1968, Vol. III, pp. 1403-1412.

Calavan, E.C. and Carpenter, J.B. 1965. Stubborn disease of citrus retards growth, impairs quality, and decreases yields. Calif. Citrogr. 50: 86-87, 96, 98-99.

Calavan, E.C. and Christiansen, D.W., 1961. Stunting and chlorosis induced in young-line citrus plants by inoculations from navel orange trees having symptoms of stubborn disease. Proc. 2nd., Conf. Int. Organ.

Citrus Virol., 1960. pp. 69-76.

Calavan, E.C. and Christiansen, D.W., 1966. Effects of stubborn disease on various varieties of citrus trees. Israel J. Bot. 15: 121-132.

Calavan, E.C., Roistacher, C.N., and Christiansen, D.W. 1968. Distribution of stubborn disease virus within trees of *Citrus sinensis* and *C. paradisi* at different seasons. Proc. 4th Conf. Int. Organ. Citrus Virol., 1966, pp. 145-153.

Calavan, E.C., Harjung, M.K., Fudl-Allah, A.E.-S., and Bowyer, J.W. 1974. Natural incidence of stubborn in field-grown citrus seedlings and budlines. Proc. 6th Conf. Int. Organ. Citrus Virol. 1972, pp. 16-19.

Calavan, E.C., Kaloostian, G.H., Oldfield, G.N., and Blue, R.L. 1976. Stubborn disease found in London rocket. Citrograph 61, 389-390.

Calavan, E.C., Mather, S.M., and McEachern, E.H. 1978. Registration, certification, and indexing of citrus trees. In: The Citrus Industry (W. Reuther, E.C. Calavan, and G.E. Carman, eds.), Vol. IV, pp. 185-222. University of California, Div. of Agricultural Sciences, Berkeley.

Calavan, E.C., Kaloostian, G.H., Oldfield, G.N., Nauer, E.M., and Gumpf, D.J. 1979. Natural spread of *Spiroplasma citri* by insect vectors and its implications for control of stubborn disease of citrus. Proc. Int. Soc. Citric., 1977, pp. 900-902.

Carpenter, J.B. 1959. Present status of some investigations on stubborn disease of citrus in the United States. In: Citrus Virus Diseases (J.M. Walace, ed.) pp. 101-107.

Univ. Calif. Div. Agric. Sci., Berkeley.

Cartia, G. and Calavan, E.C. 1974. Further evidence on the natural spread of citrus stubborn disease. Riv. Patol Veg. 10: 219-224.

Chapot. H. 1959. First studies on the stubborn disease of citrus in some Mediterranean countries. In: Citrus Virus Diseases (J.M. Wallace, ed.), pp. 109-117. Univ. Calif. Div. Agric. Sci., Berkeley.

Chapot, H. and Delucchi, V.L. 1964. Maladies, Troubles et Ravageurs des Agrumes au Maroc, pp. 359. Institut National de la Recherche Agronmique, Rabat.

Fawcett, H.S., Perry, J.C., and Johnston, J.C. 1944. The stubborn disease of citrus. Calif. Citrogr. 29: 146-147.

Fudl-Allah, A.E.-S.A., Calavan, E.C., and Igwegbe, E.C.K. 1972. Culture of a mycoplasma-like organism associated with stubborn disease of citrus. Phytopathology 62: 729-731.

Gumpf, D.J. and Calavan, E.C. 1981. Stubborn disease of citrus, In: Mycoplasma Diseases of Trees and Shrubs. (K. Maramorosch and S.P. Raychaudhuri, eds.), pp. 97-134. Academic Press, New York.

Igwegbe, E.C.K. and Calavan, E.C. 1970. Occurrence of mycoplasma-like bodies in phloem of stubborn-infected citrus seedlings. Phytopathology 60: 1525-1526.

Kaloostian, G.H., Oldfield, G.N., Pierce, H.D., and Calavan, E.C. 1976b. *Spiroplasma citri* and its transmission to citrus and other plants by leafhoppers. In: Leafhopper vectors and Plant Disease Agents

(K. Maramorosch and K.F. Harris, eds.), pp.
447-450. Academic Press, New York.

Lee, I.-M., Cartia, G., Calavan, E.C., and
Kaloostian, G.H. 1973. Citrus stubborn
disease organism cultured from beet leafhopper. Calif. Agr. 27 (11) : 14-15.

Liu, Hsing-Yeh, Gumpf, G.J., Oldfield, G.N.,
and Calavan, E.C. 1983. Transmission of
Spiroplasma citri by *Circulifer tenellus*.. Phytopathology 73: 582-585.

Nour-Eldin, F. 1967. A tumor-inducing agent associated with citrus trees infected with sa-
fargali (stubborn) disease in the United
Arab Republic. Phytopathology 57, 108-113.

Oldfield, G.N. and Kaloostian, G.H. 1979. Vec-
tors and host range of the citrus stubborn
disease pathogen, *Spiroplasma citri*.
Proc., R.O.C. -- United States Cooperative
Science Seminar on Mycoplasma Diseases of
Plants, 1978, pp. 119-124.

Rana, G.L., Kaloostian, G.H., Oldfield, G.N.,
Granett, A.L., Calavan, E. C., Pierce, H.
D., Lee, I.-M., and Gumpf, D.J. 1975. Acquisition of *Spiroplasma citri* through
membranes of homopterous insects. Phytopathology 65: 1143-1145.

Reichert, Il, and Perlberger, J. 1931. Little
leaf disease of citrus trees and its
causes. Hadar 4, 193-194.

Saglio, P., Laflèche, D., Bonissol, C., and
Bové, J.J. 1971. Isolement et culture *in
vitro* des mycoplasmes associees au
"Stubborn" des agrumes et leur observation
au microscrope electronique. C.R. Acad.
Sci. (Paris) Ser. D272, 1387-1390.

Saglio, P., L'hospital, M., Laflèche, D., Dupont, G., Bové, J.M., Tully, J.G., and Freundt, E.A., 1973. *Spiroplasma citri* gen. and sp. n.: a mycoplasma-like organism associated with "stubborn" disease of citrus. Int. J. Syst. Bacteriol. 23: 191-204.

Scaramuzzi, G. and Catara, A. 1968. Studies on sour orange stem-pitting in Sicily. Proc. 4th Conf. Int. Organ. Citrus Virol. 1967, pp. 201-205.

Vogel, R., and Bové, J.M. 1974. Studies of stubborn disease in Corsica. Proc. 6th Conf. Int. Organ. Citrus Virol. 1972, pp. 23.25.

Wallace, J.M. 1978. Virus and virus-like diseases. In: The Citrus Industry (W. Reuther, E.C. Calavan and G.E. Carman, eds.), Vol. IV, pp. 67-184. University of California Div. of Agri. Sci. Berkeley.

18

LITTLE LEAF DISEASE OF EGGPLANT

by

D.K. Mitra

Nuclear Research Laboratory

Indian Agricultural Research Institute

New Delhi - 110012, India

INTRODUCTION

Little leaf, an MLO-caused disease of egg-plant, *solanum melongena* causes consider-able economic losses (Varma, *et al.*, 1969). The infected plants are characterized by stunting, excessive branching, reduction of the leaves and phyllody of flowers. In early infection, no fruiting takes place and in late infection, fruits are formed but they are deformed and the seeds are shrivelled. The disease is transmitted by grafting and through the leafhopper vector *Hishimonas phycitis*. The following is a brief resume of the studies undertaken on this disease and the results obtained.

SURVEY

The disease is very serious and causes

qualitative as well as quantitative losses. Survey showed that out of 75 varieties and selections, none escaped the disease. Maximum infection was up to 33% in seasons under study. The commercial variety Pusa Purple Long (PPL) and the selection T were found to be highly susceptible. Once the plants were infected at early stages, no flowering took place and hence no fruiting and the loss was total.

METABOLIC STUDIES

These studies were carried out from 1976 to 1984 in order to obtain pertinent information about the effect of the disease on host physiology and to evolve some biochemical parameters to detect the disease at a very early stage of infection when symptoms were not pronounced, so that screening could be done much quicker (Mitra *et al.*, 1976, 1978, 1981; Mitra and Majumdar, 1977; Mitra and Sengupta, 1980; Mitra and Bhattacharya, 1984; Mitra and Gupta, 1984). Biochemical studies showed that the respiration rate increased as a result of little leaf infection while carbohydrate content and activities of peroxidase, catalase, polyphenol oxidase and nitrate reductase were reduced. However, indole acetic acid oxidase activity increased.

Effect of little leaf disease on photosynthetic activity of the host plant was studied with the help of C. Photosynthetic activity was reduced as a result of little leaf infection. The reduction was 30% in case of mildly infected plants but 91% in case of severly infected plants. However, no reduction in chlorophyll content was observed. This is a unique situation considering the fact that reduction in photosynthetic activity is generally accompanied by reduction in chlorophyll content. No difference in the number of stomata and opening of guard cells was observed in healthy and diseased plants.

Effect of little leaf disease on metabolic processes influencing carbon balance in eggplant was studied using radioisotopes and other methods. Considerable reduction in translocation of photosynthates was observed in leaf tissues of diseased plants. Sugar content, soluble protein content and total free amino acid content were found to be much higher in leaf tissues from diseased plants than in those from healthy ones. While protease activity was much higher in diseased tissues, - amylase activity was much lower. These findings are interesting in view of increased respiration and reduced photosynthetic activity in eggplants affected by little leaf disease.

Little leaf disease affected eggplants appeared to develop resistance to translocation of photosynthates and probably develop mechanisms to convert proteins into amino acids and amino acids to sugars through various metabolic processes.

The results obtained so far with regard to metabolic studies are significant because pertinent information with regard to pathophysiology has been obtained. Diseases caused by MLOs have been recognized only recently and not much information on pathophysiology is available with regard to MLO diseases in general and little leaf disease of eggplant in particular.

TISSUE CULTURE STUDIES

Callus tissues were grown from healthy and from little leaf infected eggplant tissues in modified Murashige and Skoog's medium. Callus growth was faster and callus initiation earlier in cultures from healthy tissues compared to those from infected ones. Differentiation was observed only in calli developed from diseased tissues. Plantlets were regenerated from callus tissues developed from diseased tissues. The same callus produced two different types of plantlets--one apparently healthy and the other apparently diseased showing all characteristic symptoms of little leaf. Production of plantlets from diseased tissues has not been reported earlier in the case of MLO diseases and this work might be helpful in conducting exacting experiments under controlled conditions permitting maintenance of MLO cultures in plant tissues grown *in vitro*.

Disease-free eggplant explants have been developed from stem tissue segments of the variety Pusa Kranti and several plantlets were regenerated. The plantlets were transferred from medium to pots in a greenhouse and they did not develop the disease symptoms for a long time. These plants and control plants were grafted with scions from severly infected plants. The regenerated plants did not exhibit any symptom of little leaf even 2-1/2 months after grafting whereas control plants showed symptoms within two to three weeks after grafting. The possible origin of these resistant plants could be from somaclonal resistant variant cells.

DETECTION STUDIES

A simple technique has been developed, following Namba, *et al.*, 1981, for rapid detection of viral and MLO infection in plants by fluorescent microscopy without using any fluorochrome or any other stain. Thin, free hand sections of leaf petioles and tender stems of a large number of virus and MLO infected plants including little affected eggplants as well as healthy plants were mounted in glycerine and examined under a fluorescent microscope with a specific exciter-barrier filter combination. While the phloem tissues in healthy leaf petioles exhibited a bright orange red to brick red autofluorescence, those from infected plants showed yellowish green to green autofluorescence. In phloem tissues from infected plants, the intensity of yellowish green to green autofluorescence was directly proportional to disease development. This characteristic yellowish green to green autofluorescence was observed much earlier than the appearance of visible symptoms. This technique might prove useful as a rapid and simple method for detection and diagnosis for viral and MLO infections of plants and also for screening against those diseases particularly in case of perennials where disease symptoms appear after a long time. This could also be useful in plant quarantine for rapid testing of viral and MLO infection in plants.

REFERENCES

Mitra, D.K., Majumdar, M. and Farkas, G.L., 1976. Effects of little leaf disease on respiration and enzymatic activities of *Solanum melangena*. Phytopath. Z., 86, p 310-313.

Mitra, D.K. and Majumdar, M., 1977. Changes in total phenolic content and polyphenol oxidase activity in brinjal (*Solanum melangena*) as a result of little leaf infection. Proc. Indian Nat. Sci. Acad., 43(b), pp. 228-231.

Mitra, D.K., Majumdar, M. and Sengupta, U.K., 1978. Effect of some viral and mycoplasmal diseases on the nitrate reductase activity in some Solanaceaeous host plants. GEOBIOS. 5, pp. 74-75.

Mitra, D.K. and Sengupta, U.K., 1980. Influence of little leaf disease on photosynthesis and chlorophyl content in brinjal leaf tissues. Annals. of Agric. Research, 1, pp. 108-111.

Mitra, D.K., Gupta, N. and Bhaskaran, S., 1981. Development of plantlets from brinjal stem tissues infected with little leaf, a mycoplasma disease. Indian J. Exptl. Biol., 19(12), pp. 1177-1178.

Mitra, D.K. and Bhattacharya, A., 1984. Influence of little leaf disease on carbon balance in infected eggplant (*Solanum melangena*) leaf tissues. Indian Phytopath., 37(3), pp. 453-457.

Mitra, D.K. and Gupta, N., 1984. A rapid technique for diagnosis of diseases by fluorescent microscopy. Science and Culture, 50, pp. 119-120.

Namba, S., Yamashita, S., Doi, Y. and Yora, K., 1981. Direct fluorescence detection method for diagnosing yellows-type virus diseases and mycoplasma diseases of plants. Ann. Phytopath. Soc. Japan, 47: pp. 258-263.

Varma, A., Chenulu, V.V., Raychaudhuri, S.P., Prakash, N. and Rao, P.S., 1969. Mycoplasma-like bodies in tissues infected with sandal spike and brinjal little leaf. Indian Phytopath., 22: 289.

19

EARLIEST HISTORICAL RECORD OF A TREE MYCOPLASMA DISEASE: BENEFICIAL EFFECT OF MYCOPLASMA-LIKE ORGANISMS ON PEONIES

by

M.Q. Wang
Department of Biology,
Fudan University, Shanghai, P.R. China

and

Karl Maramorosch
Department of Entomology and Economic Zoology
Rutgers, The State University
New Brunswick, New Jersey 08903, USA

INTRODUCTION

Tree peonies (*Poeonia suffruticosa Andr.*) were appreciated in China for many centuries as the most attractive traditional flowers. Expressions such as "natural peony color", "natural peony fragrance" and "peony, the king of all flowers" clearly describe the importance given to peonies and the emotional feelings evoked by these plants. Tree peonies were linked with "good fortune and peace" and they were symbols of "thriving and prosperity" (Yu Heng, 1980). Records of tree peony cultivation date back to the 5th century A.D. and they were grown continuously during 1400 years in different areas of China (Ou-yong Xiu, 1031; Yu Heng. 1980). Most likely peony cultivation started even earlier than the 5th century.

During the Song dynasty (960-1227) the so-called "Yao-yellow kind" peony was hailed as the best-shaped and most beautiful variety, along with the purple peony known under the name of "Queen of Wei-purple". At that time, a new variety appeared and it became known as "Ou-Jia-bi". The flowers of this new variety were green and the plants bearing such flowers soon became the most precious of all tree peonies. Every year a special tribute was being paid to the imperial court by offering these green flowers, since they were considered the most beautiful and more valuable than the "Yao-yellow" variety. Numerous statements in the literature indicate that these green flowers were ever since considered more valuable and rare than the red or yellow ones.

During the Quin dynasty (1662-1714) a "Synopsis of Sources of Classified Books" was published. In the part dealing with flowers, there appeared a statement that during the Song dynasty, a flower grower of Shan-fu succeeded in obtaining thousands of different types of tree peonies. This was achieved by modifications in the cultivation methods. During the same Song dynasty period, Ou-yong Xiu (1031) reported that "during a period of forty years a hundred variations of tree peonies were produced." At that time, as well as during the Ming and Qui dynasties, the green flowering tree peonies were always considered the most desirable and widely acclaimed of all varieties.

In May of 1962, a green flower blossomed unexpectedly at the tip of a branch of a white peony tree named "Kun-Shan-Ye-Guang", grown in the imperial garden of the Palace Museum in Beijing. In 1964 at Heze, Shan-dong province, flower growers obtained a new type of tree peony by natural hybridization and artificial selection (personal communication). They named it "Green fragrant ball". Apparently, it was not too difficult to obtain different varieties of tree peonies by means of hybridization, grafting, and natural or induced mutation.

From 1980 until 1986, one of us (M.Q.W.) studied collections of green types of tree peonies from Luo-uong, Zheng-ghou, Henan province, from Heze, Shan-dong province, and at the Botanical Garden in Shanghai. Tree peonies producing green flowers were usually less vigorous (Fig. 1). Green flowers differed from other colored flowers, having modified anthers and pistils and producing no seed. In fact, some of these plants were propagated for several centuries despite the induced sterility (Yu, 1980). Samples were collected from these plants and specimens prepared for electron microscopy examination. After fixation in glutaraldehyde, postfixing in osmium tetroxide, dehydration in graded ethanol and embedding in an epoxy resin, ultrathin sections were cut, stained with uranyl magnesium acetate, and examined with an electron microscope. The results (Fig. 2) indicated that the chlorotic, green flowers were associated with mycoplasma-like organisms. We concluded, therefore, that the change in the color of flowers which made them more desirable centuries ago was in reality the result of an MLO infection.

In addition to green peony flowers, the same method was used to examine green flowers of several other ornamental plants growing in China and it was found that their chlorosis was also directly or indirectly caused by MLO infection. This was established in green flowers of garden balsam (*Impoticus balsamina*), Chinese azalea (*Rhododendron excellsus*), Chinese monthly rose (*Rosa chinensis*), narcisus (*Narcissus tazella* var *orientalis*), *Prunus mume*, and certain orchids. Apparently, these flowering plants derived their unusual, often very attractive and desirable characteristics from MLO infections (Maramorosch, 1974; Maramorosch and Raychaudhuri, 1981).

"Broken" tulips, very popular and precious in the 17th century in Holland (Dubos, 1959) were usually quoted as the earliest record of a plant virus disease. In a way, the virus infection could be considered beneficial to tulips, since it enhanced their appeal and caused an actual craze, the "tulipomania" of that period. Due to the lack of adequate scientific knowledge and proper techniques, the demonstration that tulip breaking was due to virus infection could only be made in the 20th century (Cayley, 1928; McKay *et al.*, 1929. In 1981 Inouye and Osaki called attention to a poem composed by the Japanese Empress Koken in the year 752, and included it in the classic 8th century anthology "Manyoshi". The yellow-leaf disease of *Eupatorium chinense* var *simplicifolium*, described in this poem, constitutes the earliest record of a plant virus disease. The cause is a small isometric gemini virus, transmitted by the whitefly vector *Bemisia tabaci* (Inouye and Osaki, 1981. The discovery of MLO diseases by Doi *et al.*, (1967), Ishie *et al.*, (1967) and Nasu *et al.*, (1967) revealed that mulberry dwarf and several other diseases, earlier believed to be caused by viruses, were actually caused by mycoplasma-like organisms. Until the present study, it was believed that the earliest description of a tree mycoplasma disease was that of a mulberry dwarf,described during the Bunsei period, 1818-1831 in Japan (Ishijima and Ishiie, 1981). As shown by us now, the descriptions of green flowering tree peonies preceded that of broken tulips by at least 500 years and that of mulberry dwarf by 700 years. The famous and precious trees were known and hailed around 1119-1127 in Chinese writings but it took nearly 860 years before the cause of greening and the beneficial effect of MLOs on peonies could be documented. The MLO infection of peonies may, or may not affect the long-term survival of the trees. As in the case of broken tulips that were propagated and appreciated by many generations, the green flowering peonies were also selected and propagated for centuries, because of their strange beauty and unusual color.

The finding that green flowers of tree peonies in China are caused by MLO infection is interesting not only from a historical point of view. One wonders whether it is proper to call this MLO infection a "disease". It made the plants more desirable and their very long survival indicates that, in this instance, the MLO infection is actually beneficial to tree peony plants, as they were cared for more than ordinary peonies. Some MLO diseases kill their hosts, others cause prolonged chronic diseases, and still others, like the green peony infection, can be considered beneficial. However, all seem to induce sterility, as has been pointed out by Kunkel (1954) many years ago. In this respect "beneficial" MLO infections differ from most algal symbioses (Goff, 1983) or mycorrhiza-tree interactions (Mikola, 1980).

ACKNOWLEDGEMENT

The authors acknowledge the technical assistance of Chen Zhang-yi, Cai Tong-run, and Sun Guang-rong.

REFERENCES

Cayley, D.M. 1928. Breaking in tulips. Ann. Appl. Biol. 15: 529, 1928.

Doi, Y., Terenaka, M., Yora, K., and Asuyama, H. 1967. Mycoplasma- or PLT group-like microorganisms found in the phloem elements of plants infected with mulberry dwarf, potato witches' broom, aster yellows, or Paulownia witches' broom. Ann. Phytopathol. Soc. Japan 33, 259-299.

Dubos, Rene J. 1959. Tulipomania and the benevolent virus. In: Perspectives in Virology (I), M. Pollard, Ed. John Wiley: 291-299.

Goff, L.J. ed. 1983. Algal symbiosis, a continuum of infection strategies. 1980.

Inouye T., and Osaki, T., 1981. The earliest record of a possible plant virus disease. Abstract P22/05, 5th Int. Cong. Virology, Strasbourg: 238.

Ishijima, T. and Ishiie, T. 1981. Mulberry dwarf: first tree mycoplasma disease. In: Mycoplasma Diseases of Trees and Shrubs. K. Maramorosch and S.P. Raychaudhuri, eds. Academic Press: 147-184.

Ishiie, T., Doi, Y., Yora, K., and Asuyama, H. 1967. Suppressive effects of antibiotics of tetracycline group on symptom development of mulberry dwarf disease. Ann.Phytopathol. Soc. Japan 33, 267-275.

Kunkel, L.O., 1954. Maintenance of yellows-type viruses in plant and insect reservoirs. p. 150-163. In: The Dynamics of Virus and Rickettsial Infections. F.W. Hartman, F.L. Horsfall, and J.G. Kidd, eds. Blakiston, New York.

McKay, M.B., Brierly, P., and Dykstra, T.P. 1929. Tulip "breaking" is proved to be caused by mosaic infection. U.S. Depart. Agric. Year Book 1928: 596-597.

Maramorosch, K. 1974. Mycoplasmas and rickettsiae in relation to plant diseases. Annu. Review of Microbiology 28: 301-324.

Maramorosch, K. and Raychaudhuri, S.P., eds. 1981. Mycoplasma Diseases of Trees and Shrubs. Academic Press. p. 362.

Mikola, P. ed. 1980. Tropical mycorrhiza research.

Nasu, S., Sugiura, M., Wakimoto, T., and Iida, T. 1967. On the etiologic agent of rice yellow dwarf disease. Ann. Phytopathol. Society, Japan 33, 343-344.

Xiu, Ou-yong. 1031. Lou-yong tree peony: a record.

Yu, Heng. 1980. Heze tree peony.

Figure 1. Green flower on a tree peony at Botanical Garden, Shanghai.

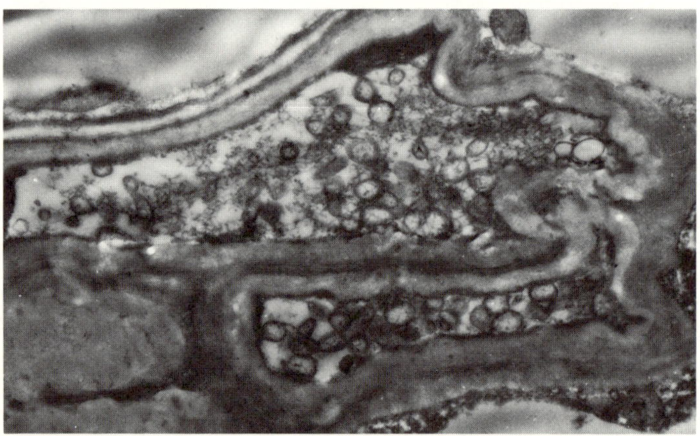

Figure 2. Electron micrograph of mycoplasma-like organisms within a sieve element of a green flower-producing tree peony.

Part IV: Chemotherapy and Other Methods of Control

20

THE PATHWAYS OF CHEMOTHERAPEUTANTS

IN THE CONTROL OF TREE DISEASES

by

V.M.G. Nair

College of Environmental Sciences

University of Wisconsin-Green Bay

Green Bay, Wisconsin, U.S.A.

INTRODUCTION

The role of systemic chemotherapeutants in the control of tree diseases is still considered as a new field of study. The major interest in the control of various plant pathogens by chemotherapy took place only in the 1940s when Florey and Chain revived Fleming's discovery of the antibiotic nature of *Penicillium* followed by the isolation of a powerful bacteriostatic antibiotic "Penicillin" and the demonstration of its therapeutic value in the control of bacterial infections in warm blooded animals and humans.

Early developments of systemic chemotherapeutants were directed towards the control of vascular wilt diseases such as "Oak Wilt" of *Quercus sp.* caused by *Ceratocystis fagacearum;* Dutch elm disease of *Ulmus sp* by

C. ulmi. Unlike surface infection, the vascular pathogens multiplied and spread through the deep-seated xylem tissues without showing any external symptoms of the disease during the course of their initial infection and spread of the pathogen within the host, and, as such, this group of vascular pathogens offered one of the greatest challenges for the development of systemic chemotherapeutants in their control.

With an initial start in the control of "wilt diseases" by systemic chemotherapeutants additional tree diseases caused by fungi controlled at varying degrees of success include, bleeding canker of maples by *Phytophthora cactorum;* decline of live oak by Cephalosporium diospyri; blister rust of white pine by *Cronartium ribicola;* Fusiform rust of slash pine by *Cronartium fusiforme;* needle cast of scots pine by *Lophiodermium pinastrai;* anthracnose of sycamore by *Gnomonia plantani;* Hypoxylon canker of aspin by *Hypoxylon pruniatum* and chestnut blight by *Endothea parasitica.*

The isolation of *Spiroplasma citri,* the causal agent of citrus stubborn disease, (Saglio *et al.,* 1971, 1973) is considered one of the major discoveries in plant pathology. In addition, researchers have now realized the role of these unique organisms in the development of plant epidemics of economically important crops, often eliminating the entire host populations, as in the case of *Paulownia* witches' broom which eliminated this superior timber from Japan and Taiwan; citrus likubin which eliminated poncan orange in Taiwan; lethal yellowing of coconut and other palm species in Florida, Jamaica and western Africa; phloem necrosis of elms in the United States of America and spike disease of sandalwood trees in India. Recently, Maramorosch (1985) has speculated on the impact of mycoplasmal diseases on the political fate of countries such as Cuba, where lethal yellowing of coconut palms destroyed all palms around 1900, forcing small landowners to sell their land to owners of huge estates. They, in turn, planted sugar cane and became immensely wealthy "sugar cane barons." Later the impoverished masses rapidly altered the economic structure of the island, providing suitable conditions for the political changes in Cuba. Similarly, the collapse of the Mayan civilization appears due to the loss of corn, *Zea mays*, the most important staple food of Mayan Indians. Corn stunt, caused by spiroplasma, significantly reduced not only the staple food, but also the lowland population of Yucatan peninsula, facilitating Spanish occupation (Maramorosch, 1985).

Research workers have now characterized spiroplasma organisms and placed them in the class Mollicutes. Spiroplasmas lack cell walls but have a pliable unit-type membrane and ribosomes consisting of RNA; a coil of DNA constitutes their genome. They require lipids and lipid precursors for membrane synthesis and exogenous plant or animal-derived cholesterol or related sterols for their growth. They have only a poorly defined nuclear area devoid of a nuclear membrane and are highly sensitive to tetracycline antibiotics (Maramorosch *et al.*, 1970; Saglio and Whitcomb, 1979).

Chemotherapeutic controls of tree diseases caused by MLOs and spiroplasmas are being developed with the additional knowledge about the organisms (a) growth and transport in phloem fluids which transport photosynthetic and metabolic products to plant parts and (b) growth in hemolymph, the fluid of arthropods in the coelomic cavity which covers the internal organs and transport metabolites from various tissues. The contol of diseases to varying degrees, caused by mycoplasma-spiroplasma include lethal yellowing of coconut palms, lethal decline of date palm, pear decline, peach yellows leaf roll, peach X disease, apple proliferation, buck skin of cherry, paulownia witches' broom, jujube witches' broom, mullberry dwarf, American elm yellows phloem necrosis, mulberry dwarf, ash yellows, and spike disease of sandalwood.

Researchers in the development of chemo-therapeutic control of tree diseases have now developed various (a) <u>injection systems</u> for dicots and monocots which include chisel-cone method trunk infusion, Mauget capsule trunk injection, gravity flow infusion, low and high pressure trunk injection, root injection, root collar and butress root injection as well as foliar sprays and (b) <u>systemic chemotherapeu-tants</u> such as benomyl (methyl-1 butylcarbamoyl) benzimidazol-2Y1 carbamate Lignasan <u>R</u> BLP (methyl benzimidazol-2Y1 carbamate H PO) Arbo-tect <u>R</u> (2- thiazol-4Y1 benzimidazol - H2PO2), carbendazim-HCL, thiabendazole, oxycarboxin, thiophanate methyl, triarimol, triforine, 2,3, 5, triiodobenzoic acid. dithane M-45, 2,3,6 trichloro-phenylacetate, 6 benzyl aminopurine, cycloheximide, Actidione, Tetracycline, oxy-tetracycline-HCL (OTC) (Terramycin <u>R</u>), dimethyl-chlortetracycline, Sulfaphenazole, Sulfaguani-dine, Amphotericin-B, Digitonin polyanithiole sulfate, 2 dimethyl-1-7-bensophonemyl methyl carbamate, s-methyl N-thio acetimidate.

Investigators working in the field of chemo-therapeutic control of tree diseases, in gene-ral, and systemic chemotherapeutants, in parti-cular, are fully aware of the complexity of the internal conducting tissues of various host spe-cies, and are trying to understand the pathways involved in the translocation of systemic chemo-therapeutants and their metabolites along with the precise nature of pathogens entry, multipli-cation and spread during initial and advanced stages of pathogenesis. In addition, various related aspects and phenomena should be taken into consideration while developing systemic chemotherapeutants as they influence the ef-ficacy of these compounds in the control of tree diseases. These include (a) solubility, phytotoxicity and immunity of these compounds, (b) efficient methods of injection, (c) path-ways and various aspects involved in the trans-

location of systemic chemotherapeutants and
their derivatives, (d) role of root grafts as
pathways of chemotherapeutants and pathogens,
(e) the effects of systemic chemotherapeutants
and their derivatives, during various stages of
host parasite interaction and host physiology
during seasonal variations.

SOLUBILITY AND PHYTOTOXICITY AND IMMUNITY OF CHEMOTHERAPEUTANTS

Good water solubility is essential for any
systemics such as systemic fungicides, insecticides, antibiotics, etc. for their uptake and
systemic distribution inside tree species. The
early use of HC1-acetone, glacial acidic acid,
etc. for solubilizing benomyl proved to be highly phytotoxic in the control of oak wilt and
Dutch elm disease (Smalley *et al.*, 1973;
Nair and Kuntz 1975). At present, good water
soluble systemics include fungicides, fungi-
cidal derivatives and antibiotics alone or in
combination such as benomyl (Methyl-(1-butylcar-
bamoyl)benzimidazol-2Y1 carbamate); chlortetracycline (Aureomycin) dimethylchlortetracycline
HCL (Ledermycin) oxytetracycline-HCL (OTC) (Terramycin R); and a mixture of Ledermycin + benomyl or terramycin + benomyl (McCoy, 1973;
McCoy *et al.*, 1976; Raychaudhuri, 1977,
1984; Raychaudhuri and Rishi, 1985; Boligala,
C.R. 1985; Nair, 1981; Carroll, 1982.
The toxicity of various chemotherapeutants
alone and/or their enzymatic, non-enzymatic and
possible conjugative derivatives or other metabolites formed in soil, plants and animals is
of prime importance and concern to researchers
as well as manufacturers. The biologically
active compounds should be very safe to man and
all the members of the ecological chain start-
ing with microorganisms. Knowledge about the
pathways of penetration of these compounds into
various organisms is essential, including res-
piratory passages and those absorbed dermally
and through membranes.

The growing list of safe systemic chemotherapeutants and some of their derivatives
include: <u>heterocyclic compounds</u>: such as sub-
stituted benzimidazoles benomyl, thiabendazole,
fuberidazole, metabolic product of benomyl ben-
zimidazole-2-Yl carbamate, triforine, Dode-
morph, chloreneb; <u>carboxylic acid anilides</u>: carboxin, oxycarboxin, pyran and furan analogues;
<u>organophosphorous compounds</u>: triamiphos, phos-
phoric esters of pyrazolopyrimidines, kitazin;
<u>ammonium compounds</u>: G.N. 64 (Dimethyl diethyl
derivatives of benzyl ammonium chloride); <u>anti-
biotics</u>: streptomycin, tetracycline, oxytetra-
cyclin, Kasugamycin.

Plant pathologists working with these various antibiotics and other systemic chemotherapeutants in the field should give careful consideration to the chances of various pathogenic
bacteria to acquire resistance or immunity to
these antibiotics, as these antibiotics are
also used in the control of human and animal
diseases Use of systemic chemotherapeutants
during the development of fruits and other
plant products should be given special attention due to the chances of biologically active
resudues of these compounds to become concentrated in storage tissues of the hosts, including fruits.

METHODS OF APPLICATION

Introduction of systemic chemotherapeutants
into the stelar tissue (a) of actively conducting xylem vessels (current annual rings of
the temperate tree species such as *Quercus
sp.* and more than one growth ring produced in
one year in the case of some tropical trees -
such as *Santalum album*) (b) of actively conducting sieve tube elements is important so
that the chemotherapeutants and/or their metabolites could reach the xylem habitat or phloem
habitat pathogens during the course of pathogenesis.

Among the various injection systems (a)

pressurized lower trunk-butress root injection system as well as (b) gravity flow by injecting solubilized systemics through drilled holes fitted with plastic "Ts" into the actively conducting xylem vessels proved to me the most effective system when large volumes of materials are required (Smalley, 1978; Kondo, 1979; Gkinis and Stennis, 1980; Nair, 1981; Nair, 1985.

Earl Swazy Company's (Winnipeg, Canada) superior plastic injector head with a leak proof fitting of plastic "Ts" into smooth drilled holes (into the xylem with a sharp drilled bit) which in turn connected to the tygon tube manifold and this manifold to the pressurized (2.1 kg/cm) stainless steel "soft drink tanks" proved to be one of the cheapest and best injector systems.

Based on the bioassay results (using *Ceratocystis fagacearum* and *Arthrobacter globiformis* as test organisms) of 3200 oaks and 300 American elms injected with various chemotherapeutants for the control of oak wilt and Dutch elm disease respectively and 22 citrus trees and 28 sandalwood trees treated with oxytetracycline (Kerala State, India) Nair and Kuntz have come to the conclusion that lower trunk area at root collar and root flair injections are superior for good distribution of systemic chemotherapeutants, probably due to the wide anastamosis of the tissues at the root collar of these dicots (Nair, 1981, 1982, 1985). In the case of monocot trunks injected with oxytetracycline for the control of lethal yellowing, McCoy (1982) came to the conclusion that the unique vascular system in *Cocos* is responsible for the good distribution of the antibiotic, though it was injected through one injection hole.

PATHWAYS OF TRANSLOCATION

The control of plant pathogens by the use of systemic chemotherapeutants will depend upon our knowledge of (1) the pathways of entering pathogens and their multiplication and spread inside the tissues of the host as well as (2) the available pathways of systemic chemotherapeutants, their derivatives and other metabolic products inside the host species.

In addition, during the introduction of systemic chemotherapeutants, the presence of solutes and the pH of xylem and phloem fluids should be taken into consideration. Xylem sap is acidic (pH 5.2-6.5) and phloem sap is alkaline (pH 8.0-8.4) in various tree species (Pate 1976). Phosphorous as phosphorycholine, sulphur as cysteine glutathionin and nitrogenous constituents are among the major constituents of the xylem sap of various tree species; the xylem sap also contains amino acids, sugars, organic acids, auxins, and cytokinins (Pate 1976). Phloem sap contains high amounts of solids especially carbohydrates (50-300 mg. ml-1) while xylem sap contains a lesser amount of solids (1-20 mg. ml-1)(Pate 1976). A word of caution is required here because the water soluble systemic chemotherapeutants and their derivatives might interact with these compounds present in the xylem and phloem lowering or increasing the therapeutic effect, which could lead to inefficient control of a disease or toxicity to host species, respectively.

The systemic chemotherapeutants are transported mainly in three stages (a) diffusion of systemic compounds into free space within living tissues, (b) entry and transport through non-living parts of the cell such as through the lumen of dead xylem vessels and tracheids called "passive apoplastic movement", (c) entry and transport through the living parts of the cell, i.e., such as the phloem sieve tubes and other cells of phloem called "active symplastic movement."

At present, the systemic chemotherapeutants are being used in the control of vascular pathogens of the xylem and phloem habitat. Xylem vessels comprise major passive conduction and the xylem sap contains far less nutrients when compared to phloem. The xylem pathogens such as *Ceratocystis fagacearum* (oak wilt) have adapted to this low nutrient rich sap in their initial multiplication and spread within the tree. In fact, endoconidia can multiply even in distilled water. In comparison, the phloem tissue translocates photosynthetic products which are rich in nutrients. The hemolymph of arthropods is also very rich in nutrients and may even contain complete metabolic systems of important pathways such as glycolysis (Saglio and Whitcomb, 1979). The microorganisms adapted to one of these fluids may be able to grow in the other and vice versa in which case microorganisms such as rickettsia-like bacteria, MLOs, spiroplasmas and viruses which inhabit the phloem or hemolymph of arthropods could be controlled by developing systemic chemotherapeutants similar in their action both in the phloem and hemolymph fluids.

A. Pathways of Entry Into Free Space and Translocation

Systemic chemotherapeutants sprayed on the aerial parts of any plant tissue including leaves enter the protoplast of epidermal cells via ectodesmata and water solubility is considered essential for this transport. Entry of solutes into free space is by diffusion. Since diffusion time for the chemicals from water in the free space is limited due to external and internal conditions such as dryness, permiability, concentration gradient, etc., the use of wetting agents such as glycerol polethyleneglycols is suggested (Goodman, 1962, Goodman and Dowler, 1958, Holly, 1956) and, in fact, the uptake of streptomycin, growth regulators, etc. is increased by the use of wetting agents.
In the second stage of uptake, the removal of compounds usually takes place from the aqueous phase in the free space of immobile ions in cell walls or protoplasm (Crowdy, 1972). In the final stage, removal of compounds would

occur metabolically. However, compounds entering into the cell vacuole are under metabolic control, hence, once they entered the vacuole, they may not be available for translocation (Crowdy, 1972).

Systemic chemotherapeutants injected into roots or taken up by roots are under the influence of the transpiration stream which prevents the accumulation in free space. However, compounds removed from the transpiration stream could get accumulated (stored) in dead or living root tissues which later could be released, as in the case of antibiotics and sulphonamides (Tanton and Crowdy, 1972)

B. Pathways of Apoplastic Movement and Transport

Apoplastic movement and transport is a passive movement through the trunk to the leaves in transpiration stream via dead xylem vessels and tracheids. In the roots, this pathway could start from the root hair cell walls and intercellular spaces of the cortex but get stopped at the suberised casparian strip of the endodermis when the solutes enter the endodermis cells where they pass through the protoplast and later enter into *xylem* vessels (Lauchli, 1976; Crowdy, 1972).

The xylem sap is acidic and it is reported that acidic and neutral substances flow with ease in the xylem vessels while basic ones get adsorbed to the negative charges of xylem vessels wall. This condition has been reported for quarterly ammonium compounds (Edington and Diamond, 1964), amino acids (Hill-Cuttingham and Lloyd-Jones, 1968), basic dyes (Charles, 1953), and basic antibiotics (Crowdy, 1972). However, addition of cations such as calcium could reverse this passive adsorption.

Since transpiration pull is one of the main forces for the apoplastic transport, stomatal closure alone and accumulation of abscisic acid along with stomatal closure, would, no doubt, interfere in apoplastic transport. Stomatal closure has been reported in over 390 white ash (*Fraxinus americana*) trees infected with ash yellows (MLOs) (Matteoni and Sinclair, 1985); *in Ulmus americana* and *U. rubra* affected with elm yellows (MLO) phloem necrosis); in green leaves of palms affected with lethal yellowing (MLO) (Basham and Eskafi, 1980) and in choke cherries (*Prunus virginiana*) infected with X disease. In addition, sudden reduction of annual rings (reduction of xylem vessels), reduction of the major pathway of apoplastic movement by 50% and leaf abscission due to the accumulation of abscisic acid has been reported in trees (Matteoni and Sinclair, 1985; Milborrow, 1974; Setter *et al.*, 1980). All of these conditions, no doubt, will contribute to the poor uptake and translocation of systemic chemotherapeutants, reducing the force of the transpiration pull and the numbers of xylem vessels that constitute the main pathway of the apoplastic translocation.

In the dicot trees, introduction of the systemic chemotherapeutants deep (a) into the non-conducting annual rings (in temperate zone trees - *Quercus sp.*) or (b) growth rings (in tropical trees - *Santalum sp.*) could cause product failure because these non-conducting areas might act as reservoirs and may not distribute these chemotherapeutants (Nair,1981, 1982, 1985).

In the case of palms and other monocot species, the stems contain many vascular bundles and these vascular tissues remain functional during the entire life span of the plants (Zimmermann, 1976). The stem of palms contains many short vessels with numerous cross connections for efficient apoplastic transport, even in the event of stem injury (Zimmermann, 1976). In palms, the unique branching of leaf traces when they leave the stem and the number of hidden leaf primodia (almost same number as the visible leaves) also help in the efficient apoplastic transport. This might have been the reason for McCoy's (1982) statement concerning the "unique vascular system" responsible for the excellent distribution of oxytetracycline through a single injection site.

The systemic chemotherapeutants and antibiotics introduced into the xylem vessels (apoplastic transport) have exhibited varying degrees of control of vascular fungal pathogens (*Ceratocystis fagacearum, C. ulmi, Cephalosporium diospyri, Verticillium dahliae, Fusarium oxysporum f. sp. perniclosum*) with Benomyl (Nair, 1981). Benomyl and its fungicidally active metabolic product MBC methyl ester of 2-benzimidazole carbamic acid and thiabendazole are transported through the apoplast (Crowdy, 1972). In addition, the oxathin derivatives, carboxin and oxycarboxin; thiabendazole, tridemorph are also transported through the apoplastic system (Nair, 1981).

The MLO and spiroplasma diseases controlled to varying degrees include sandalwood spike (Raychaudhuri and Gosh, 1984) lethal yellowing (McCoy, 1973, 1979, 1982) peach X-disease (Nyland and Sachs, 1974; Sinha, R.C., 1979) pear decline (Nyland and Moller, 1973) with OTC, jujube witches' broom (La Yong-Joon, 1982) with dimethylchlorotetracycline (Lidermycin), chlorotetracycline (Aureomycin) or a combination of benlate and Lidermycin.

C. Pathways of Symplastic Movement and Transport

Symplastic movement and translocation of systemic chemotherapeutants and their metabolites, nutrients, etc. take place in the living parts of the cell through living protoplasmic connections through plasmodesmata between uni-multiseriate ray cells, xylem parenchyma, phloem parenchyma, companion cells, sieve elements (sieve tube members in angiosperms) albuminous cells and sieve cells (in gymnosperms). This movement requires metabolic energy. In addition, specialized transfer cells with ingrowth of their walls, in order to increase the plasmalemma area, are closely associated with symplastic transport (Gunning and Pate, 1969; Gunning and Robarods, 1976).

Systemic chemotherapeutants and their metabolites transported through the symplastic system play an important role in the control of phloem restricted parasites such as viruses, MLOs and spiroplasmas. For symplastic transport, the systemic chemotherapeutants and/or their metabolites must cross the plasmalemma and enter and co-exist with the living protoplasm of the cells. The symplastic translocation system in the trees is a continuous pathway with the phloem tissues for long distance transport.

Our knowledge is still limited with regard to the phloem transport (symplastic). However, the "transfer cells" play a very important part in the transfer of molecules and solutes from apoplast to symplast and from symplast to apoplast (Gunning and Pate, 1969).

The transfer cells with wall ingrowth show loosely arranged microfibrils with interconnected channels between the microfibrils which allow the passage of solutes and water. Plasmalemma is the main barrier limiting the transport from apoplast to symplast. Hence, this ingrowth helps the transport. In addition, the wall ingrowth increases the surface area of plasmalemma thus enabling transmembrane transport. Gunning and Pate (1974) clearly explained the hypothesis of "electron osmosis" and "standing gradient osmotic flow" where the presence of solute pumps inside plasmalemma or next to it are capable of driving solutes unndirectionally from, or to, apoplast (wall ingrowth area), to or from the surrounding symplast. This, in turn, would produce an osmotic gradient which would allow water to flow. The presence of ATPases near the wall ingrowths supports this active transport of solutes across the membrane.

During phloem loading at the source region, translocated substances selectively and actively enter the sieve tube before translocation (Geiger, 1975).

The electro-osmotic or potassium theory proposed by Spanner (1975) is now regarded as a major force in the symplastic transport (phloem transport). According to Spanner, the extremely thin (1/10th thickness of polythene film) sieve plate is the electro-osmotic membrane which maintains a high level of electrical forces and at the same time reduces the hydraulic resistance. Potassium ions are the major mobile ions in the sap [companion cells could absorb potassium from apoplast and transfer to the sieve tube through the plasmodesmata] and the sieve pores of the sieve plate occluded with P-protein offer fine channels to initiate electro-osmotic forces (Spanner, 1975). According to Spanner, the P-protein is dispersed in the lumen or along the walls of the sieve tube at the time of loading of the sugar in the sieve tubes,, initiating the weak Munch flow extending downwards and this longitudenal flow soon reaches the sink region. At this time, the P-protein of the lumen moves down and slowly occludes the sieve plate pores when mobile ion (ATP) will build up above the sieve plate, triggering the metabolically energized differential uptake of KL + ion, thus reinforcing the Munch flow by electro-osmosis.

Bidirectional transport in the symplastic system (phloem) has been reported by many workers (Wardlaw, 1968; Crafts and Crisp, 1971; Eschrich, 1975) based on various phloem mobile tracers such as trailium, potassium fluorescein or isotopes P and C. According to Eschrich (1975) bidirectional means (a) two-way transport in different bundles, (b) in two different sieve tubes of the same bundle simultaneously, (c) transport towards the apex followed by reversing later. Moreover, traces or metabolites moving in two different sieve tubes in opposite directions could get exchanged if lateral sieve plates are present between them (homodromous loop path) which also would allow mixing of metabolites in one sieve tube without interferring with the directions of flow (Eschrich, 1967, 1975).

Bidirectional transport in the plasmodesmata has been reported under various situations such as the photosynthetic cells of ferns supplied by absorbed nutrients or the photosynthetic cells supplying the rhizoids with assimilates by the bidirectional transport through plasmodesmata (Fraser and Smith, 1974). Gunning *et al.*, (1974) reported similar bidirectional pathways in legume species which included phloem derived sugars carried to the bacteria containing cells of root nodules and later nitrates, the products of nitrogen fixation, back to the endodermis. However, the plasmodesmatal path containing a continuous endoplasmic reticulum might restrict the flow of large molecules in their preservation of genetic integrity (Kaufmann, 1976).

Symplastic pathways in the root of various species need further investigations. Systemic chemotherapeutants introduced into the root must take an apoplastic and/or symplastic pathway. In this case, the ions get blocked at the casparian strips once they reach them by apoplastic pathways of the cell wall (Lauchli, 1976). So, the only possibility for the ions to pass through the endodermal cells is either (a) by the uptake through plasmalemma at the outer tangential wall, after the transport through the cytoplasm is released through the plasmalemma at the inner tangential wall into the apoplast of the stele or (b) by a symplastic transport through plasmodesmata in the inner or outer tangential wall (Lauchlin, 1976). The electron microscopic investigations on the pathways of Cl transport after precipitation of Cl with organic Ag+ salts have revealed that the suberin lamella in the endodermis blocked apoplastic transport of Cl to the stelar tissue (Lauchlin, 1976). According to Gunning and Robards (1976), the endoplasmic reticulum appears continuous through the plasmodesmata for the intercellular symplastic transport and the ion transport is through this symplasmic pathway which extends to the xylem parenchyma surrounding the vessels which, in turn, release the ions to the xylem vessels. Ion secretory process has been reported in the xylem parenchyma cells and such ions could be released to the xylem vessels through the pits on the vessel walls (Lauchli, 1976). However, metabolic inhibitors such as abscisic acid cytokinins could inhibit ion transport.

Obligate parasites such as viruses or rusts, which do not possess their own metabolic system, must co-exist with the living protoplasm of the host through which symplastic pathways of systemic compounds and their derivatives take place.

MLOs or spiroplasma pathogens of tree diseases are also closely associated with the sieve element, multiply within sieve tubes, and the largest MLOs (1.2 in diameter) reported, could pass without squeezing through the functioning sieve element pores (McCoy, 1979). However, the sieve pore diameter may play an important role in host susceptibility or resistance because tree species with small sieve pore sizes may restrict the passage of mycoplasmas.

Obligate parasites, as well as MLOs and spiroplasmas which are closely associated with the symplastic system of host species, can only be affected if the systemic chemotherapeutic compounds, their derivatives or antibiotics possess phloem mobility (McCoy, 1979). In this respect, longevity of sieve elements should be taken into consideration. In the secondary phloem of dicots, depending upon the species, the longevity and conduction of sieve elements vary, i.e., in *Vitis*, *Ulmus*, *Quercus* and *Acer* for two years and in *Telia* one to five years (Lamoureux, 1975). In several conifers, most sieve elements function for one growing season though some overwinter and function during the early spring of the following year (Lamoureux, 1975) while in perennial monocots metaphloem sieve elements of palms function up to 50 years (Parthasarathy, 1975).

Extreme caution should be taken while introducing chemotherapeutants into sieve elements for symplastic transport. Any type of rupture to the sieve elements could cause the release of hydrostatic pressure, spreading to other areas of the sieve elements. Consequently, sieve pores could become plugged with P-protein, mitochondria, starch grains, etc., which in turn could cause poor distribution of systemic chemotherapeutants. However, antibiotics (Chloromycin, Terramycin, Lidermycin) dyes (crystal violet, acridine orange) some growth regulators (indoleacetic acid, indolebutyric acid, gibberellic acid, Kinetin) have inhibited or controlled various virus and plant mycoplasma diseases (Sinha, 1978, Raychaudhuri, 1977, Plavsic *et al.,*, 1987). Introduction of systemic compounds into the xylem vessels is suggested with the hope that these compounds would reach the symplastic system, without rupturing any part of the sieve elements, passing through the continuous network of protoplasmic connections between xylem and phloem tissues.

ROOT GRAFTS AS PATHWAYS OF CHEMOTHERAPEUTANTS AND PATHOGENS

In a mature forest stand of a single species, in trees of a single species planted closely under urban conditions as street or park trees (oak, maple, linden, birch, elm, etc.) or trees of the same species planted in rows under orchard conditions, a single tree should be regarded as a member of the community of trees because the surrounding trees usually establish connecting root grafts. Such intraspecific root grafts are most common in oaks, elms and citrus, while interspecific root grafts are very rare (Nair and Kuntz, 1975). However, many phanerogamic parasites such as *Striga* or *Balanophora* are known to establish interspecific root grafts. *Santa-*

lum album, a root parasite, is capable of establishing root grafts with more than 100 species of unrelated hosts. These root grafts usually support interdependency between the species acting as pipelines transporting nutrients between them. Unfortunately, these root grafts can also transport vascular pathogens such as *Ceratocystis fagacearum*, the cause of oat wilt, *C. ulmi*, the cause of Dutch elm disease, etc.

The role of inter and intraspecific root grafts in the transport of MLOs is being investigated. The pathogens of apple proliferation and pear decline were transmitted to healthy apple and pear trees by root grafting (Schaper and Seemuller, 1982.) In addition, Braun and Sinclair (1976) have observed the presence of MLOs in the functional sieve tubes of roots of apple injected with apple proliferation, pear with pear decline, and elm with elm necrosis. Moreover, overwintering of MLOs in the roots was reported for mulberry dwarf (Tahama, 1975), X-disease of peach (Stoddard, 1947), phloem necrosis of elm (Braun and Sinclair, 1976; Rosenberger and Jones,1977) and for peach decline and apple proliferation (Schaper and Seemuller, 1982)

Systemic chemotherapeutants such as benomyl when pressure injected at the root collar area of a non-dominant oak, were transported to a large extent to three neighboring large oaks (revealed by bioassay) which might have been due to a greater transpiration pull exhibited by these large dominant oaks (Nair, 1985). In addition, under orchard conditions, seven citrus trees also revealed the presence of benomyl and/or its fungicidally active MBC (by bioassay) 48 hours after pressure injection of two citrus trees at the root collar area, which suggested the presence of root grafts and the apoplastic pathways (Nair, 1985).

Release of antibiotics and sulphonamides, once accumulated or stored in the living or dead root tissues could contribute to the control of pathogens transported through the apoplastic or symplastic pathways of roots at a later stage. (Tanton and Crowdy, 1972).

THE EFFECTS OF SYSTEMIC CHEMOTHERAPEUTANTS AND THEIR DERIVATIVES ON HOST PARASITE INTERACTIONS AND HOST PHYSIOLOGY

Systemic chemotherapeutants or their metabolites transported through the apoplastic or symplastic system could kill, or inactivate the multiplication of pathogenic agents in the same systems or induce the host to alter the physiology so that the host becomes resistant to the invading pathogen.

Production of fungitoxic molecules by host cells was reported by Dekker and Oort (1964) after procaine and kinetin injection, which in turn stopped the haustorial establishment of powdery mildews. Similarly, conversion of benomyl into its metabolically active fungitoxic molecule MBC (methyl 2-benzimidazole carbamate) by hydrolysis has been reported in various plant species including American elm (Clemons and Sisler, 1969; Smalley, 1981; Nair, 1981) in the control of Dutch elm disease.

Induced production of phytoalexins, neutralization of fungal toxins, and inactivation or regulation of enzyme production by the use of various chemotherapeutic agents such as phenyl alanine, 6-carboxymethyl and 2, 4-D; feroulic acid or refianic acid has been practiced in various disease control strategies (Davis and Diamond, 1956; Grossman, 1958; Tamari *et al.*, 1966; Matta *et al.*, 1967). Liao and Chen (1981) have suggested that the reason for the high antispiroplasmal activity of the antibiotics Neomycin, Tobramycin, and Gentamycin may be the presence of deoxystreptamine moiety.

In the case of vascular wilt diseases such as oat wilt, elimination of the site of multiplication and distribution of the pathogen (in xylem vessels) was achieved by Nair and Kuntz (1976a; 1976b). In pin oaks *Q. ellipsoi-*

dalis (highly susceptible to oak wilt,) this was achieved by injecting the growth regulator sodium-2, 3, 6-trichlorophenyl acetate (TCPA) through the apoplastic system. The xylem vessel production was stopped not only the year of treatment but also in the two succeeding years (Ven *et al.*, 1968; Nair and Kuntz 1976b; Nair 1981). The wood formed after treatment mainly contained xylem tracheids, parenchyma and fibers.

In the case of MLOs, sieve pore diameter as a factor of resistance has been suggested by McCoy (1979). MLOs overwintering in the root may not be able to reinvade the above ground parts if one could change the sieve pore diameter of newly forming phloem by the use of growth regulators so as to make them smaller than the overwintering MLOs.

Seasonal variations on host physiology, variations in environmental conditions, such as high and low wind velocity, cloud cover, humidity, temperature, etc. will have a profound influence on the conducting patterns (diffusion, apoplastic and symplastic patterns) in various tree species. The time required for the infusion of oxytetracycline-HCl through three holes into peach trees (infected with MLO-X disease) was 14 days in May, and 1-3 days in September. This difference may be due to the presence of partial spring wood vessels in May and complete spring and summer wood vessels by September in their conducting pattern (Rosenberg and Jones, 1977). Similarly bioassay results of over 2,000 elm trees and 300 oaks treated with Lignasan and Arbotec have revealed that the uptake and distribution of these systemic chemotherapeutants was enhanced after full leaf expansion along with low humidity (30-35%), high temperature (80 F-86 F), and high wind velocity (20-30 MPH) during, and just prior to, treatment (Nair 1982, 1985).

CONCLUSION

Chemotherapeutic control of tree diseases by the use of systemic chemotherapeutants and/ or their derivatives will depend upon our knowledge of the internal conduction pathways of various tree species as well as the pathways of pathogens during various stages of pathogenesis. Unfortunately, our knowledge is still sparce in both of these aspects with regard to various pathogens. In vascular wilt diseases (oak wilt, Dutch elm disease, etc.) investigators agree in most instances about the pathways of the pathogens. In MLO diseases, confusion still exists with regard to the exact multiplication sites of the pathogens, their modes of systemic spread through xylem vessels, their capacity to produce antibiotic resistant strains, their dependency on plant or insect sterols for growth, and other aspects, all of which require additional research. Further research is also required to understand the exact pathways of transport of chemotherapeutants in various tree species.

One should take advantage of our knowledge of various modes of pathogen entry, spread and existance in host tree species and insect vectors in order to develop new systemic chemotherapeutants. In the case of MLOs which live in the nutrient, rich phloem sap of infected plant species as well as in the hemolymph of insect vectors (Shanta, P.R., 1985) and occurrence of combined virus and MLO infections in plants and vectors (Banttari, 1985), development of effective systemic chemotherapeutants translocated through the symplastic system of plants and hemolymph could offer efficient control. Moreover, obligate parasites have to co-exist with the living cells of the host due to their lack of independent metabolic systems. Therefore, systemics transported through symplastic pathways will have a profound influence in the control of obligate parasites such as rusts and viruses. Future development of systemic chemotherapeu should include the development of (a) antibiotics with the presence of deoxystreptamine moiety against spiroplasmas (Liao and Chen, 1981); (b) systemics to reduce the availability of plant or animal derived sterols required by MLOs; (c) growth regulators, to reduce the sieve pore size in newly forming sieve elements in order to block the reinvasion of above ground parts by mycoplasmas overwintering in roots (McCoy, 1979) or to eliminate the site of multiplication of vascular wilt pathogens (Nair, 1981); (d) systemics to induce the host to produce phytoalexins and nuteralize fungal toxins as well as inactive enzyme production by the pathogens: (e) microbial and viral insecticides (Maramorosch, 1985), including antibiotic producing living cells of *Pseudomonas syringae* (Strobel and Lanier, 1981) all of which could improve chemotherapeutic control of tree diseases.

REFERENCES

Banttari, E.E., 1987. Interactions of plant viruses and mollicutes in plants and insect vectors. In: Mycoplasma diseases of crops: Basic and applied aspects. Maramorosch, K. and Raychaudhuri, S.P., (eds.) Rutgers University Press.

Basham, H.G. and Eskafi, F.M., 1980. Xylem transport in palms with lethal yellowing. Proc. 4 Meeting of Intl. Council on Lethal Yellowing. Inst. of Food and Agri. Sciences, Univ. of Florida, Fort Lauderdale, Florida, p. 8.

Boligala, C. Raju, 1984. Chemotherapy of mycoplasma diseases of fruit trees. Proceedings 7th IUFRO conference on mycoplasma diseases. Agriculture-Forestry Center, University of Alberta, Edmonton, Alberta, Canada, p. 14.

Braun, E.J. and Sinclair, W.A., 1976. Histopathology of phloem necrosis in *Ulmus americana*. Phytopathology 66: 598-607.

Carroll, V.J., 1982. Role of industry in developing a product for control of mycoplasmalike plant diseases. Rev. Infec. Dis. 4 (Suppl.), 162-166.

Charles, A., 1953. Uptake of dys into cut leaves. Nature, London, 171: 435-436.

Clemons, G.P. and Sisler, H.D., 1969. Formation of a fungitoxic derivative from benlate. Phytopathology 59: 705-706.

Crafts, A.S. and Crisp, C.E., 1971. Phloem transport in plants. Freeman and Co., pp. 481.

Crowdy, S.H., 1972. Translocation in Systemic Fungicides. R.W. Marsh, (ed.), Longman, London, pp. 92-115.

Davis, D. and Dimond, A.E., 1956. Site of disease resistance induced by plant growth regulators in tomato. Phytopathology 46: 551-552.

Dekker, J. and Oort, A.J.P., 1964. Mode of action of 6-azauracil against powdery mildew. Phytopathology 54: 815-818.

Edgington, L.V. and Dimond, A.E., 1964. The effect of adsorption of organic cations to plant tissue on their use as systemic fungicides. Phytopathology 54: 1193-1197.

Eschrich, W., 1967. Bidirectional Transport, *Planta* 44: 37-74.

Eschrich, W., 1975. Bidirectional transport in phloem transport. Arnold, S., Dainty, J., Gorham, P.R., Srivastava, L.M., and Swanson

C.A., (eds.) Plennum Press, N.Y. and London, p. 401-416.

Fraser, T.W., Smith, D.L., 1974. Young gametophytes of the fern *Polypodium vulgare L..* An Ultrastructural Study. Protoplasma 82: 19-32.

Geiger, D.R., 1975. Phloem loading in transport in plants I. Phloem Transport Encyclopedia of Plant Physiology, New Series. Vol. I. Zimmerman, M.H. and Milburn, J.A. (eds.) p. 395-431.

Gkinis, A. and Stennis, M., 1980. How to inject elms with systemic fungicides. Minn. Agr. Ext. Folder, 504, pp. 5.

Goodman, R.N., and Dowler, 1958. The absorption of streptomycin as influenced by growth regulators and humecants. Pl. Dis. Rept. 42: 122.

Goodman, R.N., 1962. The impact of antibiotics upon plant disease control. Adv. Pest Control Res. 5: 1-46.

Grossman, 1958. Uber die Hemmung pektolytischer enzyme von *Fusarium oxysporum f. lycopersici* durch Gerbstoffe. Naturwiss. 45: 113.

Gunning, B.E.S. and Pate, J.S., 1969. Transfer cells. Plant cells with wall ingrowths specialized in relation to short distance transport of sollutes. Their occurrence, structure and development. Protoplasma 68: 107-109.

Gunning, B.E.S., Pate, J.S., Minchin, F.R. and Marks, I., 1974. Qualitative aspects of transfer cell structure in relation to vein loading in leaves and solute transport in legume nodules. Symp. Soc. Biol. 28:87-126.

Gunning, B.E.S. and Pate, J.S., 1974. In: Dynamic Aspects of Plant Ultrastructure, Robards, A.W., (ed.) McGraw-Hill, Maidenhead (UD) p. 441-480.

Gunning, B.E.S. and Robards, A.W., 1976. Plasmodesmata and symplastic transport. In: Transport and Transfer Processes in Plants. Wardlane, I.F. and Passioura, J.B. (eds.), Academic Press, New York, pp. 15-41.

Hill-Cottingham, D.G. and Lloyd-Jones, C.P., 1968. Relative mobility of some nitrogenous compounds in the xylem of apple shoots. Nature, London. 220: 389-390.

Holly, K., 1956. Penetration of chlorinated phenoxyacitic acids into leaves. Ann. Appl. Biol. 44: 195.

Kaufmann, M.R., 1976. Water transport through plants: current perspectives in transport and transfer processes in plants. Wardlaw, I.F. and Passioura, J.B., (eds.) Academic Press, N.Y. p. 313-328.

Kondo, L.S., 1979. Root flair and root injection techniques. Proc. of Symposium on Systemic Chemical Treatments in Tree Culture, Kielbaso, J.J. (ed.), Braun-Brumfield, Inc., Ann Arbor, MI, pp. 133-140.

Lamourex, C.H., 1975. Phloem tissue in angiosperms and gymnosperms. In: Phloem Transport. Aranoff, S., Dainty, J., Gorham, P. R., Srivastava, L.M. and Swanson, C.A. (eds.) pp. 1-20, Plenum, New York.

La, Yong-Joon, 1982. Mycoplasma diseases of woody plants and their control in Korea. Proceedings 5th IUFRO Conference on Mycoplasma Diseases, University of Wisconsin-Green Bay, WI, USA, p. 12.

Lauchli, a., 1976. Symplastic Transport and ion Release to the Xylem. In: Transport and Transfer Process in Plants, Wardlaw, I.F. and Passioura, J.B., (eds.), Academic Press, N.Y., p. 101-102.

Lauchli, A., 1976. Apoplastic transport in tissues. In: Encyclopedia of Plant Physiology, New Series, Vol. 2., Transport in Plants 2, Part B. Luttge, V., and Pitman, M.G. (eds.), Springer, Berlin-Heildebert, N.Y., p. 3-34.

Laio, C.H. and Chen, T.A., 1981. *In vitro* susceptibility and resistance of spiroplasma to antibiotics. Phytopathology 71: 442-445.

Maramorosch, K., Grandados, R.R. and Hirumi, H., 1970. Mycoplasma diseases of plants and insects. Advances Virus Research 16: 135-193.

Maramorosch, K., 1985. Non-chemical control of mycoplasma diseases. Proceedings, Indo-U.S. Workshop on Mycoplasma Diseases of Crops: Basic and Applied Aspects. Forest Inst., Dehran Dun, India, p. 21.

Maramorosch, K., 1985. Spiroplasma, vectors, and history. Proceedings Indo-U.S. Workshop on Mycoplasma Diseases of Crops: Basic and Applied Aspects. Forest Inst., Dehran Dun, India, p. 21.

Matta, A., Gentile, I.A. and Giai, I., 1967. Variazioni del contenuto in feloli solubilitibili indotte del *Fusarium oxysporum f. lycopersici* in piante di pomodoro susceptibili e resistenti. Ann. Fac. Sci. Agri. Della Univ. Degli Studi di Torino 4: 17.

Matteoni, J.A. and Sinclair, W.a., 1985. Role

of the mycoplasmal disease, ash yellows in decline of white ash in New York State. Phytopathology 75: 355-360.

McCoy, R.E., 1973. Effect of various antibiotics on development of lethal yellowing in coconut palms. Proc. Florida State Horto. Soc., 86: 503-506.

McCoy, R.E., 1979. Mycoplasma and yellows diseases In: The Mycoplasma, Vol. III, Whitecomb, R.F. and Tully, I.G., (eds.) Academic Press, Inc., New York, pp. 229-264.

McCoy, R.E., 1982. Use of tetracycline antibiotics to control yellows diseases. Plant Dis. 66: 539-542.

McCoy, R.E., Carroll, V.J., Poucher, C.P., and Gwin, G.H., 1976. Field control of coconut lethal yellowing with oxytetracycline hydrochloride. Phytopathology 66: 1038-1042.

McNabb, H.S., Heybroek, H.M. and MacDonald, W. L., 1970. Anatomical factors in resistance to Dutch Elm Disease. Meth. J. Pl. Pathol. 76: 196.

Milborrow, B.V., 1974. The chemistry and physiology of abscisic acid. Annu. Rev. Plant. Physiol. 25, 259-307.

Nair, V.M.G., 1981. Control of tree diseases by chemotherapy. In: Mycoplasma Diseases of Trees and Shrubs. Maramorosch, K. and Raychaudhuri, S.P., (eds.) Academic Press, Inc., New York, pp. 325-349.

Nair, V.M.G., 1982. Injection systems of systemic chemotherapeutants and electron microscopic investigations of the pathways of systemic compounds. Proceedings, 5th IUFRO Conference on Mycoplasma Diseases, University of Wisconsin-Green Bay, Wisconsin, U.S.A., p. 9.

Nair, V.M.G., 1985. Electron microscopic investigations in the translocation of systemic chemotherapeutants in woody dicots and monocots and improved injection systems of systemic compounds. Proceedings. Indo-U.S. Workshop in Mycoplasma Diseases of Crops: Basic and Applied Aspects. Forest Res. Inst. Dehran Dun, India, pp. 25-26.

Nair, V.M.G. and Kuntz, J.E., 1975. Recent advances in oak wilt research. In: Advances in mycology and plant pathology. Raychaudhuri, S. P., Anupam Varma, Bhargava, K.S. and Mehrota, B.H., (eds.) M/S Harish Kumar at Sagar Printers, 8 Jantar Mantar Road, New Delhi, India, pp. 231-240.

Nair, V.M.G. and Kuntz, J.E., 1976a. Physiological control of vascular wilt diseases of trees. XVI IUFRO World Congress (IUFRO/ FAO), Norway, pp. 3.

Nair, V.M.G. and Kuntz, J.E., 1976b. Anatomical changes in xylem tissues associated with oak wilt. XVI IUFRO World Congress (IUFRO/ FAO), Norway, pp. 2.

Nyland, G. and Moller, W.J., 1973. Plant Disease Reporter 57: 634-637.

Nyland, G. and Sachs, R.M., 1974. Collog. Inst. Natl. Sante Rech. Med. 33: 235-242.

Parthasarathy, M.V., 1975. Discussion of phloem tissues in angiosperms and gymnosperms. In: Phloem Transport. Arnoff, S., Dainty, J., Gorham, P.R., Srivastava, L.M. and Swanson, C.A., Plenum Press, N.Y.., p. 21-31.

Pate, J.S., 1976. Nutrients and metabolites of fluids recovered from xylem and phloem:

Significance in relation to long distance transport in plants. In: Transport and Transfer processes in plants. Wardlaw, I.F., and Passioura, J.B., (eds.) Academic Press, N.Y., p. 253-281.

Plavsic, B., Krivokapic, K., and Eric, Z. 1987. Kinetin treatment of stolbur-diseased plants. In: Mycoplasma diseases of crops: Basic and applied aspects. Maramorosch, K. and Raychaudhuri, S.P. (eds.) Rutgers University Press.

Raychaudhuri, S.P. and Rishi, N., 1985. Chemotherapy--Basic approaches of research. Proceedings, Indo-U.S. Workshop on Mycoplasma Diseases of Crops: Basic and Applied Aspects, Forest Inst. Dehran Dun, India, pp. 22-23.

Raychaudhuri, S.P., 1977. A manual of virus diseases of tropical plants. MacMillan Company of India, Ltd. New Delhi, India. pp. 299.

Raychaudhuri, S.P. and Ghosh, S.K., 1984. Possible chemotherapeutic control of sandal spike disease in India. Proceedings, IUFRO 7th World Conf. on Mycoplasma Diseases, Agri-Forestry Center, Univ. of Alberta. Edmonton, Alberta, Canada, p. 15.

Rosenberger, D.A. and Jones, A.L., 1977. Seasonal variation in the infectivity of inoculum from X-diseased peach and chokecherry plants. Plant Dis. Reptr. 671, 1022-1024.

Saglio, P. and Whitcomb, R.F., 1979. Diversity of wall-less prokaryotes in plant vascular tissue, fungi and invertebrate animals, The Mycoplasmas III. Plant and Insect mycoplasmas. Whitcomb, R.F. and Tully, J.G., (eds.) Academic Press, N.Y., p. 1-31.

Saglio, P., Lafleche, D., Bonissol, C., and Bove, J.J., 1971. Isolement et culture *in vitro* des mycoplasmes associees au Stubborn des agrumes et leur observation au microscope electronique. C.R. Acad. Sc. (Paris) Ser D272, 1387-1390.

Saglio, P., L'hospital, M., Lafleche, D., Dupont, G., Bove, J.M., Tully, J.G., and Freundt, E.A., 1973. *Spiroplasma citri gen. and sp.*: a mycoplasma-like organism associated with stubborn disease of citrus. Int. J. Syst. Bacteriol. 23: 191-204.

Schaper, V. and Seemuller, E., 1982. Conduction of the phloem and persistance of MLO associated with apple proliferation and pear decline. Phytopathology 72: 736-742.

Setter, T.L., Brown, W.A. and Brenner, M.L., 1980. Effect of obstructed translocation on leaf abscisic acid and associated stomatal closure and photosynthesis decline. Plant Physiol. 65: 1111-1115.

Shanta Prem Rau, 1985. Insect vectors of plant mycoplasma diseases. Proceedings, Indo-U.S. Workshop on Mycoplasma Diseases of Crops: Basic and Applied Aspects, Forest Inst. Dehra Dun, India, P. 18.

Sinha, R.C., 1978. Chemotherapy of mycoplasmal plant diseases. In: The Mycoplasmas III. Plant and Insect Mycoplasmas, Whitcomb and Tully, J.G., (eds.) Academic Press, N.Y. p. 310-334.

Sinha, R.C., 1979. Chemotherapy of mycoplasmal plant diseases. In: The Mycoplasmas III Plant and Insect Mycoplasmas, Whitcomb and Tully, J.G., (eds.) Academic Press, N.Y. p. 309-335.

Smalley, E.B., 1978. Systemic chemical treatments for protection and therapy. Dutch Elm Disease: Perspectives after 60 years. Search 8, 34-39.

Smalley, E.B., 1981. Chemotherapy of vascular wilt diseases of trees. University of Wisconsin, Madison, Wisconsin, Special Paper, pp. 12.

Smalley, E.B., Meyers, C.J., Johnson, R.N., Fluke, B.C. and Vieau, R., 1973. Benomyl for practical control of Dutch elm disease. Phytopathology 63: 1239-1252.

Spanner, D.C., 1975. Electro-osmotic flow. In: Transport in Plants I. Phloem Transport. Zimmermann, M.H. and Milburn, J.A., (eds.) Springer-Verlay, Berlin Heidelberg, N.Y., p. 301-327.

Stoddard, E.M., 1947. The X-disease of peach and its chemotherapy. Conn. Agri. Exp. Stn. Bull., (New Haven), 506. pp. 19.

Tahama, Y., 1975. Electron microscopic examination on mycoplasma-like organisms in the overwintered mulberry trees. Bull. Hiroshima Agric. Coll. 5: 151-158.

Tanton, T.W. and Crowdy, S.H., 1972. Water pathways in higher plants II, J. Exp. Bot. 23: 600.

Tamari, K., Ogasawara, N., Kaji, J. and Togashi, K., 1966. On the effect of a piricular-indetoxifying substance, ferulic acid, on tissue resistance of rice plants to blast fungal infection. Ann. Phytopathol. Soc. Japan 32: 186.

Venn, K.I., Nair, V.M.G. and Kuntz, J.E., 1968. Effects of TCPA on oak sapwood formation and the incidence and development of oak wilt. Phytopathology 58, 1071.

Wardlaw, L.F., 1968. The control and pattern of movement of carbohydrates in plants. Bot. Rev. 34: 79.

Zimmermann, M.H., 1976. The study of vascular patterns in higher plants In: Transport and Transfer Process in Plants. Wardlaw, I.F., and Passioura, J.B., (eds.) Academic Press, N.Y. pp. 221-235.

21

CHEMOTHERAPY

BASIC RESEARCH APPROACHES

by

S.P. Raychaudhuri, A-61, Shivalik Apt.,
Alakananda Kalkaji, New Delhi

Sushma Rishi

Department of Pharmacology
Haryana Agricultural University

Narayan Rishi

Department of Plant Pathology
Haryana Agricultural University
Hisar-125005

INTRODUCTION

Despite wide interest in chemotherapy of plant mycoplasma diseases, no effective mycoplasmacidal agent that can cure the disease completely, has yet been found. Most of the compounds reported to affect plant mycoplasma diseases are mycoplasmastatic. Only in few cases, the complete remission of the disease was observed. Therefore, the basic approach in chemotherapy requires a more thorough screening of drugs that affect plant mycoplasma diseases. This may give a lead in developing a more potent chemotherapeutant. Once an effective mycoplasmacidal agent is obtained, it is essential to establish its efficacy, stability, an effective mode of application, uptake, translocation and persistence in the host so that it can reach the target site of action in optimal concentration.   Such an approach requires a thorough knowledge about the compounds known to inhibit plant mycoplasma diseases, their mode of action, route of application, uptake and translocation, stability and persistence and interaction with hosts.

COMPOUNDS AFFECTING PLANT MYCOPLASMA DISEASES

The era of chemotherapy of plant mycoplasma diseases started in 1967 when Ishiie and his co-workers reported the suppressive effects of tetracycline and chlortetracycline on mulberry dwarf disease.   Afterwards, elaborate work on chemotherapeutic effects of tetracyclines on yellows diseases was carried out by a Japanese group of workers (Asuyama and Iida, 1973). They reported the mycoplasmastatic effects of tetracyclines when applied to plants immediately before or after inoculation with the disease agent.   The preventive effects of tetracyclines against mulberry dwarf infection and eastern X-disease of peach were also reported by Asuyama and Iida (1973) and Sands and Walton (1975).   Several other antibiotics were also reported to affect plant MLO and spiroplasma diseases (Bowyer and Calavan, 1974; Jao and Finland, 1967; Slotkin *et al.*, 1967; Davis, 1981).   Some were found to be effective in suppressing the disease for extended periods but none was able to cure the disease completely.

Several other chemotherapeutants, besides antibiotics, were also found to affect mycoplasma diseases in plants and animals. They include sulpha drugs, organophosphorus compounds, insecticides and few others. For detailed information, readers are referred to other reviews on chemotherapy of plant mycoplasma diseases (Sinha, 1979; Raychaudhuri and Rishi, 1981).

Thus, a large number of chemotherapeutants have been reported for their mycoplasmastatic effects in plants and animals but none was found to be mycoplasmacidal. Only a few reports are available where permanent remission of the disease was obtained. Compound BP 101, developed by Hindustan Antibiotics Ltd., Poona, India was found to cause complete remission of citrus greening and grassy shoot disease of sugar cane (Thirumalachar, 1978). Similarly, nystatin was also reported for complete remission of the fungal Dutch elm disease in the field (Costouis *et al.*, 1978). However, these findings need further exploration.

In order to find a suitable mycoplasmacidal agent, it is essential to study the possible mechanism of action of various chemotherapeutants that are reported to have effects on mycoplasma diseases.

MECHANISM OF ACTION OF ANTIBIOTICS AND OTHER CHEMOTHERAPEUTANTS EFFECTIVE IN CONTROLLING PLANT MYCOPLASMA DISEASES

The objective of a mode of action study on an antimicrobial agent is to relate its biological effects on sensitive cells to the interaction between inhibitor and its biochemical target in the cells. This requires the interaction to be explained in molecular terms. Therefore, besides understanding the chemical structure and properties of inhibitor molecule, a detailed knowledge of the target site is also needed. In spite of considerable advances in molecular biology, the study of macromolecules of target sites in chemical terms is still in its infancy. The molecular biology of mycoplasmas is also not well studied. Difficulty in culturing plant mycoplasma in artificial media poses another problem in carrying out studies *in vitro* and establishing the mode of action of various chemotherapeutants. Only on the basis of information available on cell biology of plant spiroplasmas and other animal mycoplasmas as well as their resemblance in ribosome functions with other prokaryotes, the mechanism of action of various chemotherapeutants that were found to be effective against mycoplasma diseases, can be given.

Drugs Acting on Cell Membrane

The role of sterols in the activity of certain polyene antibiotics on *Mycoplasma gallisepticum* and *Acholeplasma laidlawii* was reported by Weber and Kinsky (1965) and Feingold (1965). *M. gallisepticum* has an absolute requirement of sterol for growth (Lampen *et al.*, 1963) while *A. laidlawii* can grow almost equally well in the presence or absence of exogenous sterol e.g. cholesterol. *Spiroplasma citri* has also been reported to incorporate cholesterol from growth medium into membrane lipid (Freeman *et al.*, 1976).

Amphotericin B methyl ester and Amphotericin B demonstrated anti-micoplasmal activity against *S. citri.*, *M. gallisepticum* and *A. laidlawii* (Goldstein *et al.*, 1976).

Filipin, nystatin and amphotericin B were found to inhibit the growth and cause lysis of *A. laidlawii* cells which had been cultured in the presence of cholesterol. These polyene antibiotics are known to interact with the sterol contents of cell membrane. One molecule of the drug binds with one molecule of sterol in the membrane, thus, altering permeability of cell membrane. This causes leakage of intracellular solutes like K^+/Na$^+$ ions and small molecules up to the size of glucose. The chemical structures of these polyene antibiotics (Fig. 1) show that they are roughly rectangular in shape, one surface being hydrophobic while other having axial hydrophilic groups, is hydrophilic. At one end, the mycosamine sugar group and carboxyl group form a zwitterionic assembly with highly polar properties. Some of these molecules (about 7) assemble together in such a manner that they form a cylinder-like structure having hydrophobic groups outside and hydrophilic groups inside. The sterol molecule is tucked between two such molecules of antibiotic.

In this way, they form pores just like those present in a cell membrane. Such a structure would accommodate itself in the membrane with the polar head group at the surface and the lipophilic groups in the lipid interior. Two such structures end to end would span the membrane and provide observed peameability which can allow the free passage of cations and other small molecules. Such pore formation of polyenes is dependent on the presence of sterols.

Filipin, by complexing with cholesterol, distributes the physical state of membrane lipids which is responsible for the growth of cholesterol requiring mycoplasmas (Bittnan and Rottan, 1976). The physical state of membrane lipids governs its fluidity which affects the optimal activities of membrane such as permeability and transport.

Drugs Affecting Nucleic Acid and Protein Synthesis

The protein synthesizing system of mycoplasmas are sensitive to usual inhibitors of prokaryotic protein synthesis. Some of the antibiotics like tetracyclines, chloramphenicol and erythromycin which specifically inhibit protein synthesis on 70 S ribosomes, also inhibit mycoplasma protein synthesis whereas cycloheximide has no effect (Tourtellotte, 1969; Stanbridge and Doorsen, 1978). The presence of formyl-methionine-tRNA in almost all mycoplasmas, suggest typical prokaryotic initiation of peptide chains (Feldman and Falter, 1971; Walker and Raj Bhandary, 1975, 1978). Messenger-RNA is short-lived in mycoplasmas (Tourtellotte, 1969). Thus, mycoplasma ribosomes appear to function in protein synthesis in a manner similar to other prokaryotes. It is possible, therefore, that various inhibitors of nucleic acid and protein synthesis that affect mycoplasma diseases may act similarly as in other prokaryotes.

(i)   Inhibitors of Nucleic Acid Synthesis

Novobiocin which inhibits mycoplasmas and spiroplasmas (Davis, 1981), is an inhibitor of DNA-gyrase enzyme. This enzyme is found in prokaryotic cells and is essential for DNA-replication and cell division. It couples the hydrolysis of ATP for the introduction of negative superhelices into DNA which are required for the interaction of DNA with specific proteins associated with replication, transcription and recombination. In this way, this antibiotic affects these functions of DNA. It also inhibits the incorporation of precursors into RNA, DNA and protein synthesis (Smith and Davis, 1967). All these effects may be attributed to the ability of antibiotic to complex with MG -ions and thus deprive a number of enzymes of an essential cofactor (Brock, 1967).

(ii)   Inhibitors of Protein Synthesis

A large number of antibiotics that are known to affect protein synthesis in prokaryotic organisms, are also reported to inhibit mycoplasma diseases in plants. These include tetracyclines, chloramphenicol, streptomycin, erythromycin, oleandomycin, tylosin, carbomycin and lincomycin. The mechanism of action of these antibiotics in prokaryotic cells is schematically represented in Fig. 2.
Tetracyclines inhibit the binding of aminoacryl-tRNA into A site on smaller subunit of 70 S and 80 S ribosomes (Cundliffe and McQuillen, 1967; Cundliffe, 1972). The 70 S ribosomes of mycoplasmas are similar to those of prokaryotes and dissociate similarly into 30 S and 50 S particles (Kirk and Morowitz, 1969; Johnson and Horowitz, 1971). It is possible, therefore, that in mycoplasmas also, the tetracyclines inhibit the same step of protein synthesis. The differences in the activities of various analogs of this group of antibiotics are due to the differences in substituents which affect

the stability of the compound. The most active analogue demethyl-chlortetracycline is most stable.

Chloramphenicol inhibits the peptide bond formation by inhibiting the enzyme peptidyl synthetase (Cundliffe and McQuillen, 1967). It also inhibits the movement of ribosomes along with mRNA.

Erythromycin and oleandomycin are macrolide antibiotics and inhibit protein synthesis in bacteria by inhibiting the translocation dependent release of deacylated-tRNA from ribosomes (Taubman *et al.*, 1963; Igarashi *et al.*, 1969). Tylosin and carbomycin cause the breakdown of polyribosomes (Ennis, 1970). Carbomycin also inhibits the binding of peptidyl-tRNA to transferase region of P site on ribosome (Cerna *et al.*, 1969.

Streptomycin and dihydrostreptomycin are aminoglycoside antibiotics. They interact with a specific protein on 30 S subunit, inhibiting the binding of aminoacyl-tRNA and peptidyl-tRNA (Modolell and Davis, 1970). They inhibit the binding of aminoacyl-tRNA to 30 S -mRNA complex, thus causing ambiguity in the translocation system (Kaji *et al.*, 1966).

Lincomycin causes degradation of polyribosomes and also inhibits peptide chain initiation (Cundliffe, 1969).

Various other antibiotics exhibit no effect on mycoplasma diseases: penicillin, bacitracin, neomycin, kanamycin, vancomycin and fusidic acid. They are known to act on cell wall of bacteria. Since mycoplasmas do not possess cell walls, these antibiotics do not affect these organisms.

Slow recovery of diseased plants affected with clover club leaf disease following treatment with penicillin, may be explained on the basis that rickettsia-like organisms have been reported to contain thin cell walls (Windsor and Black, 1973).

Sulpha Drugs

A large number of sulfa drugs, namely sulfanilamide, sulphadiazine, sulfisoxazole, sulfisomidine and sulfaphenazole were tried against plant and insect mycoplasma diseases. Suppression of aster yellow disease by sulpha drugs (Maramorosch, 1970; Klein *et al.*, 1973) and grassy shoot disease by sulfaphenazole (Rishi and Raychaudhuri, 1981) was reported. These drugs are known to inhibit dihydropteroate-synthetase enzyme by competing with the usual substrate p-aminobenzoic acid, thus inhibiting the synthesis of tetrahydrofolate in bacteria (Brown, 1962, 1964, 1967). The lack of this C1 compound affects protein synthesis through deficiency of methionine, glycine and formyl-group of formylmethionine-tRNA and affects nucleic acid synthesis through lack of purines and pyrimidines. The most striking effect is the deprivation of thymine causing thymine-less death of the cell. The importance of thymine was demonstrated in the growth of *S. citri* by studying C-thymine uptake (Townsend *et al.*, 1977).

The combination of sulpha drugs with tetracycline, oxytetracycline or chloramphenicol was reported to suppress the disease symptoms in grassy shoot disease for longer periods (Rishi and Raychaudhuri, 1981). This may be explained on the basis of the additive effects of two types of drugs.

Organophosphates and Insecticides

Various organophosphorus compounds such as O,O-diethyl-O-(2-pyrazinyl)-phosphorothioate, O,O-diethyl-S-phosphorodithioate, and O,O-dimethyl-S-phosphorothioate, were reported in controling tobacco yellow dwarf disease (Paddlick and French,1964). The effects of systemic oxime carbamate insecticide 2,3-dihydro-2-2-dimethyl-7-benzofuranyl-methylcarbamate and S-methyl-N-thioacetimidate on mycoplasma in tobacco fields have been reported (Paddlick *et al.,* 1971). Raychaudhuri *et al.,* (1972) reported the treatment of sandal spike disease with the systemic fungicide benlate, (methyl-1-butylcarbomyl)-2-benzimidazole carbamate. This fungicide was effective when applied to the trunk after removing a portion of the bark. These systemic organophosphorus and oxime carbamate insecticides reduced the incidence of yellow dwarf on tobacco, not only because of vector control, but also by disease remission in tobacco plants. Clemons and Sisler (1969) reported that systemic fungicides, such as benlate, are converted by hydrolysis to fungitoxic molecules of methyl-2-benzimidazole carbamate, which affect the separation of chromosomes during mitosis. In mycoplasma, also, the drug may act in the same way.

Hormone Therapy

Maramorosch (1957) found that stunting of plants infected with corn stunt and aster yellows could be overcome by applying gibberellic acid. However, this hormonal treatment is effective only in overcoming stunting of the plants. Hormonal treatment of stolbur is described by Plavsic *et al.,* in the present volume.

METHODS OF APPLICATION

The degree of therapeutic effectiveness of chemotherapeutic agents against mycoplasma diseases depends on the manner in which the drugs are applied to plants, and the stage of disease development in plants at the time of application. Common methods by which a drug is applied to the plant are foliar spray, root, shoot or leaf immersion, hydroponic culture, and in the case of trees, direct injection of drugs into the trunk or root. The details of these methods, as well as their modifications, were described by Sinha (1979) and Raychaudhuri

and Rishi (1981) in their reviews on chemotherapy of plant mycoplasma diseases.

UPTAKE, TRANSLOCATION AND PERSISTENCE
OF CHERMOTHERAPEUTANTS

An effective uptake and translocation of
systemic chemotherapeutants or their derivatives is essential both in protective and cura-
tive treatments of various host plants against
mycoplasma pathogens. However, proof is
lacking in most cases to confirm whether the
actual injected compound is responsible for the
disease control or its metabolite. The various
steps in the translocation process are
apoplastic movement, symplastic movement and
movement in free space, that is diffusion into
cell walls. In apoplastic movement, the main
bulk of movement is a mass flow from roots to
leaves in transpiration stream via xylem
vessels, but the chemotherapeutant must cross
the plasmolemma and enter the protoplasm. So
the nature and type of xylem wall pitts, pit
membranes, xylem plates etc. will play a roll
in the translocation process. Benomyl and its
fungicidally active metabolite, methyl-2-benzi-
midazole carbamate are transported by the apoplastic movement (Clemons and Sisler, 1969).
Symplastic movement occurs within the living
parts of the cell, protoplast and functional
sieve tubes. Therefore, the chemotherapeutant
or its metabolites must coexist with the living
protoplasm where plasmodesmata play an important role. Our knowledge is sparce with regard
to the movements of chemotherapeutants and
their metabolites in the phloem tissue from the
site of application. Since mycoplasmas are restricted to sieve tubes of host plants, further
investigations in this field will reveal a better understanding of phloem movement and hopefully should aid in the development of systemic
chemotherapeutants especially designed for symplastic movement as bark applicators. In foliar sprays, ectodesmata of cuticle act as path-

ways.   The details of uptake and translocation of antibiotics effective against mycoplasma diseases were described by Sinha (1979).

First studies on the acquisition and retention of tetracycline hydrochloride by healthy and MLO-diseased plants were carried out by Frederick *et al.,* (1971) and by Klein *et al.,* (1972).   Subsequently, Peterson and Sinha (1977) determined the uptake, translocation and persistence of four tetracycline antibiotics: Chlortetracycline, doxycycline, oxytetracycline and tetracycline, in various plant species after root immersion treatment of 100   g/ml of each antibiotic for 24 hrs.   They observed the difference in uptake, translocation and persistence of these antibiotics with different species of plants as well as with different parts of the same plant.   Oxytetracycline and tetracycline were more persistent in all the plants than chlortetracycline and doxycycline.   They explained that these differences in persistence of various antibiotics presumably reflex the rate at which they may be metabolized in different plants.   However, their data also shows that persistence of these antibiotics in different plants is directly proportional to their uptake.

EFFECTS OF CHEMOTHERAPEUTANTS ON HOST PHYSIOLOGY AND HOST-PATHOGEN INTERACTIONS

The systemic compounds used in the control of plant mycoplasma diseases may kill the pathogen, retard or arrest its multiplication or growth, or alter the physiology of the host in such a way that the host may exhibit resistance to the pathogen. Hirumi and Maramorosch (1972) and Rishi *et al.,* (1973) observed natural degeneration of mycoplasma in yellows and grassy shoot diseases of aster and sugar cane.   However, the factors responsible within the plant to induce mortality of mycoplasma have not been investigated. The production of toxins and toxic metabolites from *S. citri* that were harmful to the host plant were also reported (Sagio *et al.*, 1973; Daniels and Meddins, 1974). Interaction of mycoplasma with hormone balances controlling growth and development of a host plant was studied by Lemcke (1972). He pointed out that the pathways of biosynthesis of gibberellins and carotenoids (synthesized by acholeplasmas) diverge from a common precursor pathway and it might be possible that the carotenoid synthesis by the pathogens might deprive the plant of gibberellin precursors. The stunted growth in diseased plants is attributed to the decrease in gibberellin content which can be remitted upon treatment with extraneously applied gibberellic acid (Maramorosch, 1959).

FUTURE STRATEGIES

The studies on the chemotherapy of mycoplasma diseases are still in a primitive stage. Better understanding of the mechanisms involved in the host-pathogen interactions during pathogenesis will facilitate the development of improved methods of chemotherapy. An appraisal of the work done in this area provides guidelines for future chemotherapy of plant mycoplasma diseases. These include (a) extensive screening of additional chemotherapeutic agents, (b) design of a potent mycoplasmacidal agent, (c) improved methods of application, (d) improved solubility and persistence of chemotherapeutants in host plants, (e) study of environmental influences during and after chemotherapeutic treatment, (f) understanding of the time of application, and (g) induction of resistance in the hosts against the pathogens.

Extensive screening of drugs against mycoplasma diseases requires *in vitro* studies. Until now, most of the effects observed in controlling plant mycoplasma diseases have been carried out *in vivo*. The results did not prove that these agents have a direct effect on mycoplasma. *In vitro* screening is helpful in establishing direct effects of chemotherapeutants on pathogens. Since no plant mycoplasma other than spiroplasmas, has been successfully cultured in artificial media, spiroplasmas can be used as a tool for screening various chemotherapeutic agents *in vitro*. However, *in vitro* tests do not necessarily prove the effectiveness of a chemotherapeutant in vivo as its efficacy may be affected by several factors such as its stability and translocation in plants. Compounds known to interfere with various metabolic processes in prokaryotes should be screened. These include selective inhibitors of cytoplasmic membrane functions of nucleic acid and protein synthesis and of specific enzymes related to mycoplasma replication. Since the knowledge regarding the nature and specificity of such enzymes in mycoplasma is scarce, screening of compounds known to affect such processes in other prokaryotes should be carried out. The knowledge gained along these lines on molecular biology of animal mycoplasmas and spiroplasmas can be explored.

Quinlan and Maniloff (1972) observed that the DNA growing point in *Mycoplasma gallisepticum* is membrane bound and they described a membrane-bleb-infrableb complex at the site of DNA-replication (Maniloff and Quinlan, 1974). Any interference with this complex might result in failure of DNA replication. The synthesis of an actin-like protein by *Mycoplasma pneumoniae* which is responsible for rapid and reversible changes in cell form during growth, reproduction and motility (in some species) was reported by Neimark (1977). Effects on mycoplasma growth and vitality of various macrolide antibiotics such as monactin, diactin, triactin and nonactin should be studied. Another inhibitor of the replication process, 6-(P-hydroxyphenylazo)-uracil blocks semi-consentrative DNA-replication (Brown, 1971). However, nalidixic acid which is also a blocker of some processes was found ineffective in *S. citri* (Davis,

1981).   In bacteria, phenethyl alcohol inhibits the activity of DNA-polymerase *in vitro* (Zahn, *et al.*, 1966). This compound affects the permeability properties of *Bacillus subtilis* cell membrane. Other inhibitors of cell membrane structure and function should also be screened.   The membrane of mycoplasma contains sterol as one of its components. Triforine, miconazole and diclobutrazole fungicides that interfere with sterole biosynthesis of fungal cell membranes should also be screened.

It was observed that ionophores affect the vertical disposition of mycoplasma membrane proteins on the cell surface. Exposure of *Acholeplasma laidlawii* cells to valinomycin which dessipates potassium-gradient, or to carbonyl-cyanide-m-chlorophenyl hydrazone which causes collapse of proton-gradient, resulted in a rapid drop of iodination values of proteins exposed on the cell surface (Amar *et al.*, 1978).   Changes in proton-gradient causes conformational changes in membrane proteins or enzymes (Radnick *et al.*, 1976; Plaff and Klinberger, 1968, Jagendorf, 1975). The effects of valinomycin and other ionophore antibiotics like monensin, nigericin etc. should be tested on plant mycoplasmas.

The externally located glycoproteins on cell membranes of all mycoplasmas were thought to be responsible for their specific binding to host cells (Razin, 1978) as observed in viruses (Kahane and Tully, 1976; Kahane *et al.*, 1977).   The receptors on respective host cells are neuraminidase sensitive and contain sialic acid.   Therefore, compounds that interfere with the conformation and functions of membrane glycoproteins of mycoplasma or neuraminidases (that will affect host-cell receptors), may inhibit the bindings of mycoplasma to host cell. Oligosaccharides containing neuraminic acid should be tested.

Most of the inhibitors of protein synthesis that inhibit mycoplasma disease were found to be specific for 70 S ribosomes. Additional inhibitors of 70 S ribosomes should be screened. The effects of drugs that deplete the cell supply of nutritional precursors should be tested either alone or in combination with some protein synthesis inhibitors.

The effect of a chemotherapeutant depends upon its solubility and stability in the host tissues. To obtain its best effect, it should be highly water soluble at a low pH. It should efficiently enter into the vascular strand and in the case of trees, into the xylem vessels of the current year's annual ring for a very wide distribution into the plant. Nair (1979) observed that uptake and distribution of a systemic chemotherapeutant was enhanced in elm trees after full leaf expansion periods because of the presence of large spring wood vessels. Therefore, knowledge regarding the production of growth rings is essential before developing efficient injecting systems for controlling diseases of tropical trees.

The time of chemotherapeutant application, is important as observed in the case of tetracycline treatment of peach X-diseased trees of peach and cherry (Sands and Walton, 1975; Rosenberger and Jones, 1977). The effects of various environmental factors such as humidity, temperature, wind velocity and cloud cover on uptake and distribution of chemotherapeutants in trees was studied by Nair (1979). He concluded that high temperature and low humidity, along with high wind velocity during the days immediately before treatment increased uptake and distribution of chemotherapeutants.

In view of these studies, it is desirable to try a wide range of chemicals and their combinations and various methods of application during different periods of the year. Several mycoplasma diseases have been cured by heat therapy (Kunkel, 1941, 1945, 1952; Nariani *et al.*, 1973; Singh, 1977). More extensive studies on heat therapy should be made and where heat therapy alone is not successful, its combination with chemotherapy should be exploited.

The possibilities to induce resistance in host plants against mycoplasma diseases by treatment with chemicals should not be excluded. Cruickshank (1966) suggested the development of systemic compounds which induce the production of phytoalexins for the control of plant diseases. The role of plant growth regulators in induction of disease resistance in tomato plants was studied by Davis and Diamond (1953, 1965).

Design of a Potent Mycoplasmacidal Agent

A rational approach to the design of mycoplasmacidal agents depends on a judicious selection of biochemical pathways that are susceptible to the effects of chemical agents. The most apparent mechanisms are those which are specific for mycoplasma replication and are not required for host-cell metabolism. Elucidation of the mode of action of available agents affecting mycoplasma diseases and recent advances in molecular biology of mycoplasmas could suggest susceptible mechanisms to be exploited in search for new mycoplasmacidal agents. Although the studies on molecular biology of plant mycoplasma are still in their infancy, approaches to design a suitable chemotherapeutic agent can be made from the available information on animal mycoplasmas and spiroplasmas. According to Paul Ehrlich, who created the concept of chemotherapy, a compound must be able to bind itself to the receptor site in the host for producing its effect. It must have a binding site and a reactive site. Therefore, geometry of receptors should be known before designing a drug. In the case of mycoplasma, no studies have been made in this direction. Exhaustive screening of compounds and studies of their structure-activity relationship will give a lead to develop more potent chemotherapeutic agents. Structural modification of a parent compound by group substitutions in such a way that it improves its activity, stability and transport and reduces toxicity towards the host, ought to yield more promising chemotherapeutic agents. Another approach of drug design is to develop molecules that can inhibit various enzyme activities or other metabolic processes of pathogens without altering the functions of hosts. Since plant mycoplasmas are intracellular obligate parasites, such an approach is difficult. If some specific factors that differentiate them from host-cell functions were recognized, selective interference with pathogen metabolism would become possible. In spite of superficial resemblance with bacteria, these pathogens differ markedly in their components of membrane DNA, RNA and protein synthesizing machinery (Reff, *et al.*, 1977; Johnson and Horowitz, 1971). These differences can be explored for selective approach in developing mycoplasmacidal agents.

REFERENCES

Amar, A., Rottem, S. and Razin, S., 1978. Disposition of membrane proteins as affected by changes in the electrochemical gradient across mycoplasma membranes. Biochem. Biophys. Res. Commun. 84: 306.

Asuyama, H. and Iida, T.T., 1973. Effects of tetracycline compounds on plant diseases caused by mycoplasma-like agents. Ann. N.Y. Acad. Sci. 225: 509.

Bittman, R. and Rottem, S., 1976. Distribution of cholesterol between the outer and inner halves of the lipid bilayer of mycoplasma cell membranes. Biochem. Biophys. Res. Commun. 71, 318.

Bower, J.W., and Calavan, E.C., 1974. Antibiotic sensitivity *in vitro* of the mycoplasma-like organism associated with citrus stubborn disease. Phytopathology 64: 346.

Brock, J.W., and Calavan, E.C., 1974. Antibiotics, Vol., I, Mechanism of Action, (Gottlieb, D., and Shaw, P.D., (eds.), pp. 651, Springer-verlag, Berlin.

Brock, T.D., 1967. In: Antibiotics, Vol. I, Mechanism of Action, Gottlieb, D., and Shaw, P.D., (eds.), pp. 615. Springer-verlag, Berlin.

Brown, G.M., 1962. The biosynthesis of folic acid II. Inhibition by sulphonamides. J. Biol. Chem., 237: 536.

Brown, G.M., 1964. Methods for measuring inhibition by sulphonamides of the enzymic synthesis of dihydropteroic acid. Methods Med. Res., 10: 233.

Brown, G.M., 1967. Inhibition by sulphonamides of the biosynthesis of folic acid. Intern. J. Leporsy, 35: 380.

Cerna, J., Rychlik, I. and Pulkrabek, P., 1969. The effect of antibiotics on the coded binding of peptidyl-tRNA to the ribosome and on the transfer of the peptidyl residue to puromycin. Eur. J. Biochem., 9: 27.

Clemons, G.P. and Sisler, H.D., 1969. Formation of a fungitoxic derivative from benlate. Phytopathology 59: 705.

Costouis, A.C., Becker, E.I., Garret, W.T. and Julien, I.M., 1978. Evaluation of antibiotic, nystatin and other fungicides in Dutch Elm disease control. Proc. 3rd Int. Congr. Plant Pathol. Muchen, (Abstract).

Cruichshank, I.A.M., 1966. Defense mechanisms in plants. World Rev. Pest Contr. 5: 161.

Cundliffe, E., 1969. Antibiotics and polyribosomes. II. Some effects of lincomycin, spiramycin and streptomycin A *in vivo*. Biochem., 8: 2063.

Cundliffe, E., 1972. Bacterial protein synthesis: The effects of antibiotics upon the puromycin *in vitro*, In: Molecular Mechanisms of Antibiotic Action on Protein Biosynthesis and Membranes. Munoz, E., Ferrandiz, F. and Vazquez, D., (eds.) pp. 242 Elsevier, Amsterdam.

Cundliffe, E. and McQuillen, K., 1967. Bacterial protein synthesis: the effects of antibiotics. J. Mol. Biol. 30: 137.

Daniels, J.M., and Meddins, B.M., 1974. The pathogenicity of *Spiroplasma citri*. Colloq. Inst. Natl. Sante Rech. Med. 33: 195.

Davis, R.E., 1981. Antibiotics sensitivities *in vitro* in diverse spiroplasma strains associated with plants and insects. Appl. Environ. Microbiol. 41: 329.

Davis, D. and Dimond, A.E., 1953. Inducing disease resistance with plant growth regulators. Phytopathol. 43: 137.

Davis, D. and Dimond, A.E., 1956. Site of disease resistance induced by plant growth regulators in tomato. Phytopathol. 46: 551.

de Kruyff, B., van Dijck, P.W.M., Goldbach, R. W., Demel, R.A. and van Deenen, L.L.M., 1973. Influence of fatty acid and sterol composition on the lipid phase transition and activity of membrane-bound encymes in *Acholeplasma laidlawii*. Biochim. Biophys. Acta 330: 269.

Ennis, H.L., 1970. Synergistin. A synergistic antibiotic complex which selectively inhibits protein synthesis. In: Progress in Antimicrobial and Anticancer Chemotherapy, Vol. 11, Tokyo, Univ. of Tokyo Press, pp. 489.

Feingold, D.S., 1965. The action of amphotericin on *Mycoplasma laidlawii*. Biochem. Biophys. Res. Commun. 19: 261.

Feldmann, H. and Falter, H., 1971. Transfer ribuonucleic acid from *Mycoplasma laidlawii*. Eur. J. Biochem., 18: 573.

Frederick, R.J., Klein, M., and Maramorosch, K., 1971. Acquisition and retention of tetracycline-HCl by plants. Plant Dis. Reporter (U.S.) 55:223-226.

Freeman, B.A., Sissenstein, R., McManus, T.T., Woodward, J.E., Lee, I.M. and Mudd, J.B., 1976. Lipid composition and lipid metabolism of *Spiroplasma citri*. J. Bacteriol. 125: 946.

Goldstein, N.I., McIntosh, A.H., Fisher, P.B., Maramorosch, K. and Schaffner, C.P., 1976. *In vitro* anti-mycoplasmal activity of amphotericin B methyl ester. J. Antibiotics 29: 656-661.

Hirumi, H., and Maramorosch, K., 1972. Intracytoplasmic mycoplasma-like bodies in phloem parenchyma cells of aster yellows-infected *Nicotiana rustica*. Phytopathol. Z. 77: 71.

Igarashi, K., Ishitsuka, H. and Kahi, A., 1969. Comparative studies of the mechanism of action of lincoymcin, streptomycin and erythromycin. Biochem. Biophys. Res. Commun. 37: 499.

Ishiie, T., Doi, Y., Yora, K., and Asuyama, H., 1967. Suppressive effects of antibiotics of tetracycline group on symptom development of mulberry dwarf disease. Biochem. Biophys. Res. Commun. 37: 499.

Jagendorf, A.T., 1975. Mechanism of photophosphorylation, In: Bioenergetics of Photosynthesis, Govindjes, R., (ed.) pp. 414. Academic Press, New York.

Jao, R.L. and Finland, M., 1967. Susceptibility of *Mycoplasma pneumoniae* to 21 antibiotics *in vitro*. Am. J. Med. Sci., 253: 639.

Johnson, J.D., and Horowitz, J., 1971. Characterization of ribosomes and RNAs from *Mycoplasma hominis*. Biochim. Biophys. Acta. 247: 262.

Kahane, I., Greenstein, S., and Razin, S., 1977. Carbohydrate content and enzymic activities in the membrane of *Spiroplasma citri*. J. Gen. Microbiol. 101: 173.

Kahane, I. and Tully, J.G., 1976. Binding of plant lectins to mycoplasma cells and membranes. J. Bacteriol. 128, p. 1.

Kaji, A., Suzuka, I. and Kaji, H., 1966. Binding of specific soluble ribonucleic acid ro ribosomes. J. Biol. Chem., 241: 1251.

Kirk, R.G., and Morowitz, H.J., 1969. Ribonucleic acids of *Mycoplasma gallisepticum* strain A5969. Am. J. Vet. Sci. 30: 287.

Klein, M., Frederick, R.J., and Maramorosch, K., 1972. Chemotherapy of aster yellows: Tetracycline hydrochloride uptake by healthy and diseased plants. Phytopathology 61: 111-115.

Klein, M., Frederick, R.J., and Maramorosch, K., 1973. The effect of sulpha drugs. Ann. N.Y. Acad. Sci. 225: 522.

Kunkel, L.O., 1941. Heat cure of aster yellows in periwinkles. Amer. J. Bot., 28: 761.

Kunkel, L.O., 1945. Studies on cranberry false blossom. Phytopathol. 35: 805.

Kunkel, L.O., 1945. Transmission of alfalfa witches' broom to non-leguminous plants by dodder and cure in periwinkle by heat. Phytopathol. 42: 27.

Lampen, J.O., Gill, J.W., Arnou, P.M. and Magana Plazq, A., 1963. Inhibition of the pleuroneumonia-like organism *Mycoplasma gallisepticum* by certain polyene antifungal antibiotics. J. Bacteriol. 86: 945.

Lemcke, R.M., 1972. Osmolar concentration and fixation of mycoplasmas. J. Bacteriol. 110: 1154.

Maramorosch, K., 1957. Reversal of virus-caused stunting in plants by gibberellic acid. Science, 126: 651.

Maramorosch, K., Granados, R.R., and Hirumi, H., 1970. Mycoplasma diseases of plants and insects. Adv. Virus Res. 16: 135.

Modellel, J. and Davis, B.D., 1970. Breakdown by streptomycin of initiation complexes formed on ribosomes of *Escherichia coli*. Proc. Natl. Acad. Sci. U.S., 67: 1148.

Nariani, T.K., Ghosh, S.K., Kumar, D., Raychaudhuri, S.P., and Viswanath, S.M., 1973. Detection and possibilities of therapeutic control of the greening disease of citrus caused by mycoplasma. Sym. Mycoplasmal

Diseases, Chandigarh, (Abstract). pp. 24.

Nair, V.M.G., 1979. Chemotherapy of Dutch Elm disease. Res. Progr. Rep. Univ. Wis., North Central Res. Committee., pp. 32 (Diseases of Forest and Shade Trees).

Neimark, H.C., 1977. Extraction of an actino-like protein from the prokaryote *Mycoplasma pneumoniae.* Proc. Natl. Acad. Sci., U.S. 74: 4041.

Paddick, R.G., and French, F.L., 1964. Control of yellow dwarf of tobacco by treatment with systemic insecticides. Proc. Aust. Tobacco Res. Conf. pp. 304.

Paddick, R.G., French, F.L. and Turner, P.L., 1971. Control of leafhopper-borne plant diseases possibly due to direct action of systemic biocides on mycoplasmas. Plant Dis. Rep. 55: 291.

Peterson, E.A., and Sinha, R.C., 1977. Uptake, distribution and persistence by tetracycline antibiotics in various plant species susceptible to mycoplasma infection. Phytopathol. Z. 90: 250.

Plaff, E., and Klinberger, M., 1968. Adenine nucleotide translocation of mitochondria. I. specificity and control. Eur. J. Biochem., 6: 66.

Quinlan, D.C. and Maniloff, J., 1972. Membrane association of the deoxyribonucleic acid growing-point region in *Mycoplasma gallisepticum.* J. Bac teriol. 112: 1375.

Raychaudhuri, S.P., Chenulu, V.V., Ghosh, S.K., Verma, A., Rao, P.S., Srimathi, R.A., and Nag, K.C., 1972. Chemical control of spike disease of sandal. Curr. Sci. 41: 72.

Raychaudhuri, S.P. and Rishi, N., 1981. Chemotherapy of Plant Mycoplasma Diseases In:
Mycoplasma Diseases of Trees and Shrubs,
Maramorosch, K., and Raychaudhuri, S.P.,
(eds.) pp. 315. Academic Press, New York.

Razin, S., 1978. The mycoplasmas. Microbiol.
Rev. 42: 414.

Reff, M.E., Stanbridge, E.J. and Schneider, E.
L., 1977. Phylogenetic relationship between
mycoplasmas and other prokaryotes based
upon the electrophoretic behavior of their
ribosomal ribonucleic acids. Int. J. Syst.
Bacteriol. 27: 185.

Rishi, N., Okuda, S., Arai, K., Doi, Y., Yora,
K. and Bhargava, K.S., 1973. Mycoplasma-
like bodies, possibly the cause of grassy
shoot disease or sugarcane in India. Ann.
Phytopathol. Soc. Jap. 39: 429.

Rishi, N. and Raychaudhuri, S.P., 1981. Serological studies on plant mycoplasmas and
cor neform bacterium associated with ratoon
stunting disease of sugarcane. Proc. XVII
Int. Union Forestry Res. Org. World. Cong.
Japan. Orgn. II, pp. 351.

Rosenberger, D.A. and Jones, A.L., 1977. Symptom remission in X-diseased peach trees as
affected by date, method and rate of application of oxytetracycline-HCl. Phytopathol.
67: 277.

Rottam, S., Yashouv, J., Ne'eman, Z. and Razin,
S., 1973. Chosesterol in mycoplasma membranes, composition, ultrastructure and
biological properties of membranes from
Mycoplasma mycoides var. capri cells
adapted to grow with low cholesterol concentration. Biochim. Biophys. Acta. 323:
495.

Rudnick, G., Schuldiner, S., and Kaback, R.H., 1976. Equilibrium between two forms of the *lac* carrier protein in energized and non-energized membrane vesicles from *Escherichia coli*. Biochemistry. 15: 5126.

Saglio, P.O., L.Hospital., M., Lafleche, D., Dupont, G., Bove, J.M., Tully, J.G. and Freundt, E.A., 1973. *Sprioplasma citri gen* and sp. n: a mycoplasma-like organism associated with "stubborn" diseases of citrus. Int. J. Syst. Bacteriol. 23: 191.

Sands, D. and Walton, G.S., 1975. Tetracycline injections for control of eastern X-disease and bacterial sp. of the peach. Plant Dis. Rep. 59: 583.

Sinha, R.C., 1979. Chemotherapy of Mycoplasmal Plant Diseases. In: The Mycoplasmas, Vol. III, Plant and Insect Mycoplasmas, Whitcomb, R.F. and Tully, J.G. (eds.) pp. 310., Academic Press, New York.

Singh, K., 1977. Sugarcane diseases and three tier seed programme. Sugar News. 9: 81.

Slotkin, R.I., Glyde, Jr., W.A., and Denny, F. W., 1967. The effect of antibiotics on *Mycoplasma pheumoniae in vitro* and *in vivo*. Am. J. Epidemiol. 86: 225.

Smith, D.H. and Davis, B.D., 1967. Mode of action of novobiocin in *Escherichia coli*. J. Bacteriol. 93: 71.

Stanbridge, E.J. and Doersen, C.J., 1978. Some effects that mycoplasmas have upon their infected host In: Mycoplasma Infection in Cell Cultures. McGarrity, G. (ed.) pp. 119, Plenum, New York.

Taubmann, S. B., Young, F.E. and Corcoran, J. W., 1963. Antibiotic glycosides. IV. Studies on the mechanism of erythromycin resistance in *Bacillus subtilis*. Proc. Natl. Acad. Sci., U.S., 50: 955.

Thirumalachar, M.J., 1978. Two new chemotherapeutic agents for the control of MLO and RLO diseases. Proc. 3rd Int. Congr. Plant Pathol. Munchen, (Abstract).

Tourtellotte, M.E., 1969. Protein synthesis in mycoplasmas, In: The Mycoplasmatales and the L-phase of Bacteria. Hayflick, L., (ed.) pp. 451, Appleton, New York.

Townsend, R., Markham, P.G., Plaskitt, K.A. and Daniels, M.J., 1977. Isolation and characterization of a non-helical strain of *Spiroplasma citri*. J. Gen. Microbiol. 100, 15.

Walker, R.T. and Raj Bhandary, U.W., 1975. Formylatable methionine transfer RNA from mycoplasma. Purification and comparison of partial nucleotide sequence with those of other prokaryotic initiator tRNAs. Nucleic Acids, Res. 2: 61.

Walker, R.T. and Raj Bhandary, W.L., 1978. The nucleotide sequence of formylmethionine tRNA from *Mycoplasma mycoides sp., capri*. Nucleic Acids, Res. 5: 57.

Weber, M.M. and Kinsky, S.C., 1965. Effect of cholesterol on the sensitivity of *Mycoplasma laidlawii* to the polyene antibiotic filipin. J. Bacteriol. 89: 306.

Windsor, I.M. and Black, L.M., 1973. Remission of symptoms of clover club leaf following treatment with penicillin. Phytopathol. 65: 287.

22

KINETIN TREATMENT OF STOLBUR DISEASED PLANTS AND POSSIBILITY OF ITS APPLICATION IN CHEMOTHERAPY

by

B. Plavsic, K. Krivokapic and Z. Eric

Department of Biology,

Faculty of Science

University of Sarajevo

Sarajevo, Yugoslavia

INTRODUCTION

Stolbur of tomato and potato is one of the most important diseases in southeast Europe. Since 1968 many papers were published about the MLO etiology of stolbur.

Ever since the discovery of MLO agents of yellows diseases by Doi *et al.*, (1967) the prevention and control of these diseases has been investigated. Despite the general concensus that tetracyclines, first used by Ishii and co-workers (1967) induced only temporary remission of symptoms (McCoy, 1972; Raychaudhuri and Rishi, 1981), these antibiotics remained the only ones used for chemotherapy of MLO infections. Since the criteria of disease remission are subjective and sterility is a common feature of MLO diseases, Maramorosch (1974) proposed as a criterium for complete cure of an MLO disease the obtaining of viable seed from treated plants.

Taking into consideration already known symptomatology of the stolbur infection: chlorosis, starch accumulation in leaves, changes in chloroplast structure (Plavsic *et at.*, 1976), sterility, proliferation of axillary buds, hyperplastic and necrotic changes in the phloem (Mihailova, 1934; Samuel *et al.*, 1933; Cousin and Grison, 1966; Plavsic, 1967), we were wondering whether kinetin, because of its pecular features, might also influence MLO infections Kinetin inhibits the destruction of chlorophyll in cut leaves (Osborne and McCalla, 1961) stimulates the chloroplasts (Kursanov - *et al.*, 1964; Sveznikova and Hohlova, 1969), and prevents the destruction of mitochondrial membranes and membranes of the endoplasmic reticulum (Kursanov *et al.*, 1964). Kinetin affects cytokinesis and cell differentiation, induces organogenesis and participates in apical domination. Under high temperature kinetin, as other cytokinins, plays a protective role in plants. Kinetin also exerts a stimulative effect on RNA-, DNA-, and protein synthesis (Kulaeva, 1973; Letham *et al.*, 1978). Because of these features kinetin has been named "juvenile substance" (Reunov *et al.*, 1977).

MATERIALS AND METHODS

Naturally infected potato plants provided the source of stolbur-diseased material. The transmission of the disease was carried out through grafting of infected potato to healthy tomato plants. All infective as well as all healthy control plants were grown in a greenhouse under controlled conditions. Synthetic kinetin was obtained from Koch-Light Laboratories, Ltd. London, England.

The effectiveness of kinetin treatment was evaluated on the basis of its influence on (1) the MLO agent, (2) chloroplast structure, (3) the concentration of photosynthetic pigments, and (4) the general appearance of treated plants, with special attention to the reproductive organs. The parameters under (1) and (2) were investigated by electron microscopy. Leaf samples were fixed in buffered glutaraldehyde (pH 7.2) and postfixation was in buffered osmium tetraoxide at the same pH. Dehydration was carried out using ethanol and propylene oxide. For embedding we used epoxy resin (EPON 812). Thin sections were prepared with a diamond knife and an LKB ultramicrotome. The sections were contrasted in grids with uranyl magnesium acetate and lead citrate, then examined using a JEM (JEOL) 100B electron microscope. The analysis of chlorophyll (a+b) was carried out using Arnon's (1949) spectrophotometric method. Carotenoids were analyzed by spectrophotometry as described by Wettstein (1957). Kinetin-treated diseased plants and healthy treated and untreated controls were kept together with diseased untreated plants in a greenhouse for comparative observations.

Kinetin treatment was carried out in two ways. Since tomato plants are easily rooted, they were immersed into the kinetin solution. Foliar treatment with kinetin was used for potato plants, because they develop roots rather slowly. The tops of tomato plants, including 3 to 4 leaves, were cut and placed in Knopp's solution until a few roots appeared. Subsequently 0.05 mg/l of kinetin were added. After 11 days of this treatment the plants were planted in soil in a greenhouse. For foliar treatment we used an aqueous solution of kinetin at a concentration of 0.05 mg/l, with a few drops of Tween 80 to improve penetration. Plants were sprayed once a day for 5 days. Electron microscopy and biochemical investigations were carried out using the same samples for both investigations.

RESULTS AND DISCUSSION

Four weeks after grafting tomato plants showed very prominent symptoms of stolbur infection, typified by "big bud" malformation (Samuel *et al.*, 1933). Electron microscopy of phloem revealed MLO in sieve tubes. This provided direct confirmation of the MLO infection (Fig. 1, insert; Fig. 2).

The photosynthesis apparatus became significantly changec compared to that of healthy control plants. The functional surfaces of tylakoid membranes became extremely reduced. Chloroplasts were filled with starch and their membranes desintegrated (Fig. 4b). In the same plants the level of photosynthetic pigments (chlorophyll a+b) and carotenoids decreased (Table 1B). There was correlation between the structural changes and the decreased level of pigments. Enzyme chlorophylase located in chloroplasts induced the first step of chlorophyll destruction in virus infected plants (Goodman *et al.*, 1967). In tissues that suffered the biggest loss of chlorophyll, the strongest activity of chlorophyllase was demonstrated. On the contrary, tissues that lost little or no chlorophyll manifested low chlorophyllase activity. The results differed from those of Peterson and McKinney (1938) but it should be pointed out that in their experiments the alteration of chlorophyll was induced by a virus, and not by MLO.

Strong starch accumulation in chloroplasts interfered with their function. Probably this was the result of suppressed hydrolysis as well as disturbances in the phloem (Samuel *et al.*, 1933; Mihailova, 1934; Causen and Grison, 1966; Plavsic, 1967). Although we are not able to explain the michanism of the described changes in the photosynthetic apparatus of stolbur infected plants, we can state that they are in direct connection with the induction of the "yellowing" symptom, specific for MLO infection. Proliferation of axillary buds and flowers also might be a result of hormone imbalance (Krivo-kapic *et al.*, 1979). The results confirm previous findings that stolbur symptoms are induced by changes in the metabolism of organic substances (Blattny, 1956; Valenta *et al.*, 1961; Plavsic, 1967).

The same tomato plants have been investigated on the 9th day of kinetin treatment (concentration 0.05 mg/l). Electron microscopy of phloem from diseased treated plants revealed degenerative forms of MLO. The comparison with MLO of untreated plants (Fig. 2 and 3) revealed that kinetin caused destructive changes in this prokaryotic microbe. There was no detectable DNA in the prokaryon and no ribosomes in the procytoplasm. Most of MLO cells were empty. Electron microscopy of chloroplasts from diseased plants treated for 9 days showed the recovery of normal photosynthetic surfaces of tylakoid membranes. Also, there was no trace of abnormal starch accumulation. The chloroplast membrane clearly separated chloroplasts from the surrounding cytoplasm (Fig. 4D).

Biochemical investigations of leaves from treated diseased plants revealed that there was a remarkable increase in the concentration of photosynthetic pigments (Table 1, D). That increase occurred not only in diseased, but also in healthy treated plants (Table 1,C).

After 11 days of kinetin treatment, all plants were planted in soil and maintained in a greenhouse. A month and a half later, the diseased, treated plants which earlier were expressing the "big bud" symptom, started to produce normal flowers, normal fruits, and viable seeds (Fig. 6). Plants that originated from these seeds were completely healthy. Their progenies have now been observed over a period of three years, without the slightest signs of MLO infection. As far as healthy treated plants were concerned, we noticed that they were more vital and greener than healthy untreated controls. Obviously, kinetin strongly stimulated the synthesis of tylakoid membranes and the correlated synthesis of photosynthetic pigments. The stimulative effect of kinetin was present in treated diseased, as well as in treated healthy plants. As a result, very intensive greening of plants occurred, contrasting sharply with the chlorosis, the typical symptom of all "yellows" diseases.

In our experiments, kinetin exerted a destructive effect on the MLO agent. The mechanism of action is difficult to explain. Srivastava (1968) found that 12% of kinetin-8-C^{14} became incorporated in nucleic acids. We suppose that kinetin, a purine (a 6-furfuril aminopurin) is able to incorporate into MLO nucleic acids and in that way able to "desinform" the metabolism of the pathogen.

On the basis of our results, we conclude that kinetin, besides its other already known beneficial features, has therapeutic properties. Summarizing all that is known until now about this biologically active compound, we can say that the term "juvenile substance" (Reunov *et al.*, 1977) was justified, as shown in our investigation. We consider kinetin as a promising compound in future attempts to increase the photosynthetic activity of MLO-affected plants.

Electron microscopy of foliar kinetin treated potato showed all characteristics of normally developed chloroplasts (tylakoid membranes, no starch accumulation). Kinetin expressed a stimulating effect on photosynthetic pigments (Table 2).

Potato plants, after treatment, were grown in a greenhouse together with control plants (healthy untreated and diseased untreated potato). After 3 to 5 months observation, we concluded that: 1) treated plants were intensively green, 2) treated plants were as big as healthy controls, 3) tuber production of diseased treated plants was higher than of healthy controls (Fig. 7).

REFERENCES

1. Arnon, D.I., 1949. Copper enzymes in isolated chloroplast. Polyphenoloxidase in Beta vulgaris. Plant Physiol. 24, 1-15.

2. Blattny, C., 1956. Osnovni problemi stolbura. Zbornik z Vedeckej konferencije o stolbure a pribuznych bezsemennostiach. Smolenice.

3. Cousin, M.T., Grison, G., 1966. Premieres observations concernant une fluorescence anormale dans le liber interne de plusieurs Solanees infectees par le virus du stolbur et d'une Apocynacee atteinte de phyllodie. "Etudes de Virologie", Ann. Epiphyties. 17, No. hors-serie, 93-98.

4. Doi, Y., Teranaka, M., Yora, K., and Asuyama, H., 1967. The biochemistry and physiology of infectious plant diseases. Van Nostrand Company.

5. Goodman, R.,Kiraly, Z. and Zaitlin, M., 1967. The biochemistry and physiology of infectious plant diseases. Van Nostrand Co.

6. Ishiie, T., Doi, Y., Yora, K. and Asuyama, H., 1967. Suppressive effects of antibiotics of tetracycline group on symptom development of mulberry dwarf disease. Ann. Phytopathol. Soc. Japan 33, 267-275.

7. Krivokapic, K., Plavsic, B., Eric. Z., Buturovic, D., 1979. Changes in endogenous hormones in mycoplasma infected tomato (Lycopersicum esculentum L.) plants. Contributions of the Faculty of Agriculture, University of Sarajevo, Vol. 27, 31-49.

8. Kulaeva, O.N., 1973. Citokinini ih struktura i funkcija. "Nauka". Moskva.

9. Kursanov, A.L., Kulaeva, O.N., Svesnikova, I.N., Popova, E.A., Boljakina, J.P., Kljac-ko, N.L., Vorobljeva, I.P., 1964. Vosstano vlenie kletocih struktur i obmena vesestv zeltih listiev pod deistviem 6-benzilamino-purina. Fiziol. rast. Moskva, 11, 838.

10. Letham, D.S., Goodwin, P.B., and Higgins, T.J.V. 1978. Phytohormones and related compounds: A comprehensive treatise. Vol. 1,2. Elsevier, North Holand.

11. Maramorosch, K., 1974. Mycoplasmas and Rickettsiae in relation to plant diseases. Annual Review of Microbiology. Vol. 28, 301-324.

12. McCoy, R.E. 1972. Remission of lethal yellowing in coconut palm treated with tetracycline antibiotics. Plant Dis. Rep. 56, 1019-1021.

13. Mihailova, P.V., 1934. Anatomija odervene-nija plodov u pomidora. Sb. Virusn. bolezni rast. v Krimu i na Ukraine, Simferopol.

14. Osborne, D., and McCalla, D., 1961. Rapid bioassy for kinetin and kinins using senescing leaf tissue. Plant Physiology, 36, 219-221.

15. Peterson, P.D. and McKinney, H.H., 1938. The influence of four mosaic diseases on the plastid pigments and chlorophyllase in tobacco leaves. Phytopathology. 28, 329-342.

16. Plavsic-Banjac, B., 1967. The anatomical characteristic of plants affected by Stolbur virus. Rad JAZU, 345, 237-270.

17. Plavsic, B., Krivokapic, K., Butorovic, D., 1976. Some morphological and physiological aspects of Stolbur disease infected tomato plants (Lycopersicum esculentum var. Saint Piere). Proceedings of the Society for General Microbiology. Vol. 3, p. 153, Berkshire, England.

18. Raychaudhuri, S.P. and Rishi, N., 1981. Chemotherapy of plant mycoplasma diseases. 315-324. In: Maramorosch, K. and Raychaudhuri, S.P., eds. Mycoplasma Diseases of Trees and Shrubs. Academic Press, 1981.

19. Reunov, A.V., Reunova, G.D., Vasiljeva, L.A., and Reifman, F.G. 1977. Effect of kinetin on tobacco mosaic and potato virus X replication in leaves of systemic hosts. Phytopath. Z. 90. 342-349.

20. Samuel, G., Bald, J.G. and Eardley, 1933. Big bud virus disease of tomato. Phytopathology 23, 641-653.

21. Srivastava, B.I.S., 1968. Machanism of action of kinetin in the retardation of senescence in excised leaves. In: Biochemistry and Physiology of Plant Growth Substances. Ed. Runge Press Ottawa, 1479-1493.

22. Svesnikova, I.N., Hohlova, V.A. 1969. Citologiceskoe izucenie deistvija 6-benzilaminopurina i kinetina na izolirovanie semjodolni. Fiziol. Rastenii, 16, 687-691.

23. Valenta, V., Musil, M., and Misiga, S., 1961. Investigation on European yellows type viruses. I The stolbur virus. Phytopath. Z. 42, 1, 1-38.

24. Wettstein, D., 1957. Chlorophyll -- letal und der submikroscopische Formwechsel der Plastiden. Exp. Cell. Res. 12, 427-506.

TABLE 1. Changes in the content of the photosynthetic pigments (chlorophylls and carotenoids) on the ninth day of the kinetin treatment (0.05 mg/1) in healthy and diseased tomato plants.

		Chlorophylls (a+b) (mg/g fresh weight)	Total carotenoids (mg/g fresh weight)
A.	Healthy untreated control plant	1,122	0,440
B.	Diseased untreated control plant	0,074	0,050
C.	Healthy treated plant	2,028	0,460
D.	Diseased treated plant	0,952	0,243

TABLE 2. Changes in the content of photosynthetic pigments (chlorophylls and carotenoids) on the 5th day of foliar kinetin treatment (0.05 mg/1) of diseased potato plants.

	Chlorophylls (a+b) (mg/g fresh weight)	Total carotenoids (mg/g fresh weight)
Diseased untreated plant	0,088	0,050
Diseased treated plant	0.632	0,334

Electron microscopy of foliar kinetin treated potato showed all characteristics of normally developed chloroplasts (tylakoid membranes, no starch accumulation). Kinetin expressed a stimulating effect on photosynthetic pigments (Table 2).

Potato plants, after treatment, were grown in a greenhouse together with control plants (healthy untreated and diseased untreated potato). After three to five months observation, we concluded that: 1. treated plants were intensively green, 2. treated plants were as big as healthy controls, 3. tuber production of diseased treated plants was higher than of healthy controls. (Figure 7).

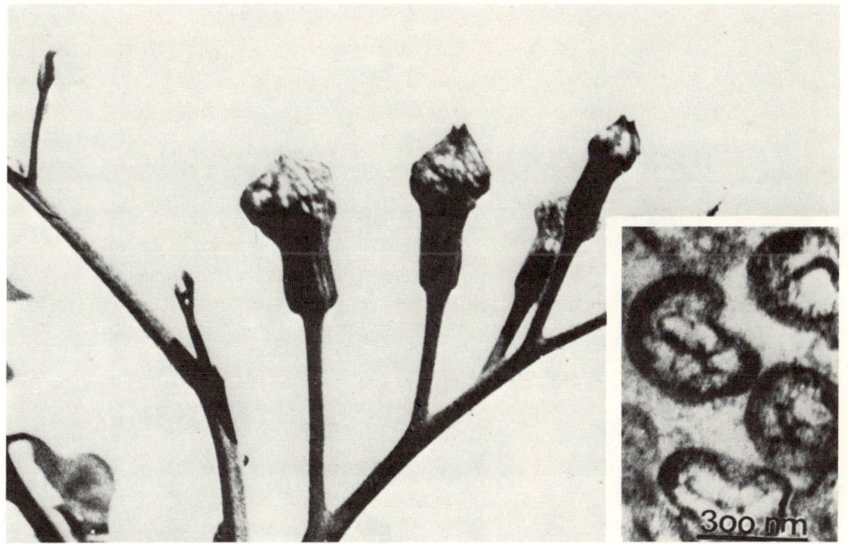

Figure 1. "Big bud" symptom and its agent (MLO).

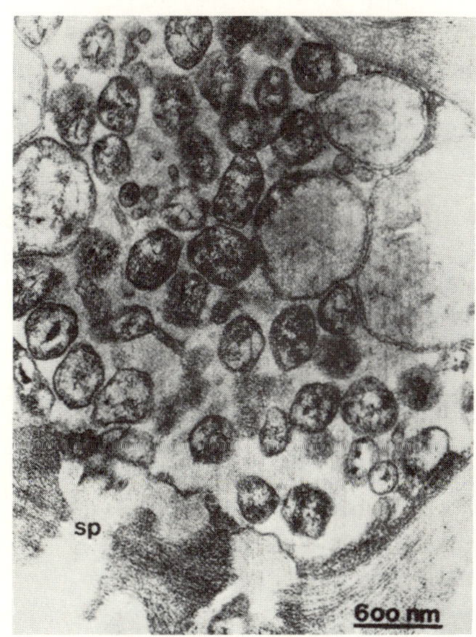

Figure 2. MLO in sieve tube of stolbur-diseased tomato.

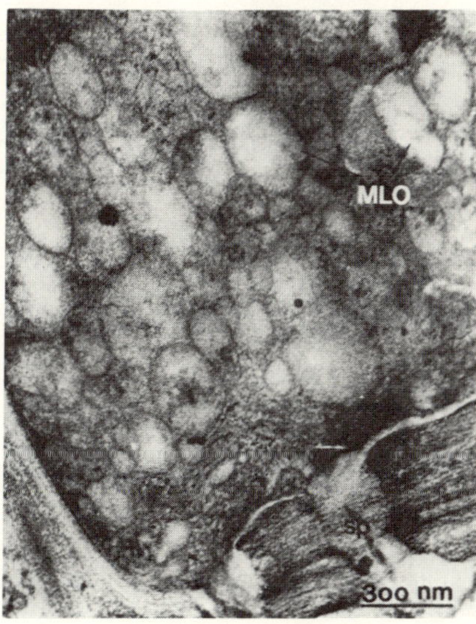

Figure 3. MLO in sieve tube of stolbur-diseased tomato treated with kinetin. sp = sieve plate.

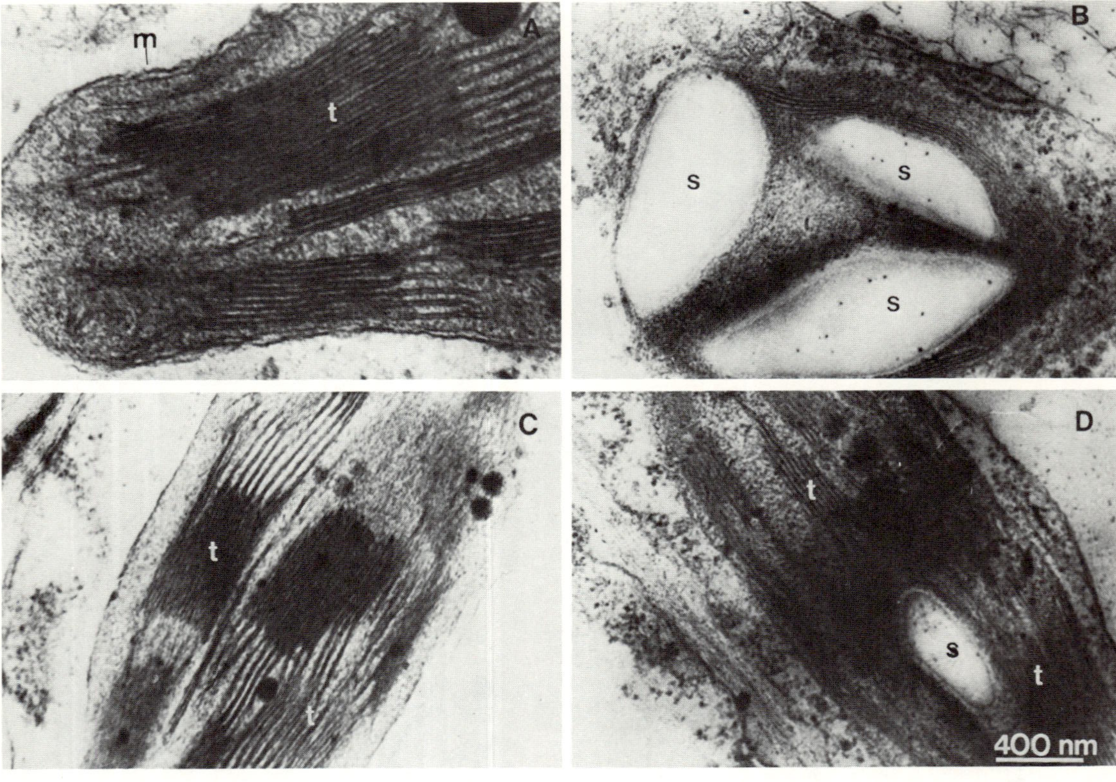

Figure 4. A, Chloroplast of healthy untreated control plant. B, Chloroplast of diseased untreated control plant. C, Chloroplast of healthy treated plant. D, Chloroplast of diseased treated plant. m = chloroplast membrane; s = starch; t = tylakoides.

Figure 5. A, Diseased treated plant. B, Diseased untreated plant.

Figure 6. A, Plant from seeds of diseased treated plant.
B, Healthy untreated control plant.

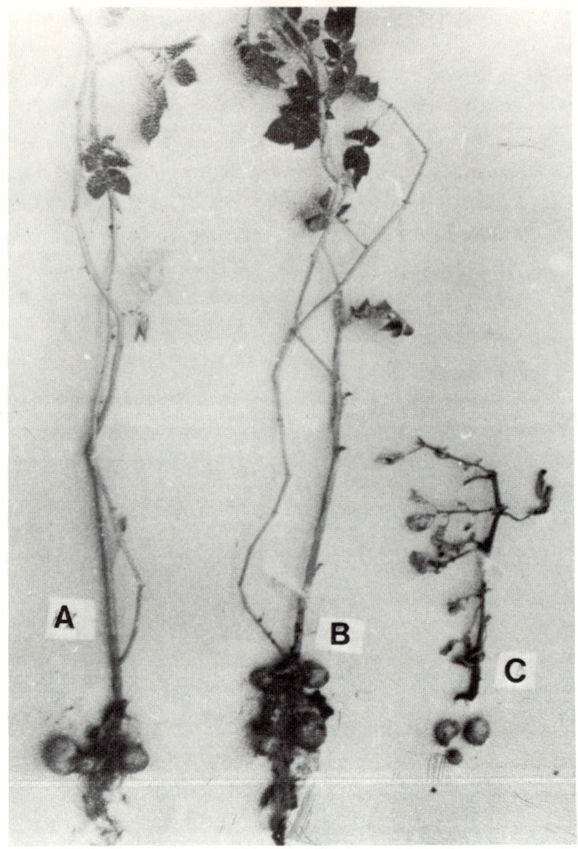

Figure 7. A, Healthy untreated potato plant. B, Diseased treated potato plant. C, Diseased untreated potato plant.

23

NON-CHEMICAL CONTROL OF

PLANT MYCOPLASMA DISEASES

by

Karl Maramorosch

Department of Entomology and

Economic Zoology

Rutgers, The State University

New Brunswick, New Jersey

INTRODUCTION

The population of the world will exceed six billion by the year 2000. Food resources that were once taken for granted and other resources, such as energy and water are gradually being exhausted or contaminated and the Earth's capacity is certainly not unlimited. Therefore, the need for restoring damaged resources becomes very urgent, and the need is obvious for protecting plants that are used for food, as well as perennial forest trees, from diseases. The primary aim of plant disease control is the reduction or elimination of losses. Additional aims include the elimination of pathogens, vectors, alternate hosts, and reservoirs of pathogens. Chemical approaches to control have been used successfully against fungal and bacterial diseases, but only rarely against viral and mycoplasmal diseases. Some of the described methods have been used successfully long before the mycoplasmal nature of plant diseases became recognized. Other control methods are now being tried or contemplated. New strategies will be needed to cope with the problems and to counterbalance the damage caused worldwide by mycoplasma diseases.

It would be desirable to develop non-chemical methods that could completely or partially control plant mycoplasma diseases. Such ultimate aims require the correct identification of the spiroplasma or MLO pathogens and an understanding of their epidemiology. Non-chemical control methods rely on prevention, restriction, or elimination of infection.

Breeding for Resistance to Mycoplasma Pathogens (MLOs)

The existence within a plant species of genetic variation to a mycoplasma pathogen provides the basis for breeding or selection programs. Since such variation exists, it can be exploited for spiroplasmas and MLOs of economic importance.

Attempts to find resistance to plant mycoplasma diseases have been made long before the mycoplasmal agents of these diseases have been recognized. Breeding for resistance was, and still remains, the most successful and cost-efficient way to protect plants from diseases. The developing of mycoplasma resistant varieties is more difficult and time consuming than the developing of plants resistant to bacteria, fungi, or mechanically transmitted viruses. All plant mycoplasma agents require insect vectors, such as leafhoppers, planthoppers, or psyllids, for their introduction into the plant phloem and for infecting susceptible plants. Work with vectors requires insectaries and it is less cost-efficient than work with mechanically transmitted pathogens. Grafting can be used for mycoplasma transmission to plants, especially in instances where vectors are still unknown.

There are but few examples of successful breeding against plant mycoplasmas, but the fact that such resistance can be obtained, indicates that this approach can be as successful as breeding for resistance to other plant pathogens.

One of the very damaging and rapidly killing MLO diseases is lethal yellowing of coconut palms. The disease has destroyed coconut palms on several Caribbean islands, including some of the Cayman Islands, and Cuba. It also destroyed many coconut palms in Florida and in West Africa, where for many years no resistance could be observed. Currently, lethal yellowing causes enormous devastations on the island of Jamaica. In addition to *Cocos nucifera*, more than 20 other palm species are susceptible to the lethal yellowing MLOs (Tsai and Maramorosch, 1987). Fortunately, one variety called in Jamaica Malayan Dwarf, has proven highly resistant and only one in 10,000 palms ever succumbs to the disease in Jamaica. The origin of the Malayan Dwarf has been traced to St. Lucia Island, where it was imported to from the Malay peninsula nearly 100 years ago. In 1963, during a worldwide survey of coconut palm diseases, I noticed that a disease closely resembling lethal yellowing, occurred sporadically in isolated areas in Malaysia (Maramorosch, 1964). In that area, as well as in Sarawak, North Borneo, very few palms succumbed and the majority remained perfectly healthy. The assumption that the disease in Malaysia is the same as lethal yellowing in America and Africa led to the speculation that lethal yellowing disease might have originated in Malaysia where eventually only resistant palms remained, survived and multiplied and where from such resistant plants were brought to the Caribbeans under the name of Malayan Dwarf (Maramorosch and Hunt, 1981). If this should prove correct, other taller varieties, producing nuts with good copra qualities, could be tested now for resistance to lethal yellowing in Jamaica. Currently, a large breeding program is underway there, crossing the Malayan Dwarf with the Jamaican Tall, to obtain resistant trees. In a few cases where such crosses became diseased, instead of being killed within the usual 4-months period, they survived for several years and some recovered completely. Since the natural mode of lethal yellowing spread and the identity of the vector are still disputed (Tsai,1980; Tsai and Thomas, 1981; Howard and Thomas, 1980), breeding for resistance has to depend on natural transmission and the palms that are being tested have to be grown in an epidemic area.

It would be much more difficult to attempt breeding for resistance against the eucalyptus witches' broom disease in India, because the disease does not seem to spread in an epidemic way, the trees are often planted at considerable distance along roads, and the insect vectors are still unknown.

The same difficulty is now being encountered with sandal spike disease in South India, where the disease occurs for many years but the vectors have not yet been determined properly (Muniyappa *et al.*, 1981). Thus no experimental infection tests can be carried out to ascertain whether a surviving tree merely escaped infection or whether it remained healthy because of resistance to the sandal spike MLOs. Sandal trees can now be propagated rapidly by tissue culture methods. This technique could provide a means for detection of resistant plants, if and when the vector of the sandal spike MLO becomes known so as to expose large numbers of tissue culture-grown seedlings to infective vectors.

In several maize varieties, natural resistance has been observed to *Spiroplasma kunkelii*, the agent of the corn stunt disease of Central America. Nearly 30 years ago, some Indian corn varieties, such as Oaxaca 40 and Chiapas 20 were found highly resistant. They were seldom, if ever, affected by corn stunt in areas where hybrid corn became decimated. Efforts are now under way to incorporate this resistance into commercial maize varieties (Ortega, 1974; Osler, 1977).

In lethal yellowing disease, an adequate source of resistance has been identified in the Malayan Dwarf, so this can be used by breeders. In other instances, such as in sandal spike, no known source of resistance has been documented as yet. Therefore, only available sandal trees can be included in any tests, but the only artificial means of inoculation of such trees is by grafting. Experimental screening of trees is very cumbersome and costly, if carried out on a wide scale. Besides, entire trees in forests are too valuable to be sacrificed for testing. Branches of presumably non-infected trees could be removed and maintained alive for prolonged periods of testing. This involves technical difficulties and unless such testing could be carried out on a large scale, its results would be meaningless. Thus, it has not yet been possible to find cultivars possessing the desirable resistance to sandal spike.

Sometimes successful screening for resistance can be carried out by relying upon natural MLO infection in the field. To succeed, it is necessary to grow a susceptible crop near a reservoir or breeding ground of the MLO vectors. Dependence on natural spread, however, is seldom reliable.

Certain MLO strains are virulent, causing severe losses, while others are mild and induced symptoms hardly recognizable. The variability among MLOs presents breeders with additional problems, because the choice of an MLO strain for use in experimental infections can be critical in attaining proper resistance.

After resistant plants have been found or selected, it may be necessary to reassure the quality and yield obtained from such plants.

Elimination of Alternate Hosts

In instances where alternate susceptible host plants within or around a crop can be identified it is sometimes possible to reduce the spread of a plant mycoplasma disease by the removal of such hosts. More often, however, it becomes difficult or even impossible to undertake such approaches, especially in instances where the alternate hosts consist of perennials or forest trees. When crops are grown near private gardens, eradication of MLO sources also might be impossible. When spread of a disease is slow, roguing within the crop can be used to control further spread. Plants, such as corn stunt-infected maize in central and tropical North America may survive a previous season's crop and thus constitute a reservior of *S. kunkelii,* from which leafhopper vectors can acquire the pathogens and carry them to a new crop. Proper cultivation can prevent further spread from such sources.

In the tropics, the continuous cultivation of rice or sugar cane might result in severe epidemics of MLO diseases, such as rice yellow dwarf or sugar cane white leaf. Unfortunately, the introduction of crop rotation or crop-free periods is seldom possible, particularly in less developed countries.

The use of mycoplasma-free planting material is of importance when plants are propagated vegetatively, or when seedlings, raised in a nursery, are later replanted in fields or forests. Mycoplasma-free plants from precious, but infected clones can be obtained by meristem tip culture and other plant tissue culture methods, discussed in this chapter, by heat therapy, or a combination of both.

Breeding for Resistance to Vectors

In 1951, Painter recognized three major types of vector resistance: tolerance, non-preference, and antibiosis. Tolerance implies the ability of a host plant to withstand the attack of a vector without severe consequences to the plant. This type of resistance is of little use in controlling the spread of MLOs, because a single feeding by an individual infective vector suffices to produces the disease. Non-preference refers to vectors landing on plants and feeding only occasionally. This incidental encounter also is sufficient to introduce MLOs into a susceptible plant and thus is of little use in finding resistance to vectors. Antibiosis, on the other hand, occurs when the growth and multiplication of an MLO vector is inhibited, thus reducing the spread of MLOs by reducing the vector population (Maxwell and Jennings, 1980).

As of now, planthopper and leafhopper resistance has been obtained in but very few instances as a secondary result of breeding programs aimed at an insect pest that, coincidentally, also happened to carry MLOs (McKelvey *et al.*, 1982). Breeding plants for resistance to leafhopper vectors can be incorporated into the general breeding plan (Maramorosch, 1980). Future approaches might give consideration to plant resistance to MLO vectors, including various types and optimal levels of resistance, genetic, environmental, and behavioral factors affecting the stability of resistance, diverse interactions between vectors and MLOs, as well as between vectors and diseased plants. One reason for omitting vector resistance from most earlier breeding programs was the attraction of vectors of mycoplasmas to diseased plants. Such attraction could be evoked by the changes in color, by bushy growth, by phytochemical and nutritional alteration of diseased plants and by other growth abnormalities, such as proliferations, or leaf curling. Hopefully, in the future, breeding for resistance to vectors will become a standard feature.

Non-chemical Control of Vectors

There are numerous theoretical ways in which vector populations can be reduced (Knipling, 1979) but since very few surviving vectors may suffice to infect numerous plants, vector control is seldom successful in case of MLO diseases. Chemical control has yielded good results in a few instances, where the specific vectors produced only one generation per year, as in cranberry false blossom, an MLO disease in the United States. More suprisingly, in the tropics, where as a rule there are several vector generations per year, a systemic insecticide, Carbofuran, has provided adequate control of *Nephotettix virescens* and *N. cincticeps*, vectors of rice yellow dwarf MLOs. Carbofuran has been declared unsafe, however, and no similar success has been obtained with any of the known biological vector control methods (Kurstak, 1981; Maramorosch and Sherman, 1985). It is difficult to predict whether biological control measures eventually will replace chemical insecticides. Perhaps integrated control of MLO vectors will be successful in the future. As for viral insecticides, none are known until now that could be used to control leafhoppers or planthoppers, the two major groups of MLO vectors (Maramorosch and Sherman, 1985). It would seem worthwhile to search for vector- pathogenic viruses by collecting planthopper and leafhopper vectors that died in nature. This search could also yield other insect pathogens that might be of interest.

Non-chemical control of vectors can sometimes be achieved by using reflective mulches. While this has been used successfully against aphid vectors of viruses, it has not been used to repell leafhoppers or planthoppers, the two groups that carry MLO pathogens. The cost of aluminum strips or plastic strips would hardly seem cost-efficient in controlling these vectors.

Oil sprays that are known to reduce the

incidence of whitefly-transmitted and aphid-transmitted viruses (Simons, 1981; 1982) have not been found effective in controlling planthopper or leafhopper vectors of mycoplasmas.

Recently, it became possible to produce commercially new varieties of *Bacillus thuringiensis* that selectively kill certain insects but remain harmless to others. In Japan, *B.t.* strains developed by Aizawa (1987), proved harmless to silkworm. Such preparations can be used for biological control of Lepidopteran pests in areas where silkworm cottage industries are thriving. Recently, a *B.t* has been produced in the United States that affects beetles selectively (Herrnstadt *et al.*, 1987). It might be possible to develop new *B.t.* by biotechnology methods (Maramorosch, 1987) that would kill certain leafhopper vectors but would not affect beneficial insects, such as silkworm. Singly or in integrated control, such future preparations could significantly reduce or entirely replace chemical insecticides, that are costly, seldom effective, and often harmful to the environment.

Surgical Procedures

Surgery is not commonly used to free plants from disease agents, because it can only succeed in instances in which the pathogens have not yet invaded the plants systemically. The classic demonstration by Baur (1904) that the virus-infected *Abutilon thompsoni* can be cured by surgery was based on the slow movement of virus in *Abutilon*. A rare instance of plant mycoplasma cure by surgery is the successful cure of papayas affected by bunchy top. Decapitation of diseased trees, that is, the removal of the bunchy top, leaves the trees free from infection. If such cured trees can be maintained so as to prevent the access of MLO-carrying *Empoasca papayae* leafhoppers, healthy leaves and fruits are formed (Maramorosch, *et al.*, 1970).

Use of Screens

For many years, screens have been used by seed producers, to prevent plant mycoplasma diseases that invariably induce sterility in plants. The demonstration by Kunkel (1929) that access to aster leafhoppers, *Macrosteles fascifrons*, can be prevented by cheesecloth screens and thus the plants can remain healthy, despite the spread of the aster yellows disease nearby, has found application in commercial enterprizes. Seed growers use screened greenhouses or screens that cover small plots, to obtain viable seed. This method cannot be used on a large area,however. Since screening plants can provide complete elimination of insect vectors and thus effectively interrupt the alternating cycle of the mycoplasma between vectors and plants, other methods that prevent vectors from feeding on plants could be attempted.

Cross-protection

Plants infected with a mild strain of a plant mycoplasma frequently will not develop severe symptoms when subsequently inoculated with a severe strain of the same mycoplasma (Kunkel, 1955; Maramorosch, 1958). This phenomenon, called cross-protection, has been known for plant viruses since 1929, when it was first described by McKinney and further developed by Thung (1936) and others. It has been assumed, earlier, that cross-protection tests can always provide evidence for virus and mycoplasma identity or strain relationships, but this is not necessarily the case. Cross-protection by related virus or MLO strains does not always occur (Gibbs and Harrison, 1976). The earlier belief, that Rio Grande spiroplasma of corn stunt is related to the MLO of the Mesa Central corn stunt (Maramorosch, 1958) was based on results of cross-protection tests. This has been challenged,and the Mesa Central disease renamed maize bushy stunt (Nault and Bradfute, 1979).

Despite such exceptions, however, the cross-protection phenomenon should be tried as a control measure in instances where the existence of mild strains can be established and the experimental transmission by proper vectors becomes feasible.

Plant Tissue Culture

Since Morel and Martin first cultured meristem tips successfully to free dahlias (1952) and potatoes (1955) from viruses, tissue culture techniques have been widely used for obtaining virus-free plants. The same techniques can be applied to free infected clones from MLO infections, but the procedures would only have practical applications for vegetatively propagated crops with desirable horticultural characteristics. While no attempts have been published concerning MLO eradication by tissue culture, diverse methods have been used successfully to eliminate viruses: callus culture, protoplast culture, culture of reproductive tissues, and meristem-tip culture (Murakishi *et al.*, 1984). The demand for pathogen-free propagative plants is on the increase and the above-mentioned techniques can be used for rapid multiplication of several vegetatively propagated crops, at the same time insuring their freedom from infection. Virus or MLO-free plantlets grown in test tubes on culture media also present a safe and convenient means for the importation of planting material, complying with strict quarantine regulations.

Plant Cell Culture to Obtain Resistant Cultivars

Cell manipulation and biotechnology present new and exciting possibilities for obtaining plants resistant to mycoplasma pathogens. Shepard *et al.*, (1980) found that plants regenerated from single-leaf cell protoplasts of a genetically stable potato variety differ greatly in their resistance to fungal infection. The finding that different cells from an individual plant can produce, upon manipulation, individual new plants of diverse resistance,could also be utilized to produce plants resistant to MLOs. It is still unclear why somaclonal variation occurs in the offspring of genetically stable plants. Probably, variation is induced during cell cultivation in tissue culture media, where chromosomes of individual cells are being altered.

Carlson *et al.,* (1972) and Melchers *et al.,* (1978) used protoplast fusion for somatic hybridization, isolating individual cells, removing their cell walls enzymatically, fusing the obtained protoplasts and regenerating the somatic hybrids into whole plants. Until now, this technique has not been used for the production of MLO resistant plants.

DNA transformation by the insertion of DNA through the intermediary of the *Ti* plasmid of *Agrobacterium tumefaciens* has been used successfully to introduce the genes of *Bacillus thuringiensis* into plants and thus make them partially resistant to damage by lepidopteran insects. In the future, this technique might allow the production of plants that would repell or kill MLO vectors. Mycoplasma resistance could also be introduced into susceptible host plants by biotechnology. Safety aspects pertaining to these exciting and novel biotechnology advances have been discussed elsewhere (Maramorosch, 1987).

Mycoplasma-infected planting material can provide a means by which MLO and spiroplasma diseases can be disseminated over great distances. The international spread of citrus stubborn disease provides a striking example. Although there is no evidence for seed transmission of the lethal yellowing MLOs or, for that matter, any MLOs, importation of coconuts from lethal yellowing-affected areas or areas close to this disease should not be attempted or permitted. Eggs of the as yet poorly determined leafhopper vectors could be on the surface of imported nuts, and even leafhopper nymphs, already infected in the country of origin, could get a "free ride" to as yet uninfested areas and cause a new outbreak. The export of coconut and oil palm seed from the Ivory Coast, a country that borders with Ghana, where lethal yellowing has been present for many years and where from the vectors can easily be blown into as yet non-infected plantations, should be scrutinized very carefully. Lethal yellowing can kill more than 20 palm species, including the oil and date palm, and the importation of propagative material from West Africa to Asia should not be permitted because of the risks involved. Not only the MLOs, but also the vectors could be carried along, but present quarantine regulations in many countries are insufficient to overcome the commercial interests of seed producers. Enforcement of existing regulations is imperative to prevent the world-wide spread of dangerous mycoplasma diseases.

SUMMARY

The primary aim of plant disease control is the reducing or eliminating of losses to affected crops. Additional aims include the elimination of pathogens, vectors, alternate hosts, and reservoirs of pathogens. All plant mycoplasma diseases require leafhopper or planthopper vectors for the transmission of the pathogens, except where transmission is by plant-parasitic plants or grafting by man. The search for vectors is undertaken so as to control, eventually, the natural spread. Vector control can be attempted by chemical insecticides but environmental, health, and even emotional factors presently curtail chemical vector control. Although non-chemical vector control is still in its infancy, its future appears bright. Microbial and viral insecticides are being studied in several laboratories, but from a practical and economic viewpoint, their future use will most likely be in combination with small amounts of chemical insecticides, that is, in integrated control.

Breeding plants for resistance to plant mycoplasmas, as well as resistance to specific vectors has been successfully carried out for virus diseases, and this approach will also yield good results in certain plant mycoplasma diseases. Heat therapy has been used successfully in but few instances. Its wider application, combined with tissue culture propagation of mycoplasma-free stock is also feasible. Surgery has worked well in papaya bunchy top disease, because of the limited spread of MLO in the plants, but it has not been useful in other instances. Screens to prevent vectors from feeding on healthy plants have been used successfully on small plots where seed production was the aim. Cross-protection has been attempted in several instances and preliminary results are encouraging. New strategies should be directed to control vectors during inter-epidemic periods when vector populations are low and foci of disease are restricted. Opportunity for successful management of mycoplasma diseases and their vectors is considerably diminished during epidemic outbreaks.

REFERENCES

Aizawa, K., 1987. Strain improvement of insect pathogens. In: Biotechnology Advances in Invertebrate Pathology and Cell Culture. Maramorosch, K., (ed.) Academic Press (in press).

Baur, E., 1904. Zur Aetiologie der infektiosen Panaschierung. Ber. Deut. Bot. Gaz. 22: 453-460.

Carlson, P.S., Smith, H.H., and Dearing, R.D., 1972. Parasexual interspecific plant hybridization. Proc. Nat. Acad. Sci. U.S. 69: 2292-2294.

Gibbs, A.J. and Harrison, B.D., 1976. Plant Virology: The Principles. Edward Arnold, London.

Herrnstadt, C., Gaertner, F., Gelernter, W. and Edwards, D.L., 1987. A *Bacillus thuringiensis* isolate with activity against Coleoptera. In: Biotechnology Advances in Insect Pathology and Cell Culture. Maramorosch, K., (ed.) Academic Press (in press).

Howard, F.W. and Thomas, D.L., 1980. Transmission of palm lethal decline to *Veitchia merrillii* by a planthopper *Myndus crudus*. J. Econ. Entomol. 73: 715-717.

Kassanis, B., 1950. Heat inactivation of leafroll in potato tubers. Ann. Appl. Biol. 37: 339-341.

Kassanis, B., 1957. The use of tissue culture to produce virus-free clones from infected potato varieties. Ann. Appl. Biol. 45: 422-427.

Knipling, E.F., 1979. The basic principles of insect population suppression and management. Agriculture Handbook, USDA, Washington, D.C., pp. 659.

Kunkel, L.O., 1929. Wire screen fences for the control of aster yellows. Phytopathology 19: 100.

Kunkel, L.O., 1936. Heat treatment for the cure of yellows and other virus diseases of peach. Phytopathology 26: 809-830.

Kunkel, L.O., 1937. Effect of temperature on ability of *Cicadella sexnotata* (Fall) to transmit aster yellows. Amer. J. Botany 24: 316-327.

Kunkel, L.O., 1941. Heat cure of aster yellows

in periwinkles. Amer. J. Botany 28: 761-769.

Kunkel, L.O., 1943. Potato witches' broom transmission by dodder and cure by heat. Proc. Amer. Phil. Soc. 86: 470.

Kunkel, L.O., 1945. Studies on cranberry false blossom. Phytopathology 35: 805.

Kunkel, L.O., 1952. Transmission of alfalfa witches' broom to non-leguminous plants by dodder and cure in periwinkle by heat. Phytopathology 42: 27-31.

Kurstak, E. (ed.) 1981. Handbook of Plant Virus Infections. Elsevier/North-Holland Biomedical Press, Amsterdam. pp. 943.

Maramorosch, K., 1958. Cross-protection between two strains of corn stunt virus in an insect vector. Virology 6: 448-459.

Maramorosch, K., 1964. A survey of coconut diseases of unknown etiology. Food and Agriculture Organization, Rome. 38 pp. + 32 color figs.

Maramorosch, K., 1980. Insects and plant pathogens. In: Breeding Plants Resistant to Insects. Maxwell F. and Jennings, P., (eds.) Wiley-Interscience: 137-155.

Maramorosch, K., 1982. Control of vector-borne mycoplasmas. In: Pathogens, Vectors, and Plant Diseases. Harris, K.F., Maramorosch, K., (eds.) Academic Press: 299-329.

Maramorosch, K., (ed.) 1987. Biotechnology Advances in Insect Pathology and Cell Culture. Academic Press (in press).

Maramorosch, K., Granados, R.R., and Hirumi, H. 1970. Mycoplasma diseases of plants and insects. Adv. Virus Res. 16: 135-193.

Maramorosch, K. and Hunt, P., 1981. Lethal yellowing disease of coconut and other palms. In: Mycoplasma Diseases of Trees and Shrubs. Maramorosch, K. and Raychaudhuri, S.P., (eds.) Academic Press. pp. 185-210.

Maramorosch, K. and Sherman, K.E. (eds.) 1985. Viral Insecticides for Biological Control. Academic Press. pp. 809

Maxwell, F. and Jennings, P. (eds.) 1980. Breeding Plants Resistant to Insects. Wiley-Interscience, New York.

McKelvey, J.J., Jr., Eldridge, B.F. and Maramorosch, K. (eds.) 1981. Vectors of Disease Agents: Interactions with Plants, Animals, and Man. Praeger, New York. pp 243.

McKinney, H.H., 1929. Mosaic diseases in the Canary Islands, West Africa and Gibraltar. J. Ag. Res. 39: 557-578.

Melchers, G., Sacristan, M.D. and Holder, A.A., 1978. Somatic hybrid plants of potato and tomato regenerated from fused protoplasts. Carlsberg Res. Com. 43: 203-218.

Morel, G.M. and Martin, C., 1952. Guerison de dahlias atteints d'une maladie a virus. C. R. Hebd. Seances Acad. Sci., 235: 1324-1325.

Morel, G.M. and Martin, C., 1955. Guerison de pommes de terre atteints de maladies a virus. Compt. Rend. Hebd. Seances Acad. Agric. 41:

Muniyappa, V., Virjayakumar, N., Subbarao, M. and Kushalappa, K.A., 1981. Technical report on studies on sandal spike disease in the forests of Karnataka State. Univ.

Agric. Sciences, Hebbal, Bangalore, India.

Murakishi, K., Lesney, M. and Carlson, P. 1984. Protoplasts and plant viruses. In: Advances in Cell Culture 3: 1-55.

Nault, L.R., and Bradfute, O.E., 1979. Corn stunt: involvement of a complex of leafhopper-borne pathogens. In: Leafhopper Vectors and Plant Disease Agents. Maramorosch, K., and Harris, K.F., (eds.) Academic Press: 561-586.

Ortega, O., 1974. Proc. World Wide Maize Improv. 70s and Role of CYMMIT. El Batan, Mexico.

Osler, R.D., 1977. Proc. Maize Virus Dis. Colloq. Workshop 1976. Ohio Agric. Res. Develop. Center, Wooster: 132.

Painter, R.H., 1951. Insect Resistance in Crop Plants. Macmillan, New York.

Shepard, J.J., Bidney, D. and Shahin, E., 1980. Potato protoplasts in crop improvement. Science 28: 17-24.

Simons, J.N., 1981. Innovative methods of control for insect-transmitted plant viral diseases. In: Vectors of Disease Agents: Interactions with Plants, Animals, and Man. McKelvey, Jr., J.J., Eldridge, B.F. and Maramorosch, K., (eds.) Prager, New York: 169-178.

Simons, J.N., 1982. Use of oil sprays and reflective surfaces for control of insect-transmitted plant viruses. In: Pathogens, Vectors, and Plant Diseases. Harris, K.F.

and Maramorosch, K., (eds.) Academic Press, 71-93.

Thung, T.H., 1936. Infective principle and plant cell in some virus diseases of the tobacco plant II. 7th Ned. Ind. Nat. Congress: 486-507, Batavia, 1935.

Tsai, J.H., 1980. Lethal yellowing of coconut palms-Search for a vector. In: Vectors of Plant Pathogens. Harris, K.F., and Maramorosch, K., (eds.) Academic Press: 177-200.

Tsai, J.H., and Maramorosch, K., 1987. Lethal yellowing. Rev. Tropical Plant Pathol. 4: (in press).

Tsai, J.H. and Thomas, D.L., 1981. Transmission of lethal yellowing mycoplasma by *Mundus crudus*. In: Mycoplasma Diseases of Trees and Shrubs. Maramorosch, K. and Raychaudhuri, S.P., (eds.) Academic Press: 211-229.